Contents

Modern Statistics for the Life Sciences

Modern Statistics for the Life Sciences

Alan Grafen
Rosie Hails
University of Oxford

OXFORD
UNIVERSITY PRESS
Great Clarendon Street, Oxford OX2 6DP

Oxford University Press is a department of the University of Oxford.
It furthers the University's objective of excellence in research, scholarship,
and education by publishing worldwide in

Oxford New York

Athens Auckland Bangkok Bogotá Buenos Aires
Cape Town Chennai Dar es Salaam Delhi Florence Hong Kong Istanbul
Karachi Kolkata Kuala Lumpur Madrid Melbourne Mexico City Mumbai
Nairobi Paris São Paulo Shanghai Taipei Tokyo Toronto Warsaw

with associated companies in Berlin Ibadan

Oxford is a registered trade mark of Oxford University Press
in the UK and in certain other countries

Published in the United States
by Oxford University Press Inc., New York

First published 2002

Library of Congress Cataloging in Publication Data
(Data applied for)

ISBN 0–19–925231–9

Typeset by Graphicraft Limited, Hong Kong

Printed in Great Britain on acid-free paper by T J International Ltd., Padstow, Cornwall

Why use this book

There are five reasons you should learn to do statistics the way this book teaches it.

It teaches a language, with which you can communicate about statistics. You use it day-to-day in telling your computer which tests you wish to do; and you use the same language to discuss with statistical advisers what tests you have been doing, might do, and should do. It is the language of **model formulae**, which was developed in the 1960s by statistical theorists, and is today universally employed by statisticians. The commonly used computer packages all have commands that use these model formulae, so this powerful language can now be usefully learnt by all users of statistics. This book teaches you that language.

The language of model formulae is based on a grand conceptual scheme, called the General Linear Model or GLM. This contains within it all the usual parametric tests, including *t*-tests, analysis of variance, contrast analysis, linear regression, multiple regression, analysis of covariance, and polynomial regression. Instead of learning these as separate tests with broadly similar features but maddening differences, this book will teach you a single coherent framework. Instead of a mish-mash of eccentrically named accidents, this book presents statistics as a meaningful whole. This is intellectually satisfying, but it is also practically useful in two ways. It's all much easier to remember *and* the unifying approach allows a lot more material to be covered in the same amount of time. Learn more faster!

Statistics textbooks tend to be divided into cookbooks (for those who want to do but not understand) and spookbooks (vice versa). The problem faced by the writers is that the obvious way to explain why a test works is to give a mathematical proof. But the mathematics is not important in everyday use of statistics, and is anyway not accessible to most users. This book chooses a different conceptual plane on which to explain statistics, one that is only possible because of the new language and conceptual unity. Some of the ideas in this plane are introduced, using geometrical pictures, to explain a bit about how GLM works. The ideas and concepts we chose to explain geometrically in *this* book are those that you really do need to use statistics properly. Concentrate your learning on what really matters.

If you are a student, and you learn the old way, then you are very likely to have the following experience at some stage in your course. You plan and carry out an experiment or survey, and go to ask for help from the person who taught you statistics. Unfortunately, the course didn't actually cover anything so sophisticated as the relatively simple project you've done. The advice is either 'Do this simpler test that was covered, even though it isn't actually the right test' or 'Give the data to me, and I'll do it for you'. However welcome this second response may be, you really would be better off knowing how to do it yourself! The power of GLM, using model formulae, is such that using the basic toolkit this book provides, you have an excellent chance of being able to analyse your project yourself.

The final reason is that GLM does not cover all of statistics. But the conceptual framework you learn from this book transfers almost unchanged into General*ised* Linear Models. These cover logistic and probit regressions, log–linear models and many more. So if you just want to learn basic statistics at the moment, the advantage of learning the way this book teaches is that, if you ever do want to go further, you will be well prepared. If, on the other hand, you already possess the laudable ambition to learn Generalised Linear Models, but lack the mathematical skills or sheer technical courage needed to tackle the textbooks on the subject, then get a firm conceptual grounding in General Linear Models from this book first. You will find the extension plain sailing.

It is important to say that many kinds of test are *not* covered in this book. The main classes are the too simple and the too complex. Nonparametric tests do not belong to GLM. If your datasets are always going to be simple enough to handle this way, then you're probably better off sticking with them—but do be aware of the danger of doing simple tests, when you *should* be doing more sophisticated ones, just because you don't know how. For example, the options for statistical elimination of variables are extremely limited, and estimation in nonparametric statistics is usually based on fairly dodgy logic.

Some of the tests that are too complex have already been mentioned: those that come under the umbrella of General*ised* Linear Models. The others include factor analysis, principal component analysis and time series analysis. These branches of statistics are all based on the simple General Linear Model, and though not directly covered here, the concepts and skills you will learn from this book are a good preparation for tackling them later on.

We have taught this course to first and second year biology undergraduates at Oxford University for about ten years. Interest around the world in the lecture notes has confirmed us in our belief that this it is the right way to teach statistics today. Those notes have here been completely rewritten with simplicity and the logical order of presentation in mind. This book represents a major attempt to make the ideas of General Linear Models accessible to life sciences undergraduates everywhere.

How to use this book

If you are in doubt as to whether you have covered the basic statistics required, much of this is briefly reviewed in the revision section at the end of the book, in a form that is only useful as a reminder. If you have any problems on reading it, this material is covered comprehensively in most elementary text books (for example *Statistics for the life sciences* by Myra Samuels, Maxwell Macmillan International Editions, and other recommendations which are given in the bibliography). The aim of our text is to present the fundamental statistical concepts, without being tied to any one statistical package. It is designed to be accompanied by one of the statistical supplements, which provide all the technical information needed to conduct the same analyses in either Minitab, SAS or SPSS. In the main text there are boxes, which present a generic form of statistical output. In the supplements, some of these same analyses are presented with package specific output. All datasets are available on the web site, so that the reader may produce the same output for themselves.

One of the first general principles of statistics is that when we wish to compare two groups, it is not sufficient to simply calculate the difference between them; it is also necessary to calculate how variable the two groups are. The amount of variation is central to deciding whether two means are far enough apart to have come from two different populations. This basic principle lies behind much of the statistics discussed here. We will not always simply want to compare two groups. Experiments will frequently involve the comparison of several variables at several levels, so we need to extend this principle to encompass the comparison of all these variables at once. Why can we not just stick to the simple tests we already know (e.g. the *t*-test), and do a number of pairwise comparisons?

The answer to this lies in the meaning of the *p*-values we produced in the simple t tests (see the revision section). Every time we conduct a test and arrive at a *p*-value of less than 0.05, we would conclude that the two groups were significantly different. We are not *certain* however that they are different. It is never possible to be absolutely certain—there is always the possibility that the two samples have been drawn, by chance, from extremes of the same population. By convention, the threshold probability is 0.05: i.e. we are prepared to accept a 5% chance that we have made a mistake. This is an acceptable risk when considering one test. If we conducted a number of pairwise comparisons, however, each test would carry the 5% risk of making an error. Even when comparing three means, we would need to do three such comparisons, four means would need six comparisons etc. The risk of making an error would rapidly become unacceptable. If all these comparisons could be effectively combined in one test, it would be less effort, we would be more likely to come to a sensible conclusion, and we would be making greater use of the information available to come to this conclusion.

It will also become apparent, as we develop this theme, that by considering several variables simultaneously, we can progress from the rather simple question 'Are these two groups different?' to investigate more interesting and subtle hypotheses, and detect patterns in our data that would otherwise escape our notice. One example would be to ask not only 'Do these fertilisers influence the yield of a crop?', but also 'Does the influence of fertiliser on yield depend upon the irrigation regime?'. Another example might be not only 'Does caffeine intake during pregnancy influence birth weight of babies?', but also 'Does caffeine intake influence birth weight over and above the influence of nicotine?'. The full advantages of following this path will become apparent as we progress along it.

Exercises are found at the end of each chapter. Answers to these exercises are found at the end of the book. The *identifying names* of all datasets are given in the text in italics, and correspond to those on the website. Each dataset will contain a number of `VARIABLES` which will be printed in courier font. We also adopt the policy of presenting **key words** or **concepts** in bold when they first appear.

How to teach this text

Throughout the text material is presented in Boxes (as well as tables). The Boxes represent input and output from a statistical package represented in a generic format. In the package specific supplements, which may be found at our website (http://www.oup.com/grafenhails/), is all the explanation you require to produce the same output in Minitab, SAS and SPSS. Further languages may be added in the future, depending upon demand. These supplements may also be downloaded in PDF format. If you intend to use this book as the basis for an intermediate level statistics course, then each chapter covers sufficient material for one lecture and one practical (so about 2 hours teaching time).

Acknowledgements

The datasets used in this book have diverse histories. Sources include the distribution of Minitab (*trees, merchantable timber*, *peru*, *grades*, *potatoes*), J.F. Osborn (1979, *Statistical Exercises in Medical Research*, Blackwell Scientific) (*antidotes*), N. Draper and H. Smith (1981, *Applied Regression Analysis* 2nd edition, Wiley Interscience) (*specific gravity*), *The Correspondence of Charles Darwin* Volume 6 (editors F. Burkhardt and S. Smith, Cambridge University Press, 1990) (*Darwin*), M.L. Samuels (1989, *Statistics for the Life Sciences*, Maxwell-Macmillan International) (*seeds*). The source of some datasets has unfortunately been obscured over the years. Many have not previously appeared in print.

We take this opportunity to thank those involved in the development of the book, and in the course on which it is based. Robin McCleery took an active role in the development of the course, in statistical, organisational and computing terms, and believed in it during the occasionally difficult early years. All the demonstrators and all the students who have enjoyed the lectures and practicals since 1990 have played their part and deserve our gratitude. The staff at OUP have been very helpful. We would like to single out Michael Rodgers, who suggested the book in the first place, and has nurtured it since 1994. Finally, both the authors have married since the book began, and we thank our spouses Elizabeth Fallaize and Peter Greenslade for their support.

Oxford A.G
January 2002 R.H

1 An introduction to analysis of variance

1.1 Model formulae and geometrical pictures

The approach we are going to take centres around two concepts: model formulae and geometry. The model formulae are not mathematical formulae but 'word formulae', and are an expression of the question we are asking. If we have data on the weight of 50 male and 50 female squirrels, we could hypothesise that these two groups are different. The question we are asking is, 'Is weight explained by the sex of the squirrel?' If the two groups are different, then knowing the sex of the squirrel will help you to predict its weight. The corresponding model formula would be:

```
WEIGHT = SEX
```

Variable names such as `WEIGHT` and `SEX` will be represented in a different font. On the left hand side of the equation is the data variable: i.e. the variable we wish to explain. We would like to know which factors are important in determining the magnitude of this variable. On the right hand side is the explanatory variable: i.e. the variable which we suspect to be important in determining the weight of a squirrel. So this simple model formula embodies our question. Later we will see how to develop these formulae to ask more complex questions.

It is the aim of this book to provide an understanding of the concepts behind the statistical analyses done, but not to expound the mathematical details. It is not necessary to be able to do the matrix algebra that lies behind the tests now that statistical packages are provided to do that for you. It is necessary however to have an understanding of the underlying principles. These principles will be illustrated using geometrical pictures rather than maths. These pictures will be used to illustrate different concepts in the early chapters.

1.2 General Linear Models

The combined approach of using model formulae and geometrical analogies (and this by-pass of mathematical details) has been made possible by a technique known as General Linear Modelling (GLM). Whilst this technique has been used for many years by statisticians, it is only relatively recently that it has been incorporated into user friendly packages used by non-specialists—Minitab being a notable example. GLMs are developed here in a framework applicable to a wide range of packages: most particularly Minitab, SAS, and SPSS, but others such as GENSTAT, BMDP, GLIM and SPLUS have a similar interface.

Having introduced the idea of General Linear Models, we will first of all turn our attention to the analysis of variance (ANOVA) for the rest of this chapter, followed by regression in the next chapter. It is a central message of Chapter 3 however that these two kinds of analysis are forms of GLM. Indeed, one of the great advantages of using GLMs is that a number of tests that have been traditionally considered separately, all come under one umbrella. They can all be expressed in terms of model formulae and the geometrical analogy, they are all subject to the same set of assumptions, and these assumptions can be tested using a common set of procedures (see Chapters 8 and 9).

1.3 The basic principles of ANOVA

The first and simplest problem to consider is the comparison of three means. This is done by the analysis of variance (ANOVA). The aim of this section is to look at an example in some detail. This will be done by actually working through the numerical mechanics, and relating it to the output. Once the origin of the output has been derived from first principles, it will not be necessary to do this again. This section will also provide you with your first introduction to model formulae, and the geometrical representation of the analysis you are conducting.

If we have three fertilisers, and we wish to compare their efficacy, this could be done by a field experiment in which each fertiliser is applied to 10 plots, and then the 30 plots are later harvested, with the crop yield being calculated for each plot. We now have three groups of ten figures, and we wish to know if there are any differences between these groups. The data were recorded in the *fertilisers* dataset as shown in Table 1.1.

When these data are plotted on a graph, it appears that the fertilisers do differ in the amount of yield produced (Fig. 1.1), but there is also a lot of

Table 1.1 Raw data from the *fertilisers* dataset

Fertiliser	Yields (in tonnes) from the 10 plots allocated to that fertiliser
1	6.27, 5.36, 6.39, 4.85, 5.99, 7.14, 5.08, 4.07, 4.35, 4.95
2	3.07, 3.29, 4.04, 4.19, 3.41, 3.75, 4.87, 3.94, 6.28, 3.15
3	4.04, 3.79, 4.56, 4.55, 4.53, 3.53, 3.71, 7.00, 4.61, 4.55

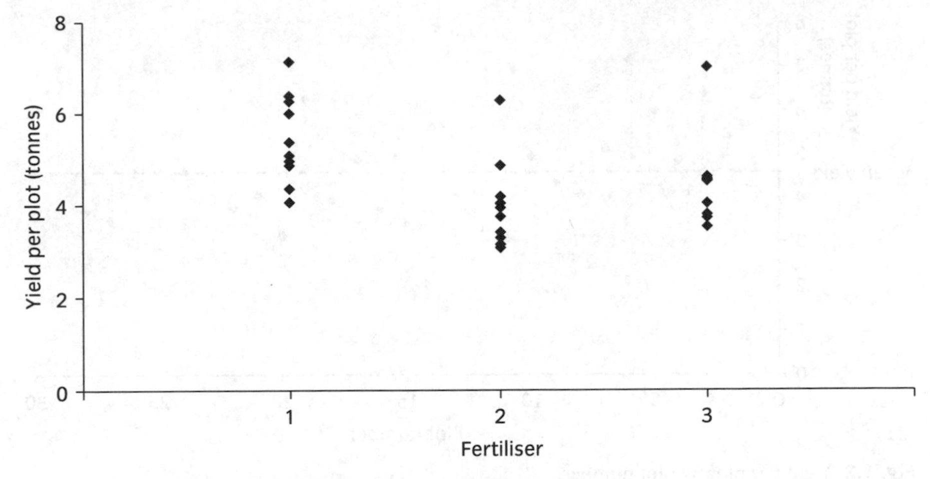

Fig. 1.1 The yield per plot for 30 plots treated with 3 fertilisers.

variation between plots given the same fertiliser. Whilst it appears that fertiliser 1 produces the highest yield on average, a number of plots treated with fertiliser 1 did actually yield less than some of the plots treated with fertilisers 2 or 3.

We now need to compare these three groups to discover if this apparent difference is statistically significant. When comparing two samples, the first step was to compute the difference between the two sample means (see revision section). However, because we have more than two samples, we do not compute the differences between the group means directly. Instead, we focus on the variability in the data. At first this seems slightly counter-intuitive: we are going to ask questions about the *means* of three groups by analysing the *variation* in the data. How does this work?

What happens when we calculate a variance?

The variability in a set of data quantifies the scatter of the data points around the mean. To calculate a variance, first the mean is calculated, then the deviation of each point from the mean. Deviations will be both positive and negative; and the sum will be zero. (This follows directly from how the mean was calculated in the first place). This will be true regardless of the size of the dataset, or amount of variability within a dataset, and so the raw deviations are not useful as a measure of variability. If the deviations are squared before summation then this sum is a useful measure of variability, which will increase the greater the scatter of the data points around the mean. This quantity is referred to as a **sum of squares** (SS), and is central to our analysis. The fertiliser dataset is illustrated in Fig. 1.2, along with the mean, and the deviation of each point from the mean. At this point, the fertiliser applied to each plot is not indicated.

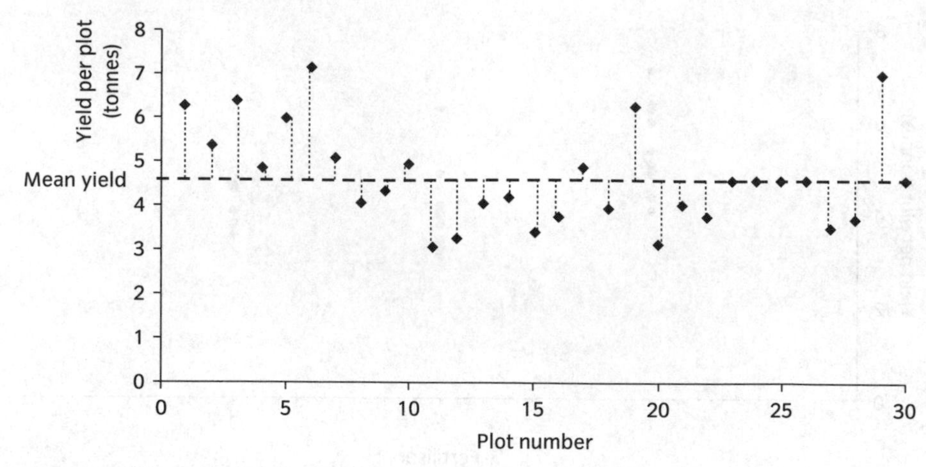

Fig. 1.2 Yield per plot by plot number.

The SS however cannot be used as a comparative measure between groups, because clearly it will be influenced by the number of data points in the group; the more data points, the greater the SS. Instead, this quantity is converted to a variance by dividing by $n - 1$, where n equals the number of data points in the group. A variance is therefore a measure of variability, taking account of the size of the dataset.

Why use n – 1 *rather than* n?

If we wish to calculate the average squared deviation from the mean (i.e. the variance) why not divide by n? The reason is that we do not actually have n independent pieces of information about the variance. The first step was to calculate a mean (from the n independent pieces of data collected). The second step is to calculate a variance with reference to that mean. If $n - 1$ deviations are calculated, it is known what the final deviation must be, for they must all add up to zero by definition. So we have only $n - 1$ independent pieces of information on the variability about the mean. Consequently, you can see that it makes more sense to divide the SS by $n - 1$ than n to obtain an average squared deviation around the mean. (A fuller explanation of this is given in Appendix 2). The number of independent pieces of information contributing to a statistic are referred to as the **degrees of freedom**.

Partitioning the variability

In an ANOVA, it is useful to keep the measure of variability in its two components; that is, a sum of squares, and the degrees of freedom associated with the sum of squares. Returning to the original question: what is causing the variation in yield between the 30 plots of the experiment? Numerous factors are likely to be involved: e.g. differences in soil nutrients between the plots, differences in moisture content, many other biotic and abiotic factors, and also the fertiliser applied to the plot. It is only the last of these that we are interested in, so we will divide the variability between plots into two parts: that due to applying different fertilisers, and that due to all the other factors. To illustrate the principle behind partitioning the variability, first consider two extreme datasets. If there was almost no variation between the plots due to any of the other factors, and nearly all variation was due to the application of the three fertilisers, then the data would follow the pattern of Fig. 1.3a. The first step would be to calculate a grand mean, and there is considerable variation around this mean. The second step is to calculate the three group means that we wish to compare: that is, the means for the plots given fertilisers A, B and C. It can be seen that once these means are fitted, then little variation is left around the group means (Fig. 1.3b). In other words, fitting the group means has removed

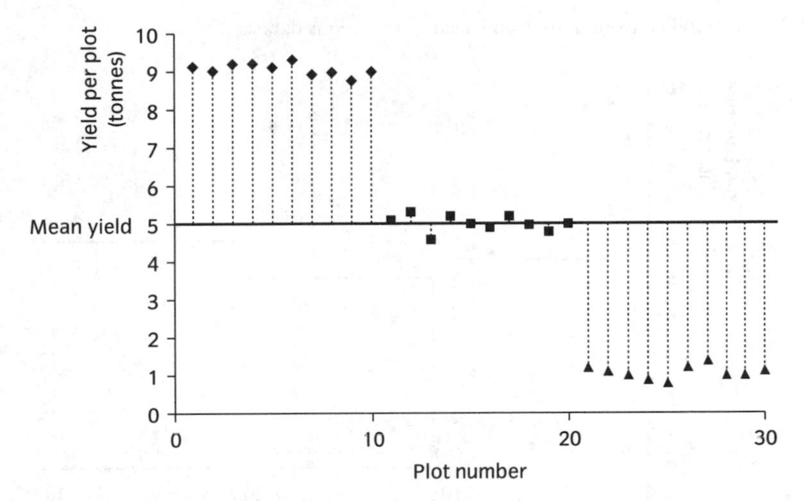

Fig. 1.3(a) Variability around the grand mean for fictitious dataset 1.

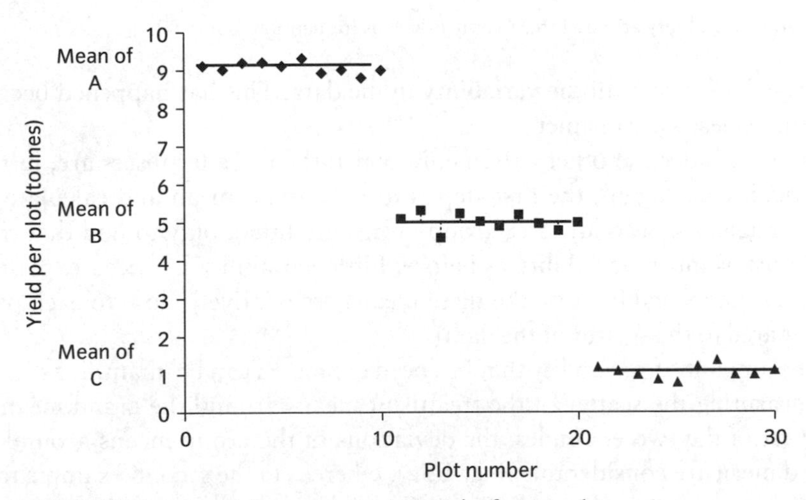

Fig. 1.3(b) Variability around three treatment means for fictitious dataset 1.

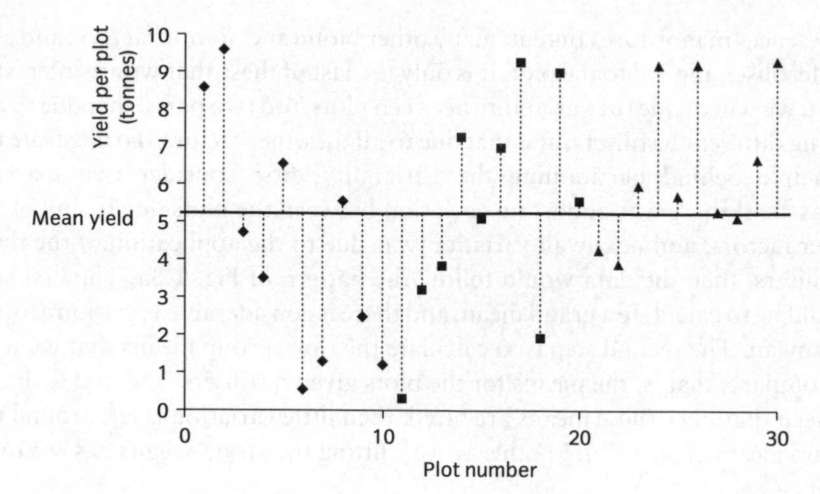

Fig. 1.4(a) Variability around the grand mean for fictitious dataset 2.

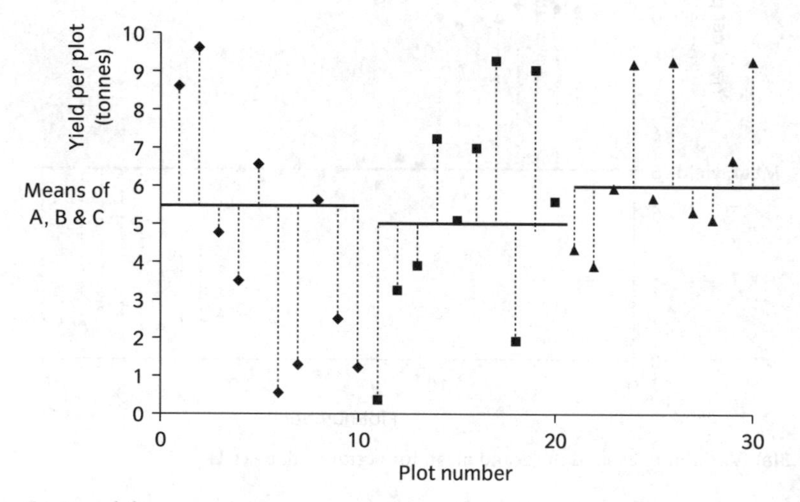

Fig. 1.4(b) Variability around three treatment means for fictitious dataset 2.

or **explained** nearly all the **variability** in the data. This has happened because the three means are distinct.

Now consider the other extreme, in which the three fertilisers are, in fact, identical. Once again, the first step is to fit a grand mean and calculate the sum of squares. Second, three group means are fitted, only to find that there is almost as much variability as before. Little variability has been explained. This has happened because the three means are relatively close to each other (compared to the scatter of the data).

The amount of variability that has been explained can be quantified directly by measuring the scatter of the treatment means around the grand mean. In the first of the two examples, the deviations of the group means around the grand mean are considerable (Fig. 1.5a), whereas in the second example these deviations are relatively small (Fig. 1.5b).

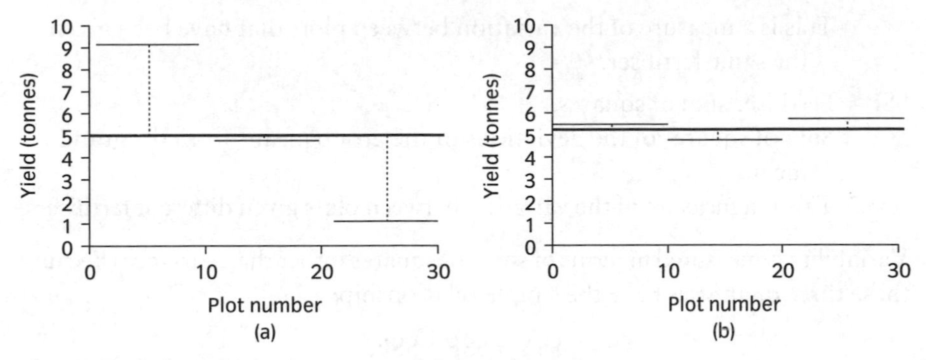

Fig. 1.5 Deviations of group means around the grand mean.

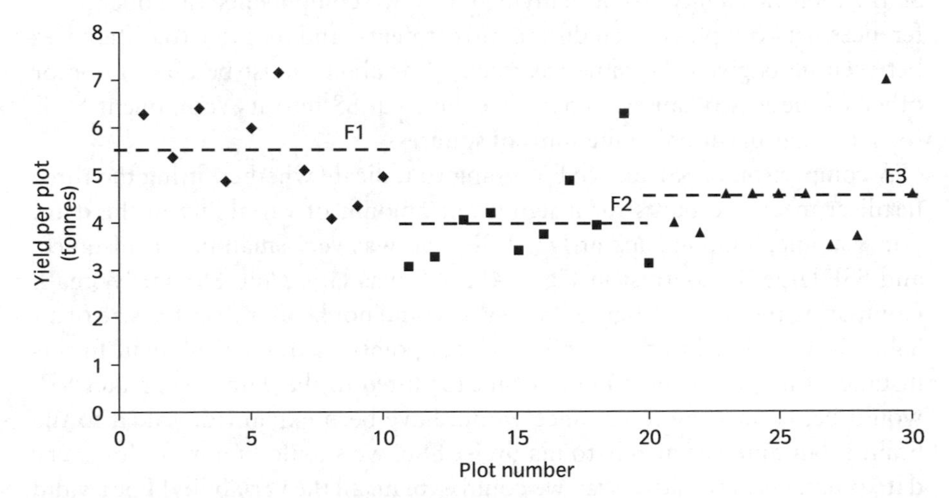

Fig. 1.6 Variability in yield around three means for the *fertiliser* dataset.

The dataset given in Table 1.1 represents an intermediate situation in which it is not immediately obvious if the fertilisers have had an influence on yield. When the three group means are fitted, there is an obvious reduction in variability around the three means (compared to the one mean) (Fig. 1.6). But at what point do we decide that the amount of variation explained by fitting the three means is **significant**? The word significant, in this context, actually has a technical meaning. It means 'When is the variability between the group means greater than that we would expect by chance alone?'

At this point it is useful to define the three measures of variability that have been referred to. These are:

SSY = Total sum of squares.
Sum of squares of the deviations of the data around the grand mean.
This is a measure of the total variability in the dataset.

SSE = Error sum of squares.
Sum of squares of the deviations of the data around the three separate group means.

This is a measure of the variation between plots that have been given the same fertiliser.

SSF = Fertiliser sum of squares.
Sum of squares of the deviations of the group means from the grand mean.
This is a measure of the variation between plots given different fertilisers.

Variability is measured in terms of sums of squares rather than variances because these three quantities have the simple relationship:

$$SSY = SSF + SSE.$$

So the total variability has been divided into two components; that due to differences between plots given different treatments, and that due to differences between plots given the same treatment. Variability must be due to one or other of these two causes. Separating the total SS into its component SS is referred to as **partitioning the sums of squares.**

A comparison of SSF and SSE is going to indicate whether fitting the three fertiliser means accounts for a significant amount of variability in the data. For example, looking back at Fig. 1.3b, SSE was very small in this instance, and SSF large. In contrast in Fig. 1.4b, SSE was large, and SSF fairly small. Comparing the two raw figures however would not be useful, as the size of an SS is always related to the number of data points used to calculate it. In this instance, the greater the number of means fitted to the data, the greater SSF would be, because more variance would have been explained. Taken to the limit, if our aim was merely to maximise SSF, we should fit a mean for every data point, because in that way we could explain all the variability! For a valid comparison between these two sources of variability, we need to compare the variability per degree of freedom, i.e. the variances.

Partitioning the degrees of freedom

Every SS was calculated using a number of independent pieces of information. The first step in any analysis of variance is to calculate SSY. It has already been discussed that when looking at the deviations of data around a central grand mean, there are $n - 1$ independent deviations: i.e. in this case $n - 1 = 29$ degrees of freedom (df). The second step is to calculate the three treatment means. When the deviations of two of these treatment means from the grand mean have been calculated, the third is predetermined, as again by definition, the three deviations must sum to zero. Therefore, SSF, which measures the extent to which the group means deviate from the grand mean, has two df associated with it. Finally, SSE measures variation around the three group means. Within each of these groups, the ten deviations must sum to zero. Given nine deviations within the group, the last is predetermined. Thus SSE has $3 \times 9 = n - 3 = 27$ df associated with it. Just as the SS are additive, so are the df.

Mean squares

Combining the information on SS and df, we can arrive at a measure of variability per df. This is equivalent to a variance, and in the context of ANOVA is called a **mean square** (MS). In summary:

Fertiliser Mean Square (FMS) = SSF/2
The variation (per df) between plots given different fertilisers.

Error Mean Square (EMS) = SSE/27
The variation (per df) between plots given the same fertiliser.

Total Mean Square (TMS) = SSY/29
The total variance of the dataset.

Unlike the SS, the MS are not additive.

So now the variability per df due to differences between the fertilisers has been partitioned from the variability we would expect due to all other factors. Now we are in the position to ask: by fitting the treatment means, have we explained a significant amount of variance?

F-ratios

If none of the fertilisers influenced yield, then the variation between plots treated with the same fertiliser would be much the same as the variation between plots given different fertilisers. This can be expressed in terms of mean squares: the mean square for fertiliser would be the same as the mean square for error: i.e.

$$\frac{\text{FMS}}{\text{EMS}} = 1.$$

The ratio of these two mean squares is the ***F*-ratio**, and is the end result of the ANOVA. Even if the fertilisers are identical, it is unlikely to equal exactly 1, it could by chance take a whole range of values. The ***F* distribution** represents the range and likelihood of all possible *F*-ratios under the null hypothesis (i.e. when the fertilisers are identical), as illustrated in Fig. 1.7.

If the three fertilisers were very different, then the FMS would be greater than the EMS, and the *F*-ratio would be greater than 1. However, Fig. 1.7 illustrates that the *F*-ratio can be quite large even when there are no treatment differences. At what point do we decide that the size of the *F*-ratio is due to treatment differences rather than chance?

Just as with other test statistics, the traditional threshold probability of making a mistake is 0.05. In other words, we accept that the *F*-ratio is significantly greater than 1 if it will be that large or larger under the null hypothesis only 5% of the time. If we had inside knowledge that the null hypothesis was in fact true, then 5% of the time we would still get an *F*-ratio that large. When we conduct an experiment however we have no such inside knowledge, and

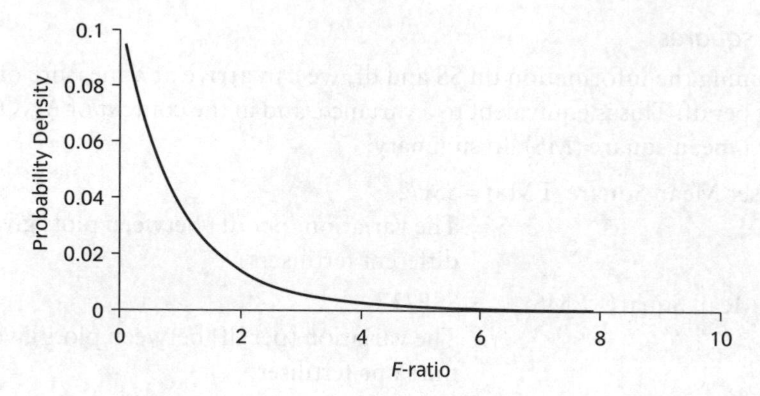

Fig. 1.7 The *F* distribution for 2 and 27 degrees of freedom (illustrates the probability of a *F*-ratio of different sizes when there are no treatment differences).

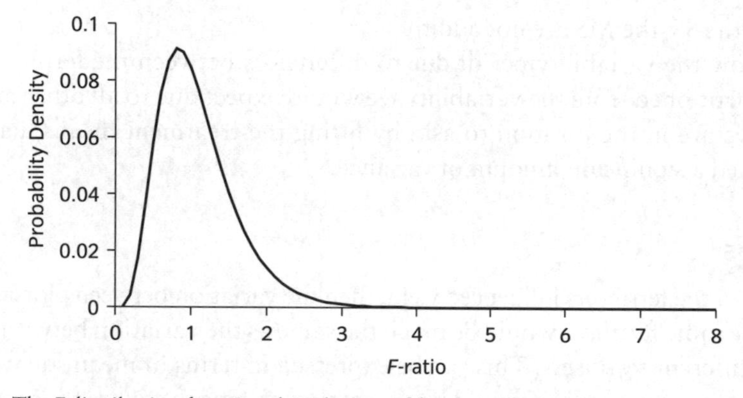

Fig. 1.8 The *F* distribution for 10 and 57 degrees of freedom.

we are trying to gather evidence against it. Our *p*-value is a measure of the strength of evidence against the null hypothesis. Only when it is less than 0.05 do we consider the evidence great enough to accept. A fuller discussion of the meaning of the *p*-value is given in Appendix 1.

It should be mentioned that the exact *F* distribution will depend upon the df with which the *F*-ratio was constructed. In this case, the df are 2 and 27, associated with the numerator and the denominator of the *F*-ratio respectively. The general shape will vary from a decreasing curve (Fig. 1.7) to a humped distribution, skew to the right (Fig. 1.8). When doing an ANOVA table in most packages the *F*-ratio, degrees of freedom and the *p*-value are provided in the output, or occasionally you are left to look up the *F*-ratio in statistical tables.

1.4 An example of ANOVA

Having explained the principles behind an analysis of variance, this section will provide an example of a one-way ANOVA. This requires two pieces of input from you.

Step 1: The data

The first point is to represent the two variables in a form that a statistical program will understand. To do this, the data should be converted from Table 1.1 to the 'samples and subscripts' form shown in Table 1.2. It can be seen here that FERTIL is represented by the subscripts 1, 2 and 3 which correspond to the

Table 1.2 Data presented as samples and subscripts

FERTIL	YIELD (tonnes)
1	6.27
1	5.36
1	6.39
1	4.85
1	5.99
1	7.14
1	5.08
1	4.07
1	4.35
1	4.95
2	3.07
2	3.29
2	4.04
2	4.19
2	3.41
2	3.75
2	4.87
2	3.94
2	6.28
2	3.15
3	4.04
3	3.79
3	4.56
3	4.55
3	4.55
3	4.53
3	3.53
3	3.71
3	7.00
3	4.61

three different fertilisers. This variable is categorical, and in this sense the values 1, 2 and 3 are arbitrary. In contrast, YIELD is continuous, the values representing true measurements. Data are usually continuous, whilst explanatory variables may be continuous (see Chapter 2) or categorical (this chapter) or both (later chapters).

Step 2: The question

This is the first use of model formulae—a form of language that will prove to be extremely useful. The question we wish to ask is: 'Does fertiliser affect yield?'.

This can be converted to the **word equation**

YIELD = FERTIL.

This equation contains two variables: YIELD, the data we wish to explain and FERTIL, the variable we hypothesise might do the explaining.

YIELD is therefore the **response** (or dependent) **variable**, and FERTIL the **explanatory** (or independent) **variable.** It is important that the data variable is on the left hand side of the formula, and the explanatory variable on the right hand side. It is the right hand side of the equation that will become more complicated as we seek progressively more sophisticated explanations of our data. Having entered the data into a worksheet in the correct format, and decided on the appropriate model formula and analysis, the specific command required to execute the analysis will depend upon your package (see package specific supplements). The output is presented here in a generalised format.

BOX 1.1 Analysis of variance with one explanatory variable

Word equation: YIELD = FERTIL

FERTIL is categorical

One-way analysis of variance for YIELD

Source	DF	SS	MS	F	P
FERTIL	2	10.8227	5.4114	5.70	0.009
Error	27	25.6221	0.9490		
Total	29	36.4449			

Output

The primary piece of output is the ANOVA table, in which the partitioning of SS and df has taken place. This will either be displayed directly, or can be constructed by you with the output given. The total SS have been partitioned between treatment (FERTIL) and error, with a parallel partitioning of degrees of freedom. Each of the columns ends with the total of the preceding terms.

The calculation of the SS is displayed in Table 1.3. Columns M, F and Y give the grand mean, the fertiliser mean and the plot yield for each plot in turn.

Table 1.3 Calculating the SS and the DF

Datapoint	FERTIL	*M*	*F*	*Y*	*MY*	*MF*	*FY*
1	1	4.64	5.45	6.27	1.63	0.80	0.82
2	1	4.64	5.45	5.36	0.72	0.80	−0.09
3	1	4.64	5.45	6.39	1.75	0.80	0.94
4	1	4.64	5.45	4.85	0.21	0.80	−0.60
5	1	4.64	5.45	5.99	1.35	0.80	0.54
6	1	4.64	5.45	7.14	2.50	0.80	1.69
7	1	4.64	5.45	5.08	0.44	0.80	−0.37
8	1	4.64	5.45	4.07	−0.57	0.80	−1.38
9	1	4.64	5.45	4.35	−0.29	0.80	−1.10
10	1	4.64	5.45	4.95	0.31	0.80	−0.50
11	2	4.64	4.00	3.07	−1.57	−0.64	−0.93
12	2	4.64	4.00	3.29	−1.35	−0.64	−0.71
13	2	4.64	4.00	4.04	−0.60	−0.64	0.04
14	2	4.64	4.00	4.19	−0.45	−0.64	0.19
15	2	4.64	4.00	3.41	−1.23	−0.64	−0.59
16	2	4.64	4.00	3.75	−0.89	−0.64	−0.25
17	2	4.64	4.00	4.87	0.23	−0.64	0.87
18	2	4.64	4.00	3.94	−0.70	−0.64	−0.06
19	2	4.64	4.00	6.28	1.64	−0.64	2.28
20	2	4.64	4.00	3.15	−1.49	−0.64	−0.85
21	3	4.64	4.49	4.04	−0.60	−0.16	−0.45
22	3	4.64	4.49	3.79	−0.85	−0.16	−0.70
23	3	4.64	4.49	4.56	−0.08	−0.16	0.07
24	3	4.64	4.49	4.55	−0.09	−0.16	0.06
25	3	4.64	4.49	4.55	−0.09	−0.16	0.06
26	3	4.64	4.49	4.53	−0.11	−0.16	0.04
27	3	4.64	4.49	3.53	−1.11	−0.16	−0.96
28	3	4.64	4.49	3.71	−0.93	−0.16	−0.78
29	3	4.64	4.49	7.00	2.36	−0.16	2.51
30	3	4.64	4.49	4.61	−0.03	−0.16	0.12
DF		1	3	30	29	2	27
SS					36.44	10.82	25.62

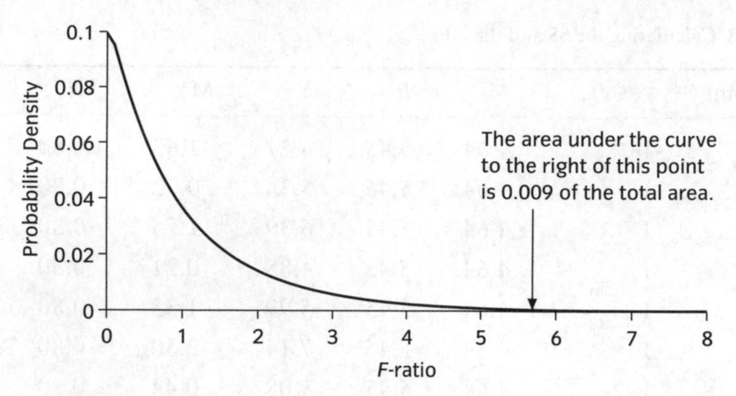

Fig. 1.9 The *F* distribution of 2 and 27 df. The area to the right of 5.7 represents the probability that the *F*-ratio is at least 5.7, and is 0.009 of the total area under the curve.

Column *MY* then represents the deviations from the grand mean for each plot. If these values are squared and summed, then the result is the total SS of 36.44. *FY* then represents the deviations from the group mean for each plot; these values squared and summed give the error SS.

Finally, *MF* represents the deviations of the fertiliser means from the grand mean; squaring and summing giving the treatment SS. Dividing by the corresponding df gives the mean square. Comparison of the two mean squares gives the *F*-ratio of 5.70. The probability of getting an *F*-ratio as large as 5.70 or larger, if the null hypothesis is true, is the *p*-value of 0.009. That is sufficiently small to conclude that these fertilisers probably do differ in efficacy.

Presenting the results

Having concluded that there is a significant difference between the fertilisers, it would be interesting to know where this difference lies. One useful way of displaying the results would be to tabulate the means for each group, and their 95% confidence intervals. What do we mean by a confidence interval, and how are they constructed?

To answer this, we need to draw together the basic principles reviewed in Revision Section 1, and apply them in the context of ANOVA. A confidence interval is an expression of how confident we are in our estimates (in this case, the three group means). For each confidence interval, we would expect the true mean for that group to lie within that range 95% of the time.

To construct a confidence interval, both the parameter estimate, and the variability in that estimate are required. In this case, the parameters estimated are means—we wish to know the true mean yield to be expected when we apply fertiliser 1, 2 or 3—which we will denote μ_A, μ_B, and μ_C respectively. These represent true population means, and as such we cannot know their exact values—but our three treatment means represent estimates of these three parameters. The reason why these estimates are not exact is because of the

Table 1.4 Constructing confidence intervals

Fertiliser	$\bar{y}$	t_{crit} with 27 df for 95% confidence	$\frac{s}{\sqrt{n}}$	Confidence interval
1	5.445	2.0518	0.3081	(4.81, 6.08)
2	3.999	2.0518	0.3081	(3.37, 4.63)
3	4.487	2.0518	0.3081	(3.85, 5.12)

unexplained variation in the experiment, as quantified by the **error variance** which we previously met as the error mean square, and will refer to as s^2. From Revision Section 1, the 95% confidence interval for a population mean is:

$$\bar{y} \pm t_{crit}\frac{s}{\sqrt{n}}.$$

The key point is where our value for s comes from. If we had only the one fertiliser, then all information on population variance would come from that one group, and s would be the standard deviation for that group. In this instance however there are three groups, and the unexplained variation has been partitioned as the error mean square. This is using all information from all three groups to provide an estimate of unexplained variation—and the degrees of freedom associated with this estimate are 27—much greater than the 9 which would be associated with the standard deviation of any one treatment. So the value of s used is $\sqrt{\text{EMS}} = \sqrt{0.949} = 0.974$. This is also called the pooled standard deviation. Hence the 95% confidence intervals are as shown in Table 1.4.

These intervals, combined with the group means, are an informative way of presenting the results of this analysis, because they give an indication of how accurate the estimates are likely to be.

It is worth noting that we have assumed it is valid to take one estimate of s and apply it to all fertiliser groups. However, consider the following scenario. Fertiliser 1 adds nitrate, while Fertiliser 2 adds phosphate (and Fertiliser 3 something else altogether). The plots vary considerably in nitrate levels, and Fertiliser 1 is sufficiently strong to bring all plots up to a level where nitrate is no longer limiting. So Fertiliser 1 reduces plot-to-plot variation due to nitrate levels. The phosphate added by Fertiliser 2 combines multiplicatively with nitrate levels, so increasing the variability arising from nitrate levels. The mean yields from plots allocated to Fertiliser 2 would be very much more variable, while those allocated to Fertiliser 1 would have reduced variability, and our assumption of equal variability between plots within treatments would be incorrect. The 95% confidence interval for Fertiliser 2 will have been underestimated.

Fortunately in this case the group standard deviations do not look very different (Table 1.5), so it is unlikely that we have a problem. In Chapter 9 we shall discuss this and other assumptions we make in doing these analyses.

Table 1.5 Descriptive statistics for YIELD by FERTIL

Descriptive Statistics for YIELD by FERTIL

FERTIL	N	Mean	Standard Deviation
1	10	5.445	0.976
2	10	3.999	0.972
3	10	4.487	0.975

1.5 The geometrical approach for an ANOVA

The analysis that has just been conducted can be represented as a simple geometrical picture. One advantage of doing this is that such pictures can be used to illustrate certain concepts. In this first illustration, geometry can be used to represent the partitioning and additivity of the SS.

The geometrical approach is actually a two-dimensional representation of multidimensional space. One dimension is represented by the position of a point on a line—one coordinate can be used to define that position. Two dimensions may be pictured as a graph, with a point being specified by two coordinates. This can be extended to three dimensions, in which the position of a point in a cube is specified by three coordinates. Beyond three dimensions it is no longer possible to visualise a geometrical picture to represent all dimensions simultaneously. It is possible however to take a slice through multidimensional space and represent it in two dimensions. For example, if a cube has axes x, y, and z, the position of three points can be specified by their x, y and z coordinates. A plane could then be drawn through those three points, so allowing them to be represented on a piece of paper (Fig. 1.10). There are still three coordinates associated with each point (and so defining that point), but for visual purposes, the three dimensions have been reduced to two. In fact, it is possible to do this for any three points, however many dimensions they are plotted in. This trick is employed by the geometrical approach.

In this case, there are as many dimensions as there are data points in the dataset (30). Each point is therefore represented by 30 coordinates. The three points themselves are the columns 3, 4 and 5 (M, F and Y) of Table 1.3.

Point Y

This point represents the data, so the 30 coordinates describing this point are the 30 measurements of yield.

Point M

This point represents the grand mean. Because we are dealing with 30-dimensional space (as dictated by the size of the dataset), this point also has 30 coordinates

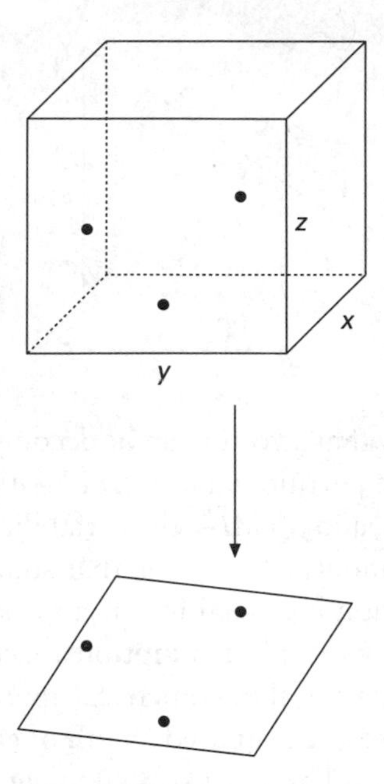

Fig. 1.10 Representing three dimensions in two dimensions.

specifying its position in multidimensional space. However, the values of these 30 coordinates are all the same (the grand mean).

Point F

This point represents the treatment means. While still 30 elements long, the first ten elements are the mean for treatment 1 (and are therefore the same value), the second ten the mean for treatment two etc. Therefore the first part of the geometrical approach is that the three **variables,** M, F and Y, are represented as **points.** These three points may be joined to form a triangle in two dimensional space as follows:

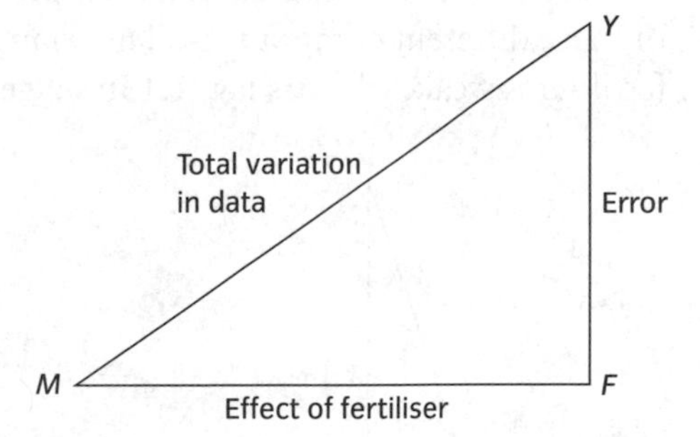

Fig. 1.11 The geometrical approach—variables represented as points, sources as vectors.

The triangle has been drawn with F at a right angle. There is a reason for this which will be explained more fully in Chapter 2. The lines joining the points are **vectors,** and these represent **sources of variability.** For example, the vector MY represents the variability of the data (Y) around the grand mean (M). In

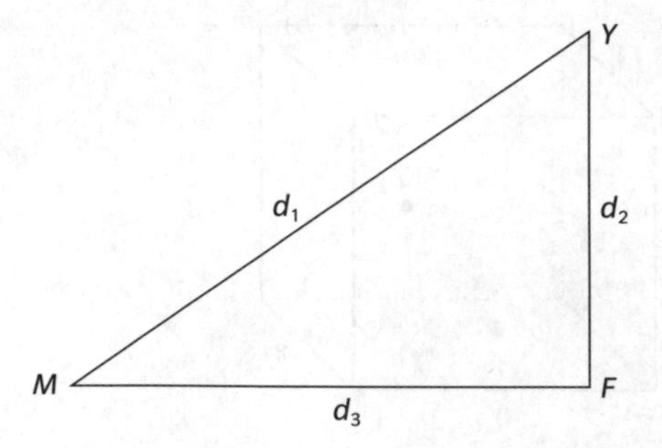

Fig. 1.12 The geometrical approach—Pythagoras theorem.

the same way that a vector can be decomposed into two components, so can the variability be partitioned into (i) *FY*—the variability of the data around their group means, and (ii) *MF*—the variability of the group means around the grand mean. The implication here is that sources of variability are **additive.** While this assumption is crucial in our approach, it is not necessarily true. Testing and correcting for this assumption are covered later (Chapters 9 and 10).

The third part of the geometrical approach relies on the fact that the triangle is right-angled. The **squared length of each vector** is then equivalent to the SS for that source. This is illustrated in Fig. 1.12.

Pythagoras states that:

$$d_1^2 = d_2^2 + d_3^2.$$

This is equivalent to:

$$\text{SSY} = \text{SSF} + \text{SSE}.$$

This illustrates geometrically the partitioning of sums of squares. It is precisely because the sums of squares can be summed in this way that they are central to the analysis of variance. Other ways of measuring variation (e.g. using variances) would not allow this, because the variances of the components would not add up to the variance of the whole.

The shape of the triangle can then provide information on the relative sizes of these different components. The triangle in Fig. 1.13a suggests the effect of fertiliser is weak, whereas Fig. 1.13b suggests that the effect is strong.

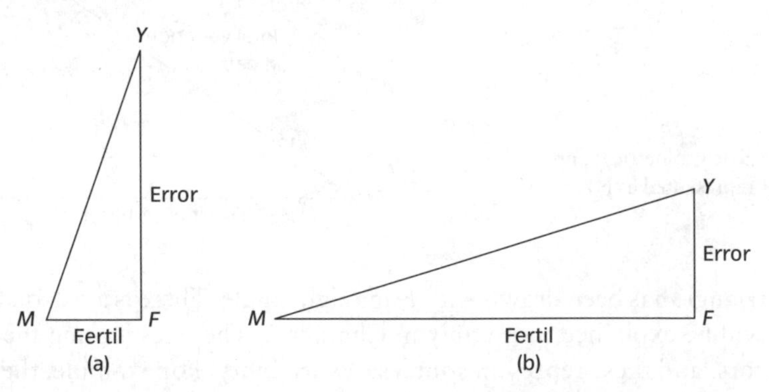

Fig. 1.13 (a) Impact of fertiliser on yield is weak; (b) Impact of fertiliser on yield is strong.

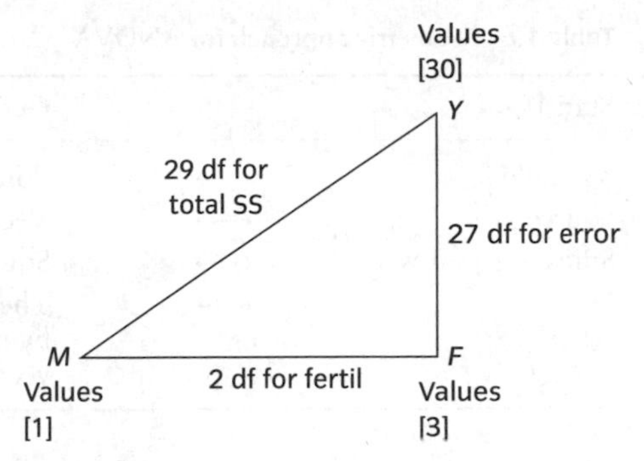

Fig. 1.14 The geometrical approach—partitioning the degrees of freedom.

It is also possible to represent the parallel partitioning of degrees of freedom in a similar manner. At each apex (Fig. 1.14), are the number of values in each variable (30 different data points, 3 treatment means, and 1 grand mean). The difference between these values gives the number of degrees of freedom associated with moving from one point to another. For example, calculating a grand mean is equivalent to moving from Y to M, and in doing so, twenty nine degrees of freedom are lost. Moving from M to F is equivalent to exchanging one grand mean for three treatment means, the difference being two degrees of freedom. These degrees of freedom are associated with the corresponding vectors and therefore with the sources represented by these vectors.

Figure 1.14 also illustrates the additivity of degrees of freedom. In moving from $M[1]$ to $Y[30]$, the total df are 29. This is true if we go directly (via vector MY), or indirectly (via MF with 2 df then FY with 27 df).

1.6 Summary

In this chapter an analysis of variance has been followed from first principles. A number of concepts have been discussed, including:

- Model formulae—a 'word equation' which encapsulates the question being asked.
- The fundamental principle behind an ANOVA—partitioning variability in data to ask questions about differences between groups.
- Degrees of freedom—the number of independent pieces of information which contribute to a particular measure.
- Three parallel partitions: sources of variability, sums of squares, degrees of freedom.
- The meaning of a p-value.
- Presenting the results of ANOVA in the form of confidence intervals.
- The geometrical approach for ANOVA, see Table 1.6.

Table 1.6 Geometric approach for ANOVA

Statistics		Geometry
Variable	$\longleftrightarrow$	Point
Source	$\longleftrightarrow$	Vector
Sums of squares	$\longleftrightarrow$	Squared length of vector
DF	$\longleftrightarrow$	The number of dimensions of freedom gained by moving from the variable at one end of the vector to the variable at the other

1.7 Exercises

Melons

An experiment was performed to compare four melon varieties. It was designed so that each variety was grown in six plots—but two plots growing variety 3 were accidentally destroyed. The data are plotted in Fig. 1.15, and can be found in the *melons* dataset under the variables `YIELDM` and `VARIETY`.

Table 1.7 shows some summary statistics and an ANOVA table produced from these data.

(1) What is the null hypothesis in this case?

(2) What conclusions would you draw from the analysis in Table 1.7?

(3) How would you summarise the information provided by the data about the amount of error variation in the experiment?

(4) Calculate the standard error of the mean for all four varieties.

(5) How would you summarise and present the information from this analysis?

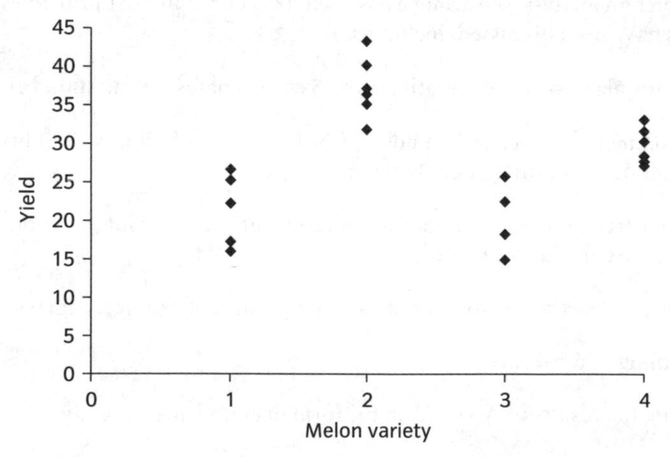

Fig. 1.15 Melon yields.

Table 1.7 ANOVA for *melons*

VARIETY	N	Mean
1	6	20.490
2	6	37.403
3	4	20.462
4	6	29.897

One-way analysis of variance for YIELDM

Source	DF	SS	MS	F	P
VARIETY	3	1115.3	371.8	23.80	0.000
Error	18	281.2	15.6		
Total	21	1396.5			

Dioecious trees

A plant species is dioecious if each individual produces all male flowers or all female flowers. The dataset *dioecious trees* contains data from 50 trees of one particular dioecious species, from a ten hectare area of mixed woodland. For each individual, the SEX was recorded (coded as 1 for male and 2 for female), the diameter at breast height in millimetres (DBH), and the number of flowers on the tree at the time of measurement (FLOWERS). This dataset will be revisited several times over the following chapters.

(1) Test the null hypothesis that male and female trees produce the same number of flowers.

(2) Show graphically how the number of flowers differs between the sexes.

Technical guidance on the analysis of these datasets is provided in the package specific supplements. Answers are presented at the end of this book.

2 Regression

2.1 What kind of data are suitable for regression?

A regression analysis looks for a linear relationship between two variables. The explanatory variable (or X variable) may be **continuous,** or an **ordinal categorical**—though it is usually continuous. The response variables (Y variables —containing the data we wish to explain) considered here will be continuous. Examples of the two types of explanatory variables are:

Continuous

Can the volume of wood in a tree (Y) be predicted from the diameter of its trunk (X)?

Ordinal categorical

Do bird species that have more parasites (Y) also have brighter plumage (X)?

In this second example, brightness of plumage is measured on a categorical scale (1 to 7—the brightest plumage being 7, the dullest plumage being 1), and the order of the categories has meaning.

These two types of analysis however cannot be treated in exactly the same way. In doing any regression analysis, the relationship between Y and X can be tested at two levels. The most basic level asks 'Is there any **association** between Y and X?' In general terms, if X increases does Y tend to increase or decrease? In other words, in these kinds of analyses we are testing a hypothesis. At a more sophisticated level, we can try and **predict** the average value of Y when given a value of X. In this case we can estimate in a quantitative manner the relationship between X and Y. With a continuous X variable, it is valid to attempt to answer both of these questions (i.e. **hypothesis testing** and **estimation**), but with an ordinal categorical explanatory variable, the second question cannot usually be usefully asked. The reason for this is that regression analysis assumes that the X variables are measured on an **interval scale:** i.e. that the intervals on the X axis have the same absolute meaning all the way along the scale. This is clearly the case for a variable such as diameter (the absolute difference between 50 cm and 55 cm is the same as between 100 and 105 cm), but not for a variable such as 'brightness of plumage' (the difference between categories 1 and 2 would be hard to quantify, let alone to compare with categories 6 and 7).

In short, in regression analysis it is legitimate to test hypotheses in both cases (interval data and ordinal categorical data), but with ordinal categorical data only the hypothesis 'slope = 0' can be tested, while other slopes can be tested with continuous *X*. Estimation is only legitimate with interval *X*.

A relationship between *X* and *Y* could take many different forms, yet a simple regression analysis fits a straight line. This is the best starting point for a number of reasons:

- It is simple to do and answers the basic question 'Is there a trend?'
- A smooth curve over a short range will approximate to a straight line.
- The theory is well developed and significant departures from linearity can be detected at a later stage.

Later chapters will discuss fitting relationships more complex than a straight line.

2.2 How is the best fit line chosen?

In the simplest regression analysis, we are trying to predict one variable using another. The first step is to plot the data (for example Fig. 2.1). Fat is estimated by a skinfold measurement in millimetres, while weight is measured in kg. The question is are heavier people also fatter? A number of lines could be drawn through such a scatter plot by eye, each of which would fit the data quite well—how is the best line chosen? The criterion used is that of **least squares:** i.e. the sum of the squared vertical deviations of the points from the line is minimised.

To illustrate this, a regression analysis can be considered in two steps. The first is to fit a grand mean, just as in an ANOVA (Fig. 2.2a). The second step is to pivot this line around the mean coordinate ($\bar{x}$, $\bar{y}$), until the vertical deviations are minimised (Fig. 2.2b).

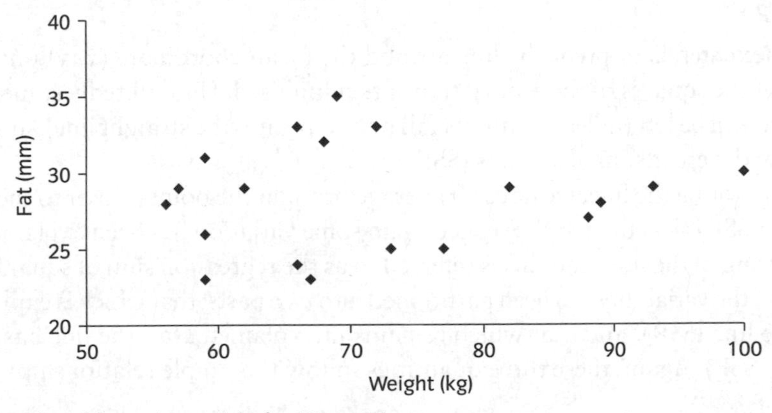

Fig. 2.1 Plot of fat against weight for 19 participants.

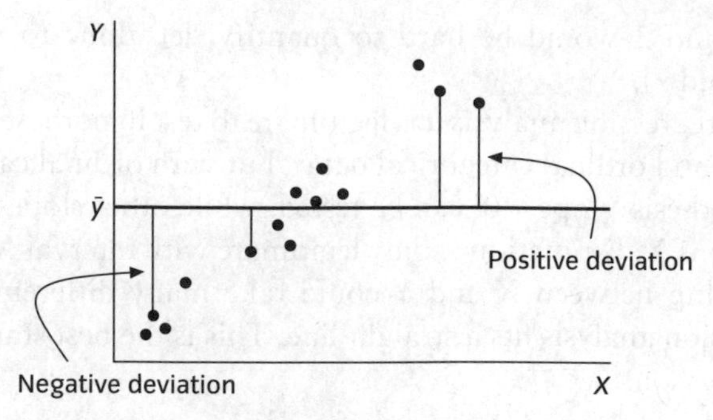

Fig. 2.2(a) Fitting the grand mean; quantifying the spread around the mean.

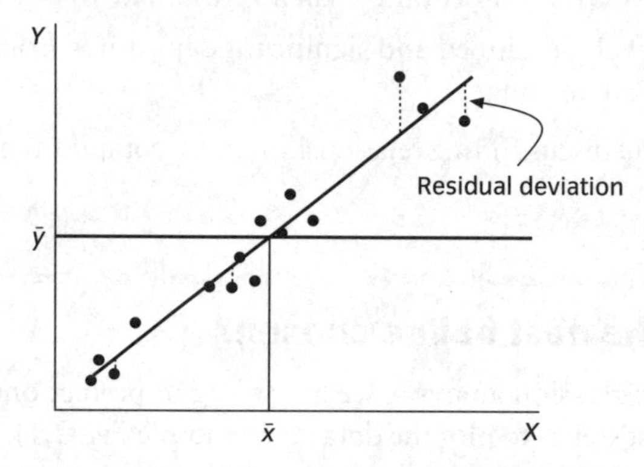

Fig. 2.2(b) Pivoting the line to minimise the sum of the squared deviations around the line.

Step 1

Fitting the grand mean of all data points will give us a line in which Y has the same value for all values of X ($Y = \bar{y}$). The deviations of all points from this line will sum to zero, and therefore once again we use the sum of the square of the deviations (SSY) to quantify the spread of the data around the mean (Fig. 2.2a).

Step 2

The next step is to pivot the line around the mean coordinate ($\bar{x}$, $\bar{y}$) until the sum of the squares of these deviations are minimised. Undoubtedly some variability will be left (unless all points fall perfectly onto the straight line), and this is now the error sum of squares (SSE).

By pivoting the line, we have on average brought the points closer to the line, so that SSE is less than SSY. So once again some variation has been explained by the fitting of the line, and this is referred to as the regression sum of squares. As before, the variability has been partitioned into two parts; that which is explained by the line (SSR), and that which remains unexplained after the line has been fitted (SSE). Again, these three quantities follow the simple relationship:

$$SSY = SSR + SSE.$$

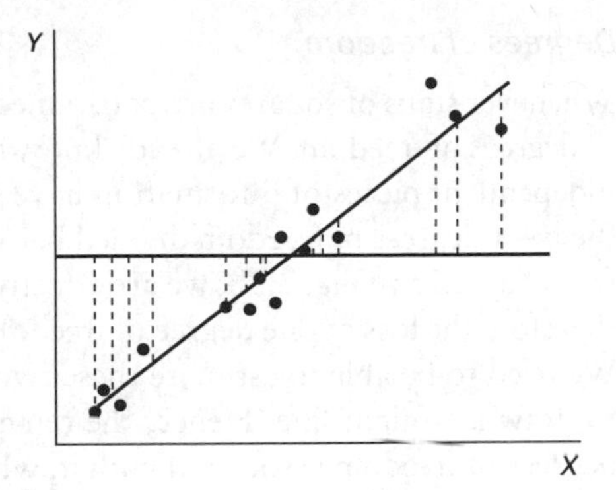

Fig. 2.2(c) SSR is the sum of the squared vertical distances between the mean and the fitted line for each datapoint.

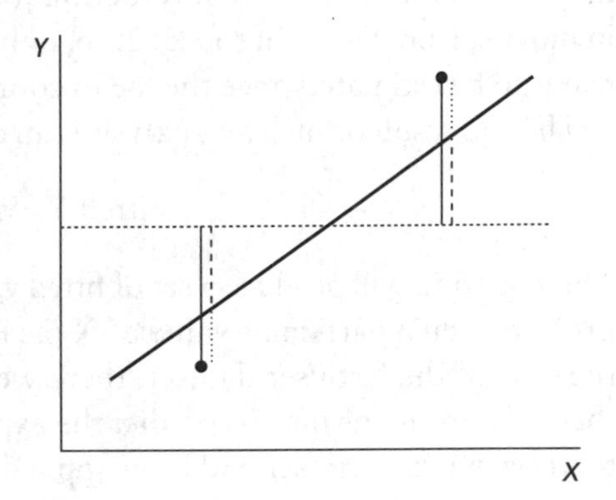

Fig. 2.2(d) The deviation of the datapoint from the grand mean is separated into two components.

The sum of squares explained by the regression, SSR, may be thought of as the sum of the squared vertical distances between the fitted line and the horizontal line through $\bar{y}$ for each data point (as illustrated in Fig. 2.2c).

To summarise the analysis we have just discussed, Fig. 2.2d illustrates how the deviations are related for two points. The solid line represents the deviation of the datapoint from the grand mean, with the long dash representing the deviation of the fitted line from the grand mean, and the short dash the deviation of the point from the fitted line. It can be seen that the solid line is composed of the sum of the long dash and the short dash.

Regression coefficients

A straight line can be expressed in general terms as:

$$Y = \alpha + \beta X$$

in which α is the intercept and β is the slope. The Greek letters are used to represent the 'true' relationship—we can only obtain estimates of this from our dataset. Hence the line from any one dataset is usually written as:

$$Y = a + bX.$$

where a and b are called the regression coefficients and are our best estimates of α and β.

Degrees of freedom

Whenever sums of squares are partitioned, there must be a parallel partitioning of degrees of freedom. We already know that with a dataset of size n, then $n-1$ independent pieces of information have been used to calculate SSY. How are the $n-1$ degrees of freedom divided between SSR and SSE? When progressing from Fig. 2.2a to Fig. 2.2b, we are effectively exchanging one grand mean (and therefore the loss of one degree of freedom) for a slope (b) and an intercept (a). We need to be able to estimate these two quantities from the data, to be able to draw a straight line. Hence, the remaining variation (SSE) now has $n-2$ degrees of freedom associated with it, whilst SSR has only 1. This latter figure may be obtained either by subtraction $[(n-1)-(n-2)]$, or by reasoning that in moving from Fig. 2.2a to Fig. 2.2b, only one additional parameter (the slope) needs to be estimated once the mean coordinate $(\bar{x}, \bar{y})$ is known.

The end result of such an analysis is an equation linking Y and X of the form:

$$\text{Fitted } Y = a + bX.$$

This equation will produce a set of **fitted values**, that is, the value of Y we would predict given a particular value of X (as if there were no error). Previously, in the case of the fertiliser dataset, there were only three distinct fitted values—the three group means. To predict the expected yield of a plot, we only needed to know which fertiliser had been applied (i.e. X category 1, 2 or 3). In the case of regression, X is continuous, so a precise value of X needs to be substituted into the equation to produce a predicted value of Y. Consequently, we could **interpolate**, and make predictions about the expected value of Y for intermediate values of X which have not actually been included in our dataset. This sounds very useful. However, we should be very cautious about **extrapolating**, and attempting to predict Y for values of X beyond our dataset—we have no evidence that the relationship continues to be linear. (We have not actually examined the dataset to see if we are justified in assuming linearity even within our dataset—but that will be done in a later chapter).

2.3 The geometrical view of regression

A regression analysis can also be represented as a geometrical picture—in fact, exactly as before with the three points, M, F and Y, forming a right-angled triangle. Y represents the data, that is, for a dataset with n points, the set of coordinates $(y_1, y_2, y_3, \ldots y_{n-1}, y_n)$. M now represents the point with coordinates $(\bar{y}, \bar{y}, \bar{y}, \ldots \bar{y})$, that is, the line drawn in Fig. 2.2a. F represents the values that have been fitted to the data: i.e. $(a + bx_1, a + bx_2, a + bx_3, \ldots a + bx_n)$. We can actually make a link between the scatter plot for which the best fit line was chosen, and the geometrical analogy.

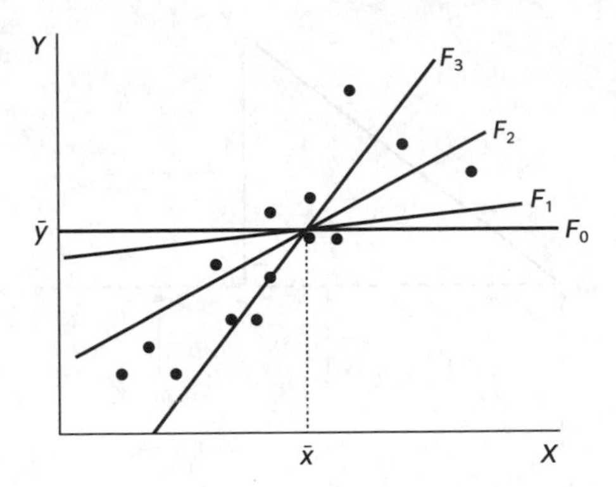

Fig. 2.3 Three of an infinite number of lines that could describe the relationship between X and Y.

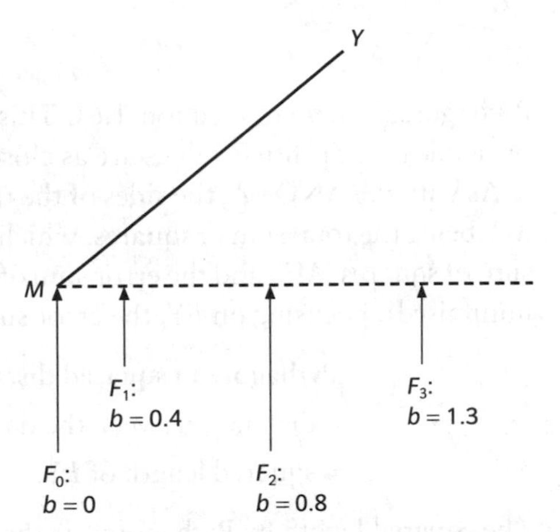

Fig. 2.4 Different lines fitted to the data represented as points in the geometrical analogy.

The best fit line must pass through $(\bar{x}, \bar{y})$ by definition. However, a number of lines do that, each of which have a different slope (b) and each of which will produce a different set of fitted values (e.g. F_1, F_2, F_3), as illustrated in Fig. 2.3.

Each set of fitted values may be represented by a point on a line that passes through M (Fig. 2.4). In fact, M itself may be considered as a fourth possible line, with the slope equaling zero. The dotted line in Fig. 2.4 is drawn as horizontal, but this is just a consequence of representing on a plane of paper, what is happening in multidimensional space.

Each of the points F_1–F_3 lies at a different distance from Y, the data. Now the distance of F_1 from Y is numerically equal to a very important quantity —namely the sum of squares of deviations between the data and the line corresponding to F_1, and we shall see this formally in a moment. But the importance now is that the distance between F_1 and Y is precisely the quantity that the criterion of least squares says we should be minimising. Thus the best fitting line must correspond to the point on the dotted line that is closest to Y. In geometrical terms this is equivalent to making the side, FY, of our triangle, as short as possible—achieved by dropping a perpendicular from Y to the line passing through M (Fig. 2.5). This ensures that MFY is a right angle—a property of this triangle that we have already made use of in applying

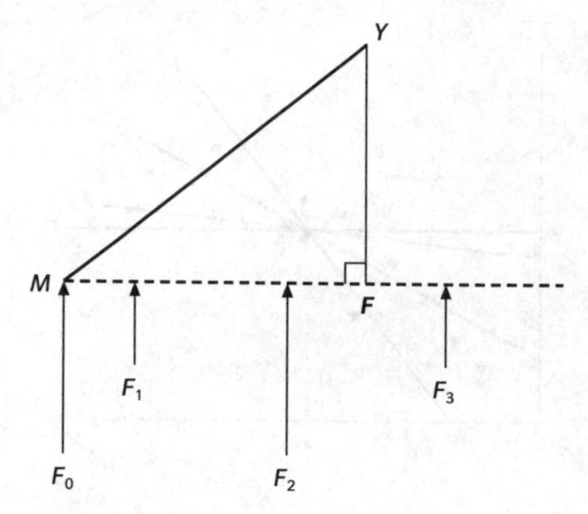

Fig. 2.5 The chosen *F* minimises the length *FY*.

Pythagoras' theorem (Section 1.6). This is a clear illustration of the general principle that the fitted values are as close as possible to the data.

As with the ANOVA, the sides of the triangle represent the sums of squares: *MY* being the total sum of squares, which is partitioned between the regression sum of squares (*MF*) and the error sum of squares (*FY*—the length we have just minimised). Focusing on *FY*, the error sum of squares

Pythagorean squared distance
= distance between the data and the fitted values
= squared length of *FY*.

The squared length by Pythagoras, is the sum of the squares of the differences in the coordinates.

For each data point, its value is represented as y_i, and its fitted value as $a + bx_i$. Substituting these into the previous equation:

$$= \Sigma_i (y_i - (a + bx_i))^2$$

This is the sum of the squared vertical discrepancies, so providing another link between the scatter plot of Fig. 2.2b and the *MFY* diagram of Fig. 2.5.

2.4 Regression—an example

Having looked at the principles behind regression, this section presents an example. A forester wished to be able to estimate the volume of merchantable timber in a tree from measuring its height. To establish this relationship, she felled 31 trees and measured both volume and height for this subsample. The dataset *trees* comprises three continuous variables, but in this example we will use just one explanatory variable to estimate the other. HEIGHT is the chosen explanatory variable, and VOLUME the variable we wish to predict.

Input

As always, it is important to correctly identify which is the response and which the explanatory variable. In this case, we wish to explain/predict `VOLUME` using `HEIGHT`. Other variables, such as `DIAMETER` of the tree, may also be a predictor of `VOLUME`, but will be ignored in this example.

Output

The details of the output will depend upon the statistical package (see supplements). The essencc of the analysis however can be summarised in two tables. The central part is common to both ANOVA and regression—the analysis of variance table. This will answer the key question 'Has more variance been explained by this line than would be expected by chance alone?' The fitted values are also often presented in the form of a table, the coefficient table, which answers the question 'What is our best estimate of the slope, and how sure are we of it?' Below, we will examine these two questions in more detail. Finally, the third component of the output is R^2, a measure of the proportion of variance explained.

Analysis of variance table

As with the analysis of variance in Chapter 2, this table compares the variation (per degree of freedom) explained by fitting the line with the variation that has been left unexplained, by constructing an *F*-ratio (details of how this is calculated are given in Section 1.4). In this case the *F*-ratio is 16.16—very much greater than 1. This leads us to conclude that the explained variation is significantly greater than zero (with a probability of less than 0.0005 of obtaining such a high *F*-ratio by chance alone). The answer to the first question is that fitting the line has accounted for a significant amount of variance—these two variables are related.

Coefficients table

The second table in Box 2.1 illustrates how the two variables are related, by providing the two coefficients which describe the line—namely the slope and intercept. This allows us to construct the fitted line, in this case given by:

$$\texttt{VOLUME} = -87.1 + 1.54\ \texttt{HEIGHT}.$$

This table however performs the additional function of testing whether the slope is significantly different from zero. A zero slope is equivalent to a horizontal line, for which X can provide no predictive information about Y. If the slope is significantly different from zero (either greater or smaller) then X can provide us with information about Y. Clearly this is a crucial question—so how is this test carried out?

The second column of that table (SECoef) gives the standard error of the coefficient—it represents the degree of uncertainty we have about that estimate. Where does this value come from? When we estimated a sample mean, the

BOX 2.1 Analysis of variance and fitted value tables for the *trees* dataset

Regression analysis

Word equation: VOLUME = HEIGHT

HEIGHT is continuous

Analysis of variance table for VOLUME

Source	DF	SS	MS	F	P
Regression	1	2901.2	2901.2	16.16	0.000
Error	29	5204.9	179.5		
Total	30	8106.1			

Giving an R^2 of 35.8%.

Coefficients table

Predictor	Coef	SECoef	T	P
Constant	−87.12	29.27	−2.98	0.006
HEIGHT	1.5433	0.3839	4.02	0.000

Giving a fitted line of VOLUME = −87.1 + 1.543 HEIGHT.

standard error was defined to be $s/\sqrt{n}$. In this case, we have estimated the slope to be 1.5433, and the standard deviation (or standard error) of that estimate to be 0.3839. (The formula for the standard error of a slope is different from the standard error of a mean, but we will not derive the exact expression for it here.) Our estimate is one of many possible estimates we could have made—if an alternative 31 trees had been chosen from the forest, undoubtedly the estimate would have been different. Our dataset is therefore just a sample of a whole population of trees that could have been measured, and hence there is a degree of uncertainty about the fitted equation. If the 'true' value of the slope is β, 1.54 is one of many possible 'guesses' at β, and the unexplained variability in the data contributes towards telling us how good that guess is likely to be. (The size of the dataset and the range of values of X also contribute to the accuracy of our guess). We can use the standard error of the slope (0.3839) in the same way, and for the same purposes, as we used the standard error of the mean.

So, to answer the question 'Is the slope significantly different from zero?', the size of the slope and the accuracy of the estimation need to be taken into account. Imagine collecting many datasets which by chance contained 31 trees of the same height as our original 31 trees, but of different volumes. These differences in measured volume would be due to measurement error, and to the range of volumes that truly exist for trees of the same height. Both of these sources of variation will cause the value of the regression slope to vary. As long

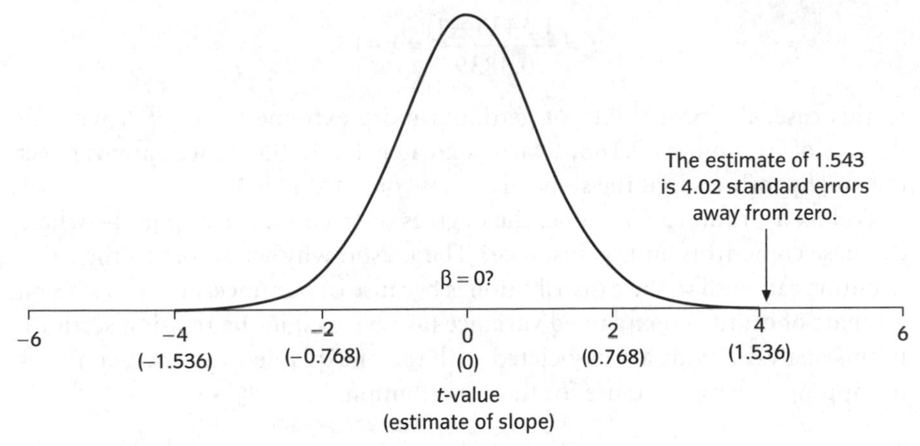

Fig. 2.6 The hypothesised distribution (with mean zero) of all possible estimates of the slope. The X axis is expressed in two units: numbers of standard errors which is equivalent to a t-value, and, in brackets, the estimate of the slope.

as our sampling was unbiased, we would hope that the average of all such estimates was the true value, β—and that all our different attempts at β are roughly normally distributed around this mean (as in Fig. 2.6). We now hypothesise that β is zero. This is the standard null hypothesis of 'no relationship'. It is usual to assume no relationship until evidence suggests that this is unlikely. If the null hypothesis is true, then our estimate of 1.54 will come from the distribution in Fig. 2.6. What is the evidence that this is the case? If the estimates follow a Normal or a t distribution, then this question is answered by looking at the number of standard errors between the hypothesised mean of zero, and the estimate 1.54—the answer to this being $1.5433/0.3839 = 4.02$. It is very unlikely that a value sampled from the distribution in Fig. 2.6 would be as far away from zero as 4.2 (in fact the probability is less than 0.0005); our estimate is clearly in the tail of this hypothesised distribution. We can therefore reject the null hypothesis that $\beta = 0$ and conclude that β is positive.

The two key questions are related—if the F-ratio of the ANOVA table is significant, then so will be the t-ratio of the coefficients table, and vice versa. (In fact, the F-ratio is the square of the t-value, and will always be so if that term has 1 df in the ANOVA table.) It is not possible to have a slope significantly different from zero, without explaining a significant amount of variance in the first place. A second point to remember is that the information given in the coefficients table can be used to test other hypotheses. We have tested the hypothesis that the true slope is zero, because that is the question most frequently of interest (i.e. do changes in X tell us anything about Y). However, other possible hypotheses include—does the relationship between `VOLUME` and `HEIGHT` have a slope significantly different from 1? This question is answered in exactly the same way—how many standard errors are there between 1.5433 and 1? This calculation would be as follows:

$$t_s = \frac{1.5433 - 1}{0.3839} = 1.415.$$

In this case, the probability of getting a more extreme value of t, with 29 degrees of freedom, is 0.168, which is greater than 0.05, so we cannot reject the null hypothesis that the slope of VOLUME on HEIGHT is 1.

To link a t-ratio to a p-value, the degrees of freedom are required—where do these come from in this instance? The reason why we resort to the t distribution rather than the z distribution is because of our uncertainty in s^2 as an estimate of σ^2, the unexplained variance (as reviewed in the revision section). In this analysis, 29 df are associated with the unexplained variance, so this is the appropriate figure to use for the t distribution.

R-*squared*

The R^2 statistic is frequently quoted for regression analyses. This is the fraction (or percentage) of variability (as indicated by Sums of Squares) explained by the line you have fitted, and as such could be calculated directly from the analysis of variance table. It could be viewed as a measure of the tightness or sloppiness of a relationship. The fraction of variability explained is calculated as:

$$R^2 = \frac{\text{Regression Sum of Squares}}{\text{Total Sum of Squares}} = \frac{\text{SSY} - \text{SSE}}{\text{SSY}}$$

$$= \frac{2901.2}{8106.1} = 0.358.$$

So 35.8% of the variability in the dataset has been explained by fitting the line. Not a very tight relationship (as might be guessed by looking at Fig. 2.7). Perhaps other variables are involved in determining the volume of wood in a tree.

These three pieces of the output form the kernel of the results. With this information, there is also the option of presenting the output in the form of confidence intervals and prediction intervals.

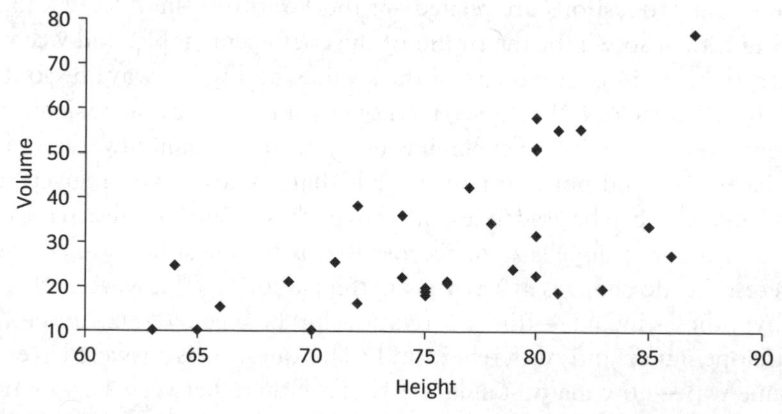

Fig. 2.7 Plot of VOLUME against HEIGHT for 31 trees.

2.5 Confidence and prediction intervals

Confidence intervals

The estimates and their standard errors allowed us to perform a *t*-test in Section 2.4, and in the same way we can construct a confidence interval for the slope and the intercept. In the revision section (Section R1.3) a general expression for the confidence interval for a parameter is derived as:

$$\text{estimate} \pm t_{\text{crit}} \times \text{standard error of the estimate.}$$

The degrees of freedom for the critical *t*-value are always the error degrees of freedom (as these represent the number of independent pieces of information we have about the error variance, s^2). For this example, and a 95% level of confidence, the interval becomes:

$$1.5433 \pm t_{\text{crit}} \times 0.3839 \rightarrow (0.758, 2.328)$$

where $t_{\text{crit}} = 2.0452$ for 29 df. This means that we are 95% confident that the true value of the slope lies between these two values.

We may however also wish to place confidence limits on the expected Y for a specific value of X, and this is rather more complex, as shown below.

Prediction intervals

One of the aims in conducting a regression is to predict a value of Y from a given value of X using the best fit equation. There are two factors which contribute to the uncertainty of such a prediction.

(1) Scatter around the line. If we initially assume that the best fit line is indeed the true line, then the error variation around the line (as measured by the error mean square) will cause some uncertainty in Y. If the expected value of y from the best fit equation is denoted $\hat{y}$, then this uncertainty can be represented as a confidence interval for values of y around the line just as with the slope above. So:

$$\text{CI} = \hat{y} \pm t_{\text{crit}} s$$

where s is the square root of the error mean square. This is therefore a confidence interval for values of y at a given value of x, if we assume that the line is correct.

(2) The second source of uncertainty is that the line fitted is not the exact true line. There will be some discrepancy between the fitted relationship and the true relationship. The accuracy with which the slope and intercept have

been estimated will influence this. Taking both sources of uncertainty into account produces a prediction interval, the formula for which is:

$$\mathrm{PI} = \hat{y} \pm t_{\mathrm{crit}} s \sqrt{\frac{1}{m} + \frac{1}{n} + \frac{(x' - \bar{x})^2}{\mathrm{SSX}}}$$

where

$$\mathrm{SSX} = \Sigma(x - \bar{x})^2.$$

This is actually a prediction interval for the mean of m values of y when $x = x'$. Whilst this formula looks complex, the principle behind it is that each source of uncertainty adds one term to the equation. The first term, $1/m$, represents the scatter around the true line—and will equal 1 when only one value of y is being predicted. With only this first term included (and $m = 1$), the prediction interval is the same as the confidence interval above. The next term, $1/n$, represents uncertainty in the intercept—the greater the number of data points, the lower this uncertainty. This term, as well as the one before, is unaffected by where, on the x-axis, we are attempting to predict our value of y. The last term represents uncertainty in the slope. If we are attempting to predict y at $\bar{x}$, then this term does not contribute to the uncertainty of our prediction. The reason for this is that all possible fitted lines would pass through the mean coordinate $(\bar{x}, \bar{y})$. However, as we move further away from $\bar{x}$ along the x axis, the greater is the numerator of this term. This is because a small change in the slope would result in relatively small changes in our predicted value of y close to $\bar{x}$, but much larger changes to $\hat{y}$ further away. The denominator of this term (SSX) quantifies the range of x values in the dataset. Clearly, if all x values are clustered close to $\bar{x}$, this will lead to a much less accurate estimate of the slope than if the x values are more widely spread along the X axis. Simply increasing the size of the dataset will also decrease uncertainty in the slope, as the size of SSX will increase.

One of the aims of regression analysis is to use the resulting equation to predict values of Y. Frequently, we will not only be interested in point estimates, but also in placing some form of confidence limits on these estimates. This discussion of prediction intervals illustrates a number of important points:

- There are a number of sources of uncertainty involved when trying to predict y: unexplained variance around the line, uncertainty in the intercept and uncertainty in the slope.
- Uncertainty in the intercept can be reduced by increasing the number of data points in the dataset.
- Uncertainty in the slope may be reduced by increasing the number of data-points, but also by collecting the data in such a way as to maximise the range of X values.
- Predicting y will always be more accurate closer to $\bar{x}$ than further away.

In summary, a 95% prediction interval around the line will take the form illustrated in Fig. 2.8.

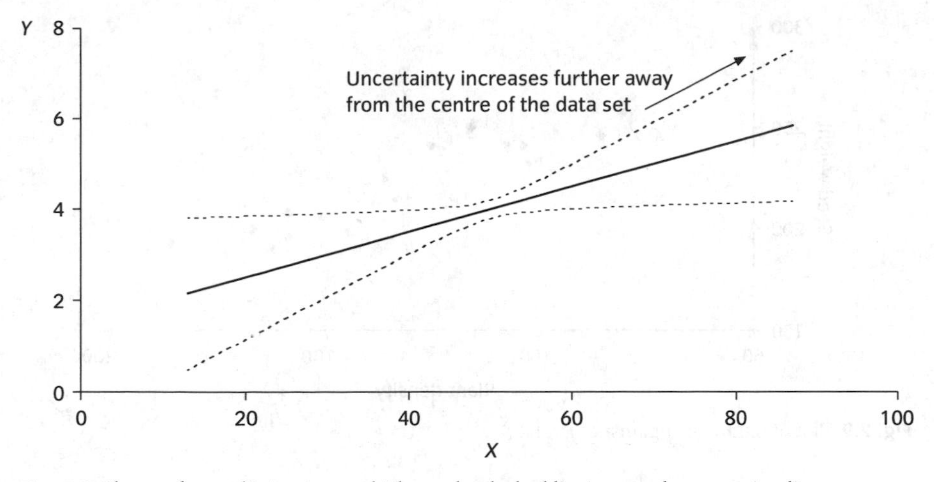

Fig. 2.8 Shape of a prediction interval (shown by dashed line) around a regression line.

2.6 Conclusions from a regression analysis

Two quite separate pieces of information result from a regression analysis—one related to the t and F-ratios, and one to the R^2 value. High t and F-ratios lead us to reject the null hypothesis of 'no relationship', and to conclude that these two variables are related. High R^2 values, on the other hand, indicate that a large proportion of the variability has been explained. It also provides some indication about whether other variables may be involved in determining the value of our Y variable. A sloppy relationship would suggest that whilst the explanatory variable in question does influence the value of Y, much is left unexplained, and other variables may also be important. These two features of the dataset do not necessarily go hand in hand. In practice, it is possible to have any one of the four possible combinations of low or high t and F-ratios, and low or high R^2 values, as is illustrated by the following four examples.

A strong relationship with little scatter

If some plants are grown at high densities, they respond by producing smaller seeds. In this first dataset, *seeds*, average seed weight (SEEDWGHT) was recorded for a random selection of plants found growing at different densities (PLANDEN). Plotting seed weight against plant density illustrates a strong negative trend (Fig. 2.9).

On performing a regression analysis, both a low p-value and a high R^2 value are obtained (Box 2.2). The conclusions drawn are that PLANDEN appears to influence average SEEDWGHT ($F_{1,18} = 11.45, p < 0.0005$). The high R^2 value indicates that we have a tight relationship—in the population of plants sampled here PLANDEN appears to be the main determinant of SEEDWGHT. The remaining

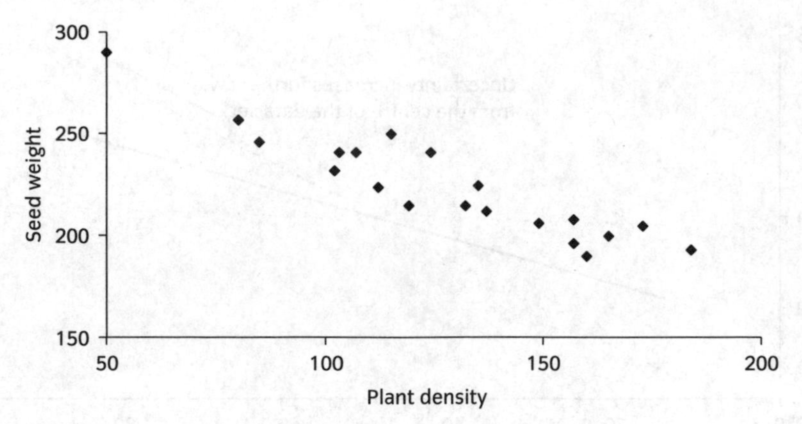

Fig. 2.9 Plot of SEEDWGHT against PLANDEN.

BOX 2.2 Regression analysis of SEEDWGHT against PLANDEN

Regression analysis

Word equation: SEEDWGHT = PLANDEN

PLANDEN is continuous

Analysis of variance table for SEEDWGHT

Source	DF	SS	MS	F	P
Regression	1	10 554	10 554	111.45	0.000
Error	18	1 705	95		
Total	19	12 259			

Giving an R^2 of 86.1%

Coefficients table

Predictor	Coef	SECoef	T	P
Constant	311.898	8.574	36.38	0.000
PLANDEN	−0.68773	0.06515	−10.56	0.000

Giving a fitted line of SEEDWGHT = 311.9 – 0.688 PLANDEN.

scatter indicates that while there may be other variables (for example, plant genotype), PLANDEN is extremely influential.

A weak relationship with lots of noise

A group of ten undergraduates sat two tests to measure their mathematical and their literary abilities. These scores, recorded as MATHS and ESSAYS were then regressed against each other (see Fig. 2.10; data to be found in the dataset *scores*; Box 2.3 for regression summary).

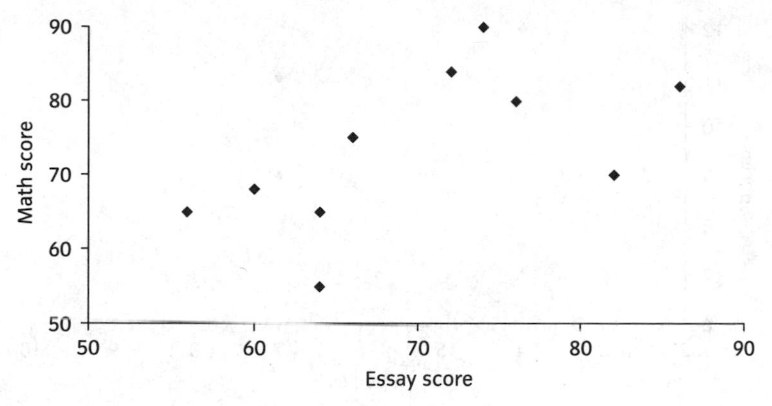

Fig. 2.10 MATHS plotted against ESSAYS for ten undergraduates.

BOX 2.3 Regression analysis of MATHS and ESSAYS

Regression analysis

Word equation: MATHS = ESSAYS

ESSAYS is continuous

Analysis of variance table for MATHS

Source	DF	SS	MS	F	P
Regression	1	360.12	360.12	4.31	0.072
Error	8	668.28	83.54		
Total	9	1028.40			

giving an R^2 of 35.0%

Coefficients table

Predictor	Coef	SECoef	T	P
Constant	27.57	22.26	1.24	0.251
ESSAYS	0.6548	0.3154	2.08	0.072

Giving a fitted line of MATHS = 27.6 + 0.655 ESSAYS.

Whilst the plot suggests there may be some trend, there is not much evidence to suggest this is significant ($F_{1,8} = 4.31$, $p = 0.072$), and only 35% of all variance has been explained. A low R^2 with low *t* and *F*-ratios indicates that the evidence of a relationship is poor, and that very little variability is explained by fitting the line. It is perfectly possible however to find that your results present you with one of the other two combinations.

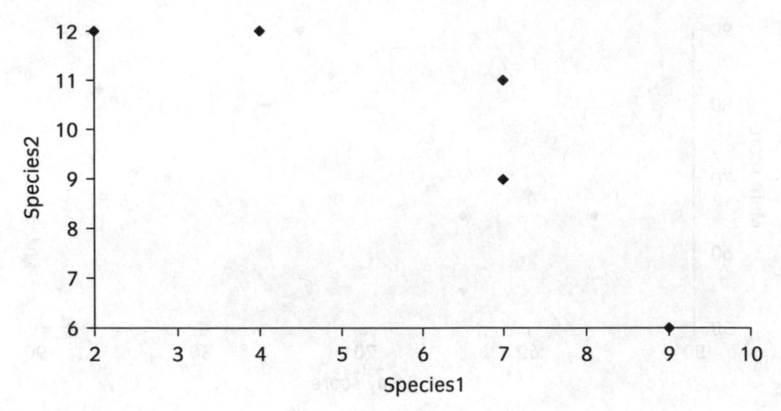

Fig. 2.11 Abundance of SPECIES1 and SPECIES2.

BOX 2.4 Regression analysis of SPECIES1 against SPECIES2

Regression analysis

Word equation: SPECIES2 = SPECIES1
SPECIES1 is continuous

Analysis of variance table for SPECIES2

Source	DF	SS	MS	F	P
Regression	1	18.701	18.701	7.69	0.069
Error	3	7.299	2.433		
Total	4	26.000			

giving an R^2 of 71.9%

Coefficients table

Predictor	Coef	SECoef	T	P
Constant	14.519	1.773	8.19	0.004
SPECIES1	−0.7792	0.2811	−2.77	0.069

Giving a fitted line of SPECIES2 = 14.5 − 0.779 SPECIES1.

Small datasets and pet theories

During the course of your PhD, you develop a theory that the population densities of two species of rodent in British woodlands are inversely correlated with each other. In your final field season, you collect data for the two species from five woodlands, store them in the *rodent* dataset and regress SPECIES1 against SPECIES2 (see Fig. 2.11). The results (see Box 2.4) are tantalising—72% of the variance is explained, but you fail to achieve significance.

This problem is particularly likely to occur with small datasets—a high proportion of variance in the dataset has been explained, but owing to the low sample size there is insufficient evidence to draw firm conclusions about relationships in the population. Given the small dataset, a *p*-value approaching significance may be worth investigation: e.g. gathering more data or following up with a manipulative experiment. However, as this analysis stands, we could not reject the null hypothesis of no relationship in the distribution of the two species. In this case it should also be pointed out that the plot suggests that if there is any relationship, it may be nonlinear in nature. In later chapters we will discuss how to follow up such a suggestion.

Significant relationships—but that is not the whole story

The earlier example of this chapter, predicting the `VOLUME` of a tree from its `HEIGHT`, is a good illustration of a significant relationship which explains a small proportion of the total variance (Box 2.1). Figure 2.12 shows the trend to be clearly present (taller trees tend to have a greater volume of wood), but noisy. It is easy to see why this might be—trees of the same height may be of very different growth forms (spindly or stubby) resulting in very different volumes. In other words, height is significant, but is not the only important variable in predicting volume; so illustrating the distinction between significance and importance. Using a legal analogy—the strength of the evidence and the seriousness of the crime are not necessarily connected.

An additional variable, such as diameter, would greatly improve our ability to calculate volume. In Chapter 4 we do exactly this—make use of two explanatory variables to predict our response variable.

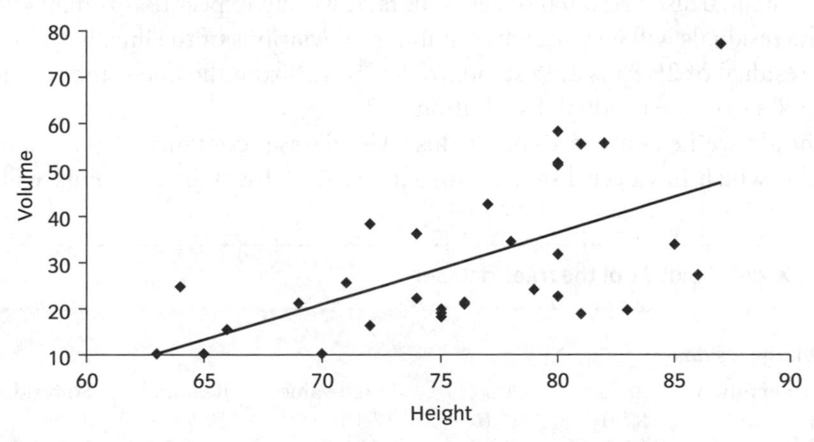

Fig. 2.12 A clear trend, but with considerable scatter of points around the line.

2.7 Unusual observations

In regression analysis, the data points may not always lie close to the fitted line. Points that are a considerable distance from the line may be indicative of errors or a poor fit. Points a long way from the mean of X may be highly influential in determining the fitted model. It is useful to be able to detect outliers of either type when they occur.

Large residuals

In the case of the *trees* dataset, datapoint number 31 has a large standardised residual (see Box 2.5 below). What exactly is meant by this, and should it concern us?

For a tree of height 87 feet, the fitted line would predict a volume of 47.15 cubic feet (the fitted value for that value of *X*). This comes from substituting 87 into the equation:

$$\text{VOLUME} = -87.12 + 1.5433\ \text{HEIGHT}$$
$$\text{VOLUME} = -87.12 + 1.5433 \times 87$$
$$= 47.15.$$

However, the observed value is 77—this datapoint lies well above the fitted line. The difference between the observed and predicted values is the residual, in this case 77 – 47.15 = 29.85. There will be a residual associated with each datapoint, and the absolute values of these residuals will depend upon the scale with which *Y* is measured. Consequently, to decide if a point lies *too* far from the line, all residuals are **standardised** (by dividing each residual by its own standard deviation, roughly the square root of the error mean square). Now, if a standardised residual is greater than 2 or less than minus 2, it lies some distance from the line. (A point exactly on the line will have a raw residual and a standardised residual of zero). In fact, we can expect that roughly 95% of the residuals will lie within two standard deviations of the line. In this case, our residual of 29.85 is 2.39 standard deviations from the line—and so it is in the 5% tail of the residual distribution.

Should we be concerned about this? The dataset contains 31 points, only one of which has been listed as an outlier. Roughly 1 in 20 points will lie

BOX 2.5 Point 31 of the *trees* dataset

A large residual

Observation	HEIGHT	VOLUME	Fitted value	Residual	St. resid
31	87.0	77.0	47.15	29.85	2.39

greater than two standard deviations from the mean. (This is exactly what we mean by a probability of 0.05.) Perhaps, then, we should not be too alarmed at this.

Influential points

In the *seeds* dataset (see Box 2.2), another type of unusual observation occurs. Datapoint number 8 (plant density 50, seed weight 90) is an outlier, even though its standardised residual is not greater than 2 or less than –2, because it is influential. This has a very specific meaning in statistics. Looking at the plot of SEEDWGHT against PLANDEN (Fig. 2.9), this point lies at the extreme left. Such points at the edges of the range of the X values will greatly influence the final fitted line—and the extent of this influence may be quantified by $(x' - \bar{x})^2$ (see the discussion on prediction intervals in Section 2.5). Again it is the standardised deviations of each x value from $\bar{x}$ on which to base a judgement of whether a data point is likely to be influential, and for this dataset it is only data point number 8 for which this value is greater than 2 or less than –2. So the important question then becomes whether this point has unduly influenced the fitted line.

The simplest way to investigate this question is to repeat the analysis omitting that one point. This produces the ANOVA table given in Box 2.6.

BOX 2.6 Repeated analysis omitting the influential point

Regression analysis

(The *seeds* dataset with point 8 omitted)

Word equation: SEEDWGHT = PLANDEN

PLANDEN is continuous

Analysis of variance table for SEEDWGHT

Source	DF	SS	MS	F	P
Regression	1	6245.8	6245.8	71.94	0.000
Error	17	1476.0	86.8		
Total	18	7721.8			

giving an R^2 of 80.9%

Coefficients table

Predictor	Coef	SECoef	T	P
Constant	302.910	9.903	30.59	0.000
PLANDEN	–0.62431	0.07361	–8.48	0.000

Giving a fitted line of SEEDWGHT = 302.9 – 0.624 PLANDEN.

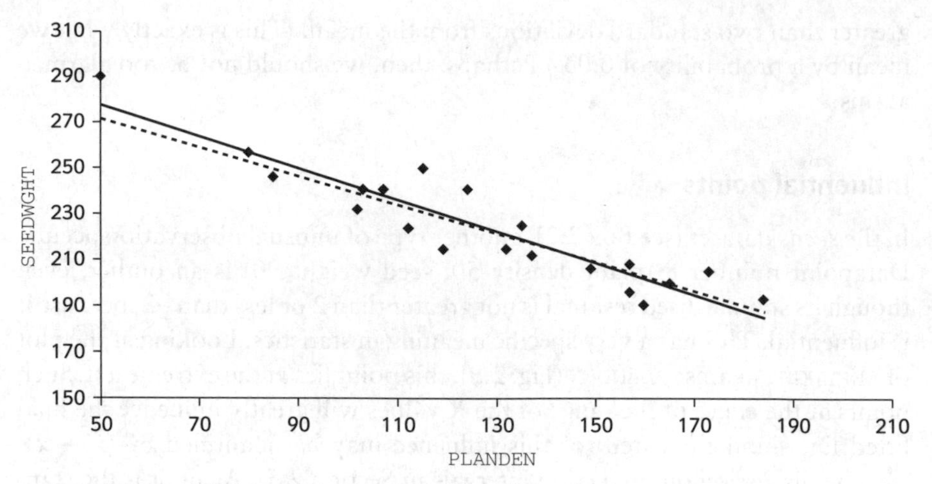

Fig. 2.13 The fitted line for the *seeds* dataset with (solid line) and without (dashed line) point 8.

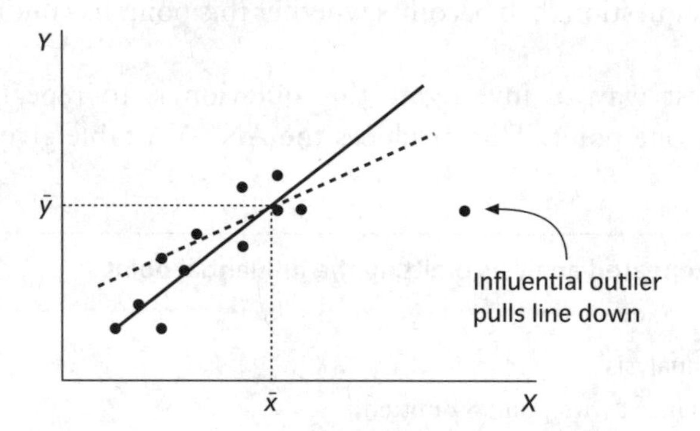

Fig. 2.14 Influential outliers may greatly influence the fitted line.

Comparing this with Box 2.2, the fitted value equation is not fundamentally different, nor are any of the conclusions altered. Figure 2.13 illustrates how the fitted line has tilted slightly in response to the omission of one point, with the solid line being the fitted values from the full dataset and the dashed line being fitted values for the reduced dataset. Do these changes matter? In later chapters we will discuss how to compare models to decide whether changes are statistically significant. If it transpires that one point quite fundamentally alters the model (and this would be more likely if the point was both an outlier and influential—see Fig. 2.14), you would be left with the quandary of how to present your results. It would not be honest to omit one point simply because it was influential, or an outlier (or both). In some cases however it may be informative to present both models, to illustrate how much difference one point has made.

2.8 The role of X and Y—does it matter which is which?

Regression involves looking at the relationship between two continuous variables—does it make any difference which variable we chose as X and which as Y? The method by which the best fit line is chosen means that in fact it does make a difference—if X and Y are interchanged, then a different best fit line results.

This can be illustrated by returning to the *trees* dataset, and doing exactly that. In Box 2.1, the regression analysis gives a best fit line of

$$\texttt{VOLUME} = -87.1 + 1.54\,\texttt{HEIGHT}.$$

This is a result of the vertical deviations being minimised. However, if volume is made the explanatory variable, and height the response variable, then the best fit equation is

$$\texttt{HEIGHT} = 69.0 + 0.232\,\texttt{VOLUME}.$$

If the results are plotted with the axes the same way round, in the second instance the fitted line is much steeper (Fig. 2.15), because in this case it is the horizontal deviations that have been minimised.

A **Model I regression** is one in which the vertical deviations have been minimised, and this is the only method that has been discussed in this chapter. We can justify our choice of method being Model I by one of two possible stances. The first of these concerns where we believe measurement error to lie. The rationale behind minimising deviations is to minimise error (so that the maximum amount of variation is explained by the line). If it is assumed that the heights are measured exactly, (and that the relationship between the two variables is linear), then the error will be reduced by minimising the vertical deviations. In other words, it is postulated that the only reason why the points do not lie exactly on a straight line is because of measurement error in volume. If however it is volume that is measured exactly, then the horizontal deviations should be minimised (in other words the axes should be swapped and a model I regression carried out). So if one variable is measured exactly, and all

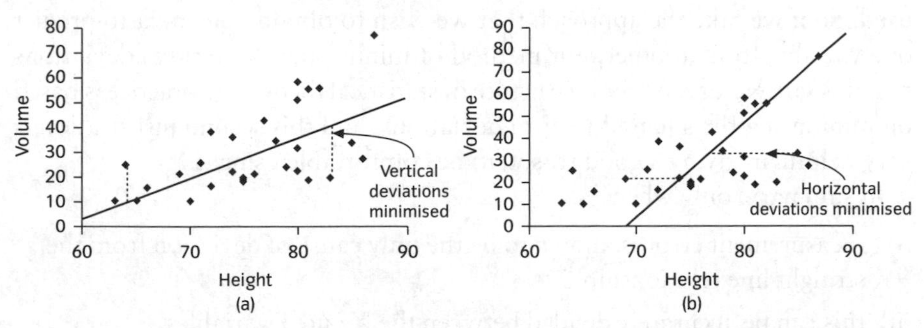

Fig. 2.15 (a) Best fit line with vertical deviations minimised; (b) Best fit line with horizontal deviations being minimised.

measurement error is in the other, then the exact variable is the explanatory one. Let us first of all consider concerns with this first stance, before considering the alternative justification.

Quite clearly in this example there is likely to be error in both height and volume measurements—does this invalidate our Model I regression? A second concern when measurement error is known to be present in both *X* and *Y*, is which of the two variables should be chosen as *X*? In such situations, sometimes a **Model II Regression** is performed. What is this and is it worth considering? Model II regression attempts to cope with those relationships in which there is clearly error in both axes by apportioning the error between the axes. This results in a best fit line being chosen which lies between the two extremes shown in Fig. 2.15. Whilst this sounds attractive, there are a number of problems with this:

(1) *The proportion of error that should be allocated to* X *and* Y *needs to be known.* There would need to be independent data that could provide some measure of this—otherwise we would be reduced to guessing.

(2) *It assumes that all deviations from the straight line relationship are due to measurement error.* In biology this would be extremely unlikely to be true. All sorts of other factors are likely to enter into play (for example, genetic and environmental differences between the trees would result in a variety of growth morphs). The omission of such relevant variables in the analysis would also cause deviation from the line, and it would be impossible to be sure that all such variables had been considered, let alone measured for each tree. The errors due to these omitted variables end up being treated as if they were measurement error in a Model II regression, and so would be divided between the two axes in some proportions, which is inappropriate when these omitted variables are not specially connected to one included variable or the other.

What then is the solution to this dilemma? Here we return to our second justification. Model I regression can also be justified by adopting the view that the fitted line is describing the relationship between '*X* as measured' and an 'expected value of *Y*'. This is not dismissing the measurement errors in *X*, but is stating that the resulting relationship is conditional on the set of *X* values being used. So if we take the approach that we wish to obtain a formula to predict one variable from another, our method of minimising the vertical deviations remains satisfactory. A whole structure of statistical theory and practice is based on minimising the squared vertical deviations, and this second justification is very helpful in giving a good reason to be comfortable using it.

In summary, only when

(i) measurement error is known to be the only cause of deviation from the straight line relationship

(ii) this can be accurately divided between the *X* and *Y* variables,

is it appropriate to perform a Model II regression.

The details will not be discussed here—but are explained more fully in 'Biometry' by R. R. Sokal and F. J. Rohlf (published by Freeman). If the first point is true, but the division between the axes is unknown, then two Model I regressions can be performed (as in the example above), giving two lines which represent the two extremes of the hypothesised relationship. If error is present in X, and there are clearly other variables involved which are omitted from the analysis, then a Model I regression between 'X as measured' and Y is perfectly valid.

2.9 Summary

In this chapter, the analysis of variance has been extended to include a continuous explanatory variable. The new concepts discussed include:

- the criterion of least squares to choose the best fit line
- fitted values, being the values of Y predicted from the best fit equation
- why the geometrical analogy involves a *right-angled* triangle
- R^2 and p-values as the key statistics from a regression analysis
- confidence and prediction intervals
- outliers and influential points
- appropriate circumstances for using Model II regressions.

2.10 Exercises

Does weight mean fat?

Can the weight of a person be used to predict how much body fat they are carrying around? In this study, total body fat was estimated as a percentage of body weight by using skinfold measurements of 19 students in a physical fitness program (stored in the dataset *reduced fats*). Weight was measured in kg.

Box 2.7 shows a regression analysis of these data and Fig. 2.16 a plot of these data.

(1) What is the best fitting straight line?

(2) What proportion of the variability in the data has been explained by fitting this line?

(3) How would you summarise the information provided by the data about the estimate of the slope?

(4) How strong is the evidence that the slope is different from zero?

(5) What would a zero slope imply about the relationship between the two variables?

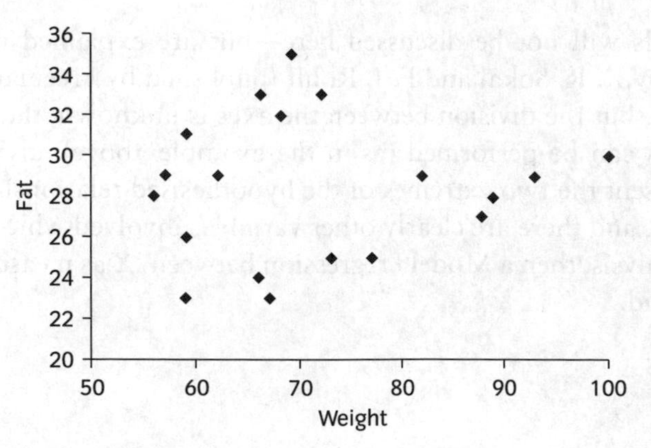

Fig. 2.16 Graph of *reduced fat* data.

BOX 2.7 Analysis of *reduced fat* data

Regression analysis

Word equation: FAT = WEIGHT

WEIGHT is continuous.

Analysis of variance table for FAT

Source	DF	SS	MS	F	P
Regression	1	1.33	1.33	0.10	0.751
Error	17	217.09	12.77		
Total	18	218.42			

Coefficients table

Predictor	Coef	SECoef	T	P
Constant	26.886	4.670	5.76	0.000
WEIGHT	0.02069	0.06414	0.32	0.751

Dioecious trees

This question returns to the *dioecious trees* dataset first used at the end of Chapter 1. The dataset contains three columns: FLOWERS, SEX and DBH (diameter at breast height).

(1) Illustrate graphically how FLOWERS and DBH are related.

(2) Using regression analysis, find the best fitting straight line predicting FLOWERS from DBH.

(3) Test the null hypothesis that the slope of the best fitting line equals 4.

3 Models, parameters and GLMs

In the last two chapters, categorical and continuous explanatory variables have been used to explain patterns in data. These two types of analysis, ANOVA and regression, use similar methods (the partitioning of variation). Here, it can be seen that they are both forms of a **general linear model**, subject to the same set of assumptions. It will no longer be necessary to consider them as two different methods of analysis—and, in fact, any combination of categorical and continuous explanatory variables can be used to extend these models.

3.1 Populations and parameters

In the revision section, the concept of a population is reviewed as an infinite number of individuals from which we take a sample, and estimate a mean. The mean is the first and simplest parameter that we meet in statistics—and the true mean of the population (μ) is a quantity that cannot be known exactly. To discover how reliable our estimate of the mean is likely to be, the sample variance was also calculated, and used in the construction of confidence intervals. The sample variance is itself an estimate of the population variance (σ^2), which also can never be known exactly. Hence, the first step in statistics involves the two parameters μ and σ^2.

In the first chapter, we progressed to estimating three means: the mean yield we would expect to obtain from a plot when using fertiliser A, B or C (μ_A, μ_B, μ_C), and also the population variance σ^2. Here, the concept of a population takes on a wider meaning. It becomes the whole set of data points that could conceivably have been obtained when the experiment was conducted. Given that it is impossible to ever obtain this complete dataset, how can the population be defined? It is defined by a model, which links together the parameters. In this case:

$$\text{YIELD} = \begin{Bmatrix} \mu_A \\ \mu_B \\ \mu_C \end{Bmatrix} + \varepsilon$$

Table 3.1 Parameters and estimates

Population parameters	Usual null hypotheses	Sample estimates
μ, σ^2	$\mu = 0$	$\bar{y}, s^2$
$\mu_A, \mu_B, \mu_C, \sigma^2$	$\mu_A = \mu_B = \mu_C$	$\bar{y}_A, \bar{y}_B, \bar{y}_C, s^2$
α, β, σ^2	$\beta = 0$	a, b, s^2

where ε is drawn, independently for each datapoint, from a Normal distribution with mean 0 and variance σ^2. The curly brackets are conditional ones: i.e. one value is chosen to predict yield, depending upon whether we are considering fertiliser A, B or C. It is the addition of ε from a continuous distribution that makes the 'population' of all possible yields an infinite one. The variance of ε, namely σ^2, is estimated from the unexplained variance of the dataset.

Then in the second chapter, a regression was conducted between the height of 31 trees and the volume of wood they produced. Again, this dataset could be considered as a sample of an infinitely large dataset. To define this population more precisely, again a model is required: in this case

$$\text{VOLUME} = \alpha + \beta.\text{HEIGHT} + \varepsilon$$

where ε is defined as before. Now, a specific value of the variable HEIGHT is required to predict a value for VOLUME. The intercept and slope in the regression equation we calculate are estimates of the 'true' intercept, α, and the 'true' slope, β, which describe the 'true relationship' present in the population. Again, it is the presence of the term ε drawn from a continuous distribution, which makes the population infinite.

All three analyses, therefore, can be viewed in terms of samples and populations, parameters and models. The null hypothesis is then phrased in terms of the population parameters, and it encapsulates the question being asked. Our sample estimates then attempt to find an answer to that question by testing the null hypothesis. See Table 3.1 for the population parameters, null hypotheses and sample estimates of the three analyses discussed.

3.2 Expressing all models as linear equations

In these early chapters, two traditionally different types of analysis (ANOVA and regression) have been performed. In both cases, the questions we have wished to pose may be expressed in the form of 'word equations'. The response variable (representing the data) is on the left, and the variables we suspect of influencing the data (explanatory variables) are on the right hand side of the formula. So both forms of analysis can be written as a linear equation, for example:

BOX 3.1 General Linear Model for the *trees* dataset

General Linear Model

Word equation: VOLUME = HEIGHT

HEIGHT is continuous

Analysis of variance table for VOLUME

Source	DF	Seq SS	Adj SS	Adj MS	F	P
HEIGHT	1	2901.2	2901.2	2901.2	16.16	0.000
Error	29	5204.9	5204.9	179.5		
Total	30	8106.1				

Coefficients table

Term	Coef	SECoef	T	P
Constant	–87.12	29.27	–2.98	0.006
HEIGHT	1.5433	0.3839	4.02	0.000

YIELD = FERTIL

VOLUME = HEIGHT.

The main difference between these two equations is that FERTIL is a categorical explanatory variable, whilst HEIGHT is continuous. Both word equations pose the question 'Is the response variable influenced by the explanatory variable?' In many statistical packages, it is no longer necessary to specify whether the analysis is an ANOVA or a regression but, by using general linear modelling, we can do these analyses by simply entering the word equation and specifying if the explanatory variable is continuous or categorical.

For the rest of this book, we will refer to both ANOVA and regression (also analysis of covariance, factorial analyses, polynomials and all such combinations) as general linear models. To assist with this general framework, we introduce two new concepts:

(1) two types of sums of squares
(2) a different way of presenting the fitted values for a model with categorical explanatory variables.

In Box 3.1 the analysis of variance table and coefficients table for the *trees* dataset is reproduced as a general linear model. The output is exactly the same as before, except that the sum of squares explained by HEIGHT now appears in two columns—once as sequential sums of squares (Seq SS) and also as adjusted sums of squares (Adj SS). These are identical because there is only one explanatory variable in our model. Quite what the terms 'sequential' and 'adjusted' mean will become apparent in the next chapter, when we move on to

BOX 3.2 General Linear Model for the *fertilisers* dataset

General Linear Model

Word equation: YIELD = FERTIL

FERTIL is categorical

Analysis of variance table for YIELD

Source	DF	Seq SS	Adj SS	Adj MS	F	P
FERTIL	2	10.8227	10.8227	5.4114	5.70	0.009
Error	27	25.6221	25.6221	0.9490		
Total	29	36.4449				

Coefficients table

Term	Coef	SECoef	T	P
Constant	4.6437	0.1779	26.11	0.000
FERTIL				
1	0.8013	0.2515	3.19	0.004
2	−0.6447	0.2515	−2.56	0.016
3	*−0.1566*			

more complex models. At present, both columns are simply the sums of squares partitioned between height and error.

Packages vary in the terms they use for sequential and adjusted sums of squares, and in where you will find them in the output. We will follow the convention of presenting the sequential and adjusted sums of squares side by side in the ANOVA table because it is particularly convenient and simple for learning. This most closely mirrors the pattern of Minitab output, but the links with the output of other packages is presented in detail in the package specific supplements.

Comparing this with a general linear model for the *fertilisers* dataset (Box 3.2), the essence of the information is the same. In both cases, the analysis of variance table tests *whether* the two variables are related. (Does HEIGHT give us any information about VOLUME? The answer is yes, with a p-value of less than 0.0005. Does FERTIL explain any differences in YIELD? The answer is yes, with $p = 0.009$). The coefficient table tells us *how* the variables are related. For the *trees* dataset, this answer is given by the fitted value equation (VOLUME = −87.12 + 1.5433 HEIGHT) as before. However, the coefficient table for the *fertilisers* dataset is presented differently. The first term in the column 'Coef' is the constant. This is exactly the same as the grand mean of the dataset in this case, $\bar{y}$, because the dataset contains the same number of plots for each fertiliser group. In more general terms, it is calculated as the unweighted mean of the three fertiliser group means. This is then followed by two values for fertilisers 1 and 2—with the value for fertiliser 3 given in italics. Why?

Table 3.2 Calculating treatment means from parameter estimates

Fertiliser	Mean
1	$\mu + \alpha_1$
2	$\mu + \alpha_2$
3	$\mu - \alpha_1 - \alpha_2$

The values given for the first two fertilisers are the deviations of the group means for fertilisers 1 and 2 from the grand mean (α_1 and α_2). Given that all three deviations must sum to zero by definition, the third deviation can be calculated from the previous two. The three treatment means may then be represented as in Table 3.2.

This may seem an odd way of presenting the results—but it has its advantages. It reflects the two steps of the model fitting process—the first being the fitting of one population mean, and the second being the fitting of three treatment means. In the second step, two *extra* parameters are calculated (and two more degrees of freedom are lost).

Each statistical package follows a different convention. The one we will follow through the book is similar to that given by Minitab, in which the grand mean of the three group means is taken as a reference point, and called the constant. However, it is also useful to consider what SAS and SPSS do when presenting the fitted values. They both take the last group (in other words, fertiliser 3) as the reference point, and provide coefficients for fertilisers 1 and 2 as the deviations from this reference point (called the intercept). The corresponding coefficients table would be structured as in Table 3.3.

If the intercept is denoted μ, and the two deviations as α_1 and α_2, then these estimates may then be converted into estimates for the three treatment means as in Table 3.4.

These different choices of reference points are referred to as **aliassing**. Other packages take group 1 as the reference point. All methods will produce the same model, and it is simply a question of how the package chooses to present the coefficients table to you. The next section is presented using Minitab's

Table 3.3 The coefficients table when fertiliser 3 is given as the reference point

Parameter	Estimate
Intercept	4.4870
FERTIL 1	0.9579
FERTIL 2	−0.4880
FERTIL 3	0.0000

Table 3.4 Calculating treatment means from the coefficients table of Table 3.3

Fertiliser	Mean
1	$\mu + \alpha_1$
2	$\mu + \alpha_2$
3	μ

method of aliassing, but full details of the same exercise for SAS and SPSS users are given in the package specific supplements.

3.3 Turning the tables and creating datasets

In an ideal world, our data would be perfectly behaved. Relationships between two variables might be linear, and data would be distributed Normally around lines and means. However, even in this ideal world, it would still not be possible to discover the 'true' values of parameters. To illustrate this, we can create a dataset in which we specify the parameters—and then analyse the data to recover estimates of those same parameters. The *fertilisers* dataset of Chapter 2 will now be 'created'.

To do this, three variables are required as a starting point: 'DUM1', 'DUM2' and F, as defined in Box 3.3. Then the 'true' parameter values are chosen (μ, α_1, α_2, σ). Finally these are combined in the model, producing the data variable Y. The random command draws values at random from a Normal distribution with

BOX 3.3 Creating a dataset

Three starting variables:

DUM1	1 for each datapoint in group 1, 0 for group 2, −1 for group 3
DUM2	0 for each datapoint in group 1, 1 for group 2, −1 for group 3
F	a code of 1 to 3 for the three fertiliser groups.

Setting the 'true' parameter values:

Let K3 = 12.2	The constant
Let K1 = 5.0	The deviation of group 1 from the constant (α_1).
Let K2 = −2.5	The deviation of group 2 from the constant (α_2).
Let K4 = 1.5	The standard deviation of the error (σ).

Specifying the model:
Random 30 error
Let Y = K3 + K1*DUM1 + K2*DUM2 + K4*error

BOX 3.4 Recovering the parameter estimates

General Linear Model

Word equation: Y = F

F is categorical

Analysis of variance table for Y

Source	DF	Seq SS	Adj SS	Adj MS	F	P
F	2	347.54	347.54	173.77	102.00	0.000
Error	27	46.00	46.00	1.70		
Total	29	393.54				

Coefficients table

Term	Coef	SECoef	T	P
Constant	12.1169	0.2383	50.85	0.000
F				
1	4.8073	0.3370	14.26	0.000
2	−2.6145	0.3370	−7.76	0.000
3	*−2.1928*			

mean 0, standard deviation 1 (precisely how to implement this command will depend upon your statistical package). This can then be converted to a Normal distribution of any variance by multiplying by a constant.

To see how this works, consider a datapoint in group 1. The model reduces to:

$$Y = K3 + K1 + \varepsilon$$

where ε is drawn at random from a Normal distribution with mean zero and standard deviation 1.5. Similarly for group 3 the model becomes:

$$Y = K3 - K1 - K2 + \varepsilon$$

which was why it was not necessary to specify a separate deviation for group 3.

Now that we have created the dataset, we can analyse it in an attempt to recover the parameter values that were used in the first place. The output shown in Box 3.4 gives an example of an analysis to do precisely this. *Code for creating such dummy datasets is given in the package specific supplements.*

In this particular case, the parameter estimates obtained seem to be moderately close to the true values (see below in Table 3.5). The standard deviation of the population (σ) was set at 1.5, but clearly as this increases, our estimates will be less reliable. This emphasises that s, our estimate of σ, is central to the accuracy of all our parameter estimates. Consequently, the error degrees of freedom (the number of independent pieces of information we have about s), determines which t distribution should be used for confidence intervals and t-tests.

Table 3.5 Comparing parameter values and GLM estimates

Parameter values		GLM estimates
μ	12.2	12.1169
α_1	5.0	4.8073
α_2	−2.5	−2.6145
$-\alpha_1 - \alpha_2$	−2.5	−2.1928
σ	1.5	1.3038

Influence of sample size on the accuracy of parameter estimates

A dataset was created using the parameter values of Table 3.5 with the exception of σ, which was set at 2. The number of replicate plots per fertiliser was varied at 5, 10, 20 and 40. For each of these sample sizes, the dataset was created and analysed ten times. Each time the commands are submitted, the 'random' command draws a different set of random numbers, and this is all that is required to generate a 'new' dataset. This provides 10 estimates of each parameter.

Figure 3.1 illustrates ten estimates of the grand mean obtained from each of four experiments, in which the number of replicate plots per fertiliser varied

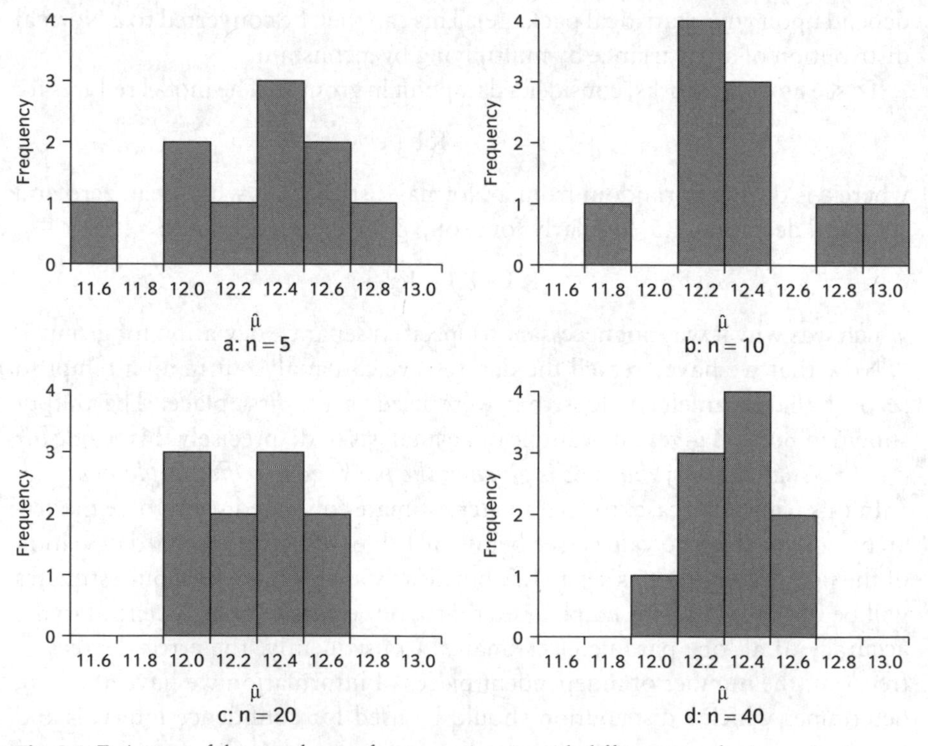

Fig. 3.1 Estimates of the grand mean from experiments with different sample sizes.

from 5 to 40. In all forty 'experiments' the true grand mean was 12.2, but the variability in the estimates obtained is far greater from experiments with only 5 replicates than from those with 40 replicates. This clearly illustrates how the accuracy of any experiment depends upon the rate of replication. In practice, replication is frequently constrained by resources.

3.4 Summary

This chapter has reviewed both ANOVA and regression as two forms of General Linear Model. The linear model is a means of defining the population in terms of parameters. To analyse a dataset, and estimate those parameters, the model can be specified in terms of a word equation, which can include either continuous or categorical variables. The next step is to include more than one variable, and to mix categorical and continuous variables. It is here that the flexibility of the General Linear Model will become apparent.

3.5 Exercises

How variability in the population will influence our analysis

Create a dataset using the parameter values of Table 3.5 with replication set at 10 plots per treatment; except alter σ to equal 2, 4, 8 and 16. Analyse each of these four datasets. Repeat this process ten times for each value of σ. Plot four histograms of the estimates of the grand mean corresponding to Fig. 3.1, for each value of σ. How does σ influence the accuracy of our parameter estimates?

4 Using more than one explanatory variable

4.1 Why use more than one explanatory variable?

In the first three chapters we used one variable to explain patterns in our data —either by looking for differences between the means, or by looking for a relationship between two continuous variables. In many situations however there will be more than one possible explanatory variable. Your first reaction may be to resist including other variables in your analysis, concerned that this may complicate the picture. However, quite the reverse may be true. If you do not include an important third variable, you could miss significant relationships between the first two variables, or even come to completely the wrong conclusion. This can be illustrated intuitively by considering two examples.

Leaping to the wrong conclusion

We could construct the hypothesis that taller children are better at maths, and test this by collecting data from a primary school. A random sample of 32 children of a range of ages sit a mathematical test, and have their height measured (the data being stored in the dataset *school children's maths*).

The corresponding General Linear Model gives a very small p-value for HGHT ($p < 0.005$) as an explanatory variable of AMA (average mathematical ability), from which we might conclude that height influences the children's numerical abilities (Box 4.1).

It would be erroneous however to conclude that this relationship implies causation (and this caution is wise with all observational data). The sample of children span a range of ages, and a logical question would be how age itself influences this relationship. This can be asked by adding a third variable to the word equation, as follows:

$$\text{AMA} = \text{YEARS} + \text{HGHT},$$

which can be interpreted as asking whether YEARS or HGHT (or both) can be used to predict mathematical ability (see Box 4.2).

Now the conclusion is that the age of the child is a strong predictor of mathematical ability ($p < 0.0005$) over and above any influence of height.

BOX 4.1 Height explaining mathematical ability

General Linear Model

Word equation: AMA = HGHT

HGHT is continuous

Analysis of variance table for AMA, using Adjusted SS for tests

Source	DF	Seq SS	Adj SS	Adj MS	F	P
HGHT	1	412.77	412.77	412.77	726.87	0.000
Error	30	17.04	17.04	0.57		
Total	31	429.81				

BOX 4.2 Years, not height, explaining mathematical ability

General Linear Model

Word equation: AMA = YEARS + HGT

YEARS and HGT are continuous

Analysis of variance table for AMA, using Adjusted SS for tests

Source	DF	Seq SS	Adj SS	Adj MS	F	P
YEARS	1	422.60	9.84	9.84	39.63	0.000
HGT	1	0.01	0.01	0.01	0.03	0.860
Error	29	7.20	7.20	0.25		
Total	31	429.81				

Also, when age is taken into account, height is unimportant ($p = 0.860$). The previous result occurred because both height and mathematical ability are highly correlated with age—so the omission of this last variable caused an artifactual association between the first two. The addition of a second explanatory variable alters the nature of the question asked by the word equation: the question has now become 'Does height influence mathematical ability once any differences due to age have been taken into account?'. The answer is 'no'. It also asks 'Does age influence mathematical ability once any differences due to height have been taken into account?'. The answer is 'yes'.

Missing a significant relationship

An experiment was carried out to determine if the amount of water given to tree saplings was vital in determining height. Forty saplings were divided between four different watering regimes. After four months, their final heights

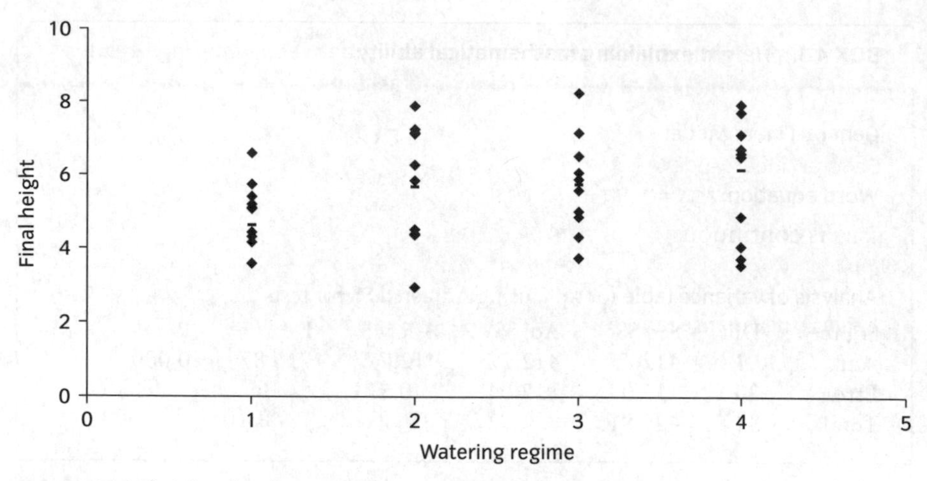

Fig. 4.1 Final height of saplings in four watering treatments.

were measured (stored in the dataset *saplings*). Figure 4.1 illustrates the final height of the saplings plotted according to the different watering regimes.

An analysis of these data to look for differences in final height (FINALHT) between the four groups suggests that there are no significant differences ($p = 0.136$ for WATER in Box 4.3(a)). However, the final height of these saplings will also be influenced to a great extent by the initial height (INITHT). As long as we have randomly allocated our saplings to the four treatments, this will not bias our results, but it will greatly add to our error variation. It would be a more sensitive analysis if the initial height of the saplings could be taken into account. This can be done by including initial height as an additional explanatory variable, and once this has been done, an insignificant result becomes a significant one ($p < 0.0005$ for WATER in Box 4.3(b)).

In summary, the first analysis of Box 4.3a asks if the final height of the saplings differs significantly between the watering regimes, whereas the second

BOX 4.3(a) Final height alone shows no differences between watering regimes

General Linear Model

Word equation: FINALHT = WATER

WATER is categorical

Analysis of variance table for FINALHT, using Adjusted SS for tests

Source	DF	Seq SS	Adj SS	Adj MS	F	P
WATER	3	12.895	12.895	4.298	1.97	0.136
Error	36	78.461	78.461	2.179		
Total	39	91.356				

BOX 4.3(b) Final height is significantly different between watering regimes when initial height is taken into account

General Linear Model

Word equation: FINALHT = WATER + INITHT

WATER is categorical, INITHT is continuous

Analysis of variance table for FINALHT, using Adjusted SS for tests

Source	DF	Seq SS	Adj SS	Adj MS	F	P
WATER	3	12.895	1.052	0.351	64.77	0.000
INITHT	1	78.272	78.272	78.272	1.4E+04	0.000
Error	35	0.190	0.190	0.005		
Total	39	91.356				

asks if final height differs, once any differences in initial height have been taken into account. Hence the second analysis detects a difference, while the first does not, as it takes into account an important explanatory variable (initial height).

These two kinds of problem (missing significant relationships, or leaping to the wrong conclusions) are very common in statistics. The simplest solution may appear to be to restrict your dataset to only include trees of a similar size (or children of a similar age). The best solution depends upon whether you become aware of this problem at the design stage or the analysis stage. If you only require an answer for one age, or one tree height, then data could be collected from a restricted set of children or trees. However, if you only become aware of this problem at the analysis stage, then by considering only a subset of the data, the size of the dataset would be drastically reduced. Instead, it would then be more efficient to include all the information available in a valid analysis.

The examples given here are transparent, but often this is not the case, and puzzling patterns result from the omission of variables. The inclusion of these third variables in your analysis allows their influence to be eliminated, and for this reason this process is called **statistical elimination**.

4.2 Elimination by considering residuals

To capture the essence of statistical elimination, consider the following analysis. If we regress HGHT against YEARS, then the residuals will tell us which children are tall or short for their age (Fig. 4.2(a)). Positive residuals (points above the line) will indicate children who are taller for their age than is predicted by the model, while negative residuals indicate children shorted than expected. Similarly, if we regress AMA against YEARS, the second set of residuals will tell

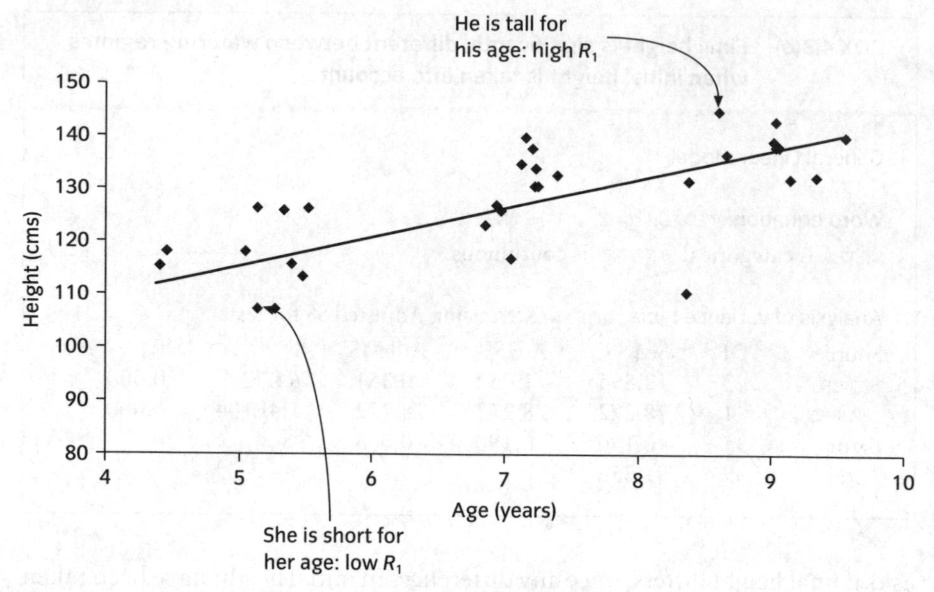

Fig. 4.2(a) HGHT against YEARS.

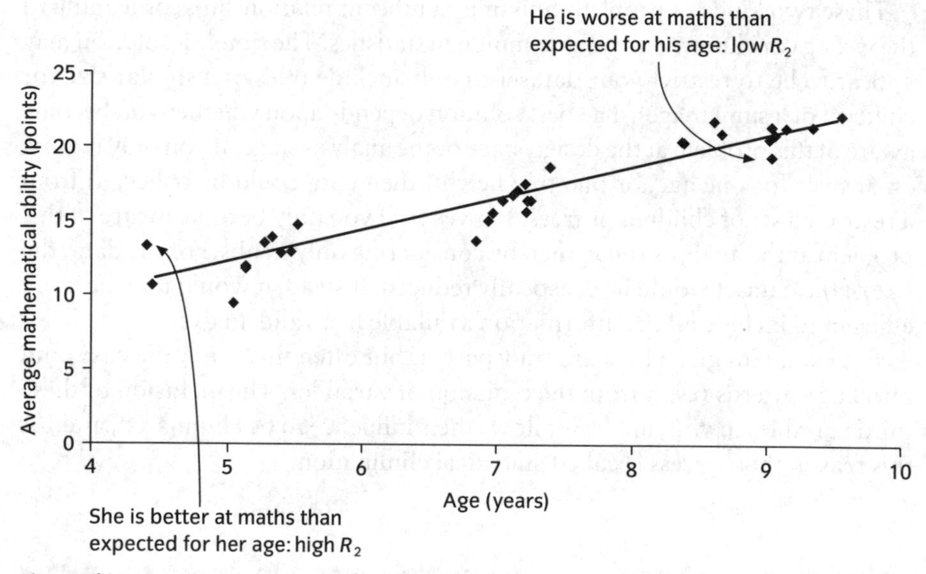

Fig. 4.2(b) AMA against YEARS.

us which children are better or poorer at maths than expected in their age group (Fig. 4.2(b)). Each of these regression plots will produce a set of residuals which will be referred to as R_1 and R_2 respectively.

By looking at the relationship between these two sets of residuals, we will actually be asking the question 'Are the children who are taller than expected for their age also better than average for their age at maths?'. So a positive

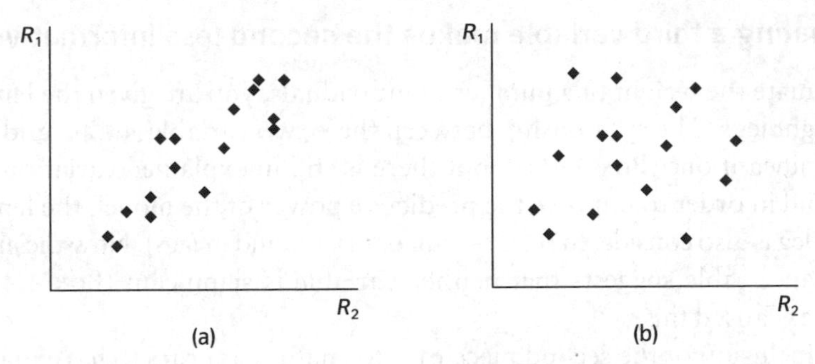

Fig. 4.3 (a) Taller (than expected) people are better at maths (than expected); (b) When age is taken into account, height isn't important.

correlation between R_1 and R_2 would suggest that this was the case (Fig. 4.3(a)), whereas no relationship would indicate that after YEARS has been taken into account, HGHT is not an important predictor of AMA (Fig. 4.3(b)).

This analysis illustrates the essence of what we want to do—namely eliminate the influence of age. It is not actually the correct way to do it. For example, by regressing R_1 against R_2 there would be too many degrees of freedom in the error $(n-2)$, given that four parameters would already have been estimated from the data in the two previous analyses of Fig. 4.2. Nevertheless, this is the general concept behind statistical elimination: that is, asking if two variables are related when a third has been taken into account.

To assist in the interpretation of our analyses when the word equation contains more than one explanatory variable, we need to consider two types of sum of squares for each variable in the model. These appeared in the last two analyses (Boxes 4.2 and 4.3) in the columns headed Seq SS for **sequential sum of squares** or **Type I sum of squares** and Adj SS for **adjusted sum of squares** or **Type III sum of squares.** We will be using the terms 'adjusted' and 'sequential' as their names very aptly describe their derivation, and so are more suitable for learning purposes. You may have noticed that we used the adjusted sum of squares to calculate the mean square, and the *F*-ratio. In these examples, that was most appropriate, but this will not always be the case. The next section examines what we mean by these two types of sum of squares.

4.3 Two types of sum of squares

To introduce the meaning of the two types of sum of squares produced by a GLM, we will consider two possible consequences of statistically eliminating a third variable. These are:

(1) eliminating the third variable makes the second less informative

(2) eliminating the third variable makes the second more informative.

Eliminating a third variable makes the second less informative

To estimate the weight of a number of individuals, you are given the length of their right legs. The relationship between these two variables (WGHT and RLEG) is a significant one (Box 4.4(a)), but there is still unexplained variation in the data, and in order to improve the predictive power of the model, the length of the left leg is also considered (LLEG—full dataset found in *legs*). Now the analysis of variance table suggests that neither variable is significant (Box 4.4(b))—what has caused this?

The inclusion of the second piece of information appears to have made the predictive power of the model much worse. The analysis of variance table has two columns for the sum of squares: sequential and adjusted. The adjusted sum of squares for RLEG is much lower than the sequential, and it is on these values that the *F*-ratios are based—hence the lack of significance. So what are sequential and adjusted sums of squares?

BOX 4.4(a) Length of right leg predicts weight of an individual

General Linear Model

Word equation: WGHT = RLEG

RLEG is continuous

Analysis of variance table for WGHT, using Adjusted SS for tests

Source	DF	Seq SS	Adj SS	Adj MS	F	P
RLEG	1	3627.7	3627.7	3627.7	125.75	0.000
Error	98	2827.1	2827.1	28.8		
Total	99	6454.8				

BOX 4.4(b) Neither RLEG nor LLEG are significant predictors of weight

General Linear Model

Word equation: WGHT = RLEG + LLEG

RLEG and LLEG are continuous

Analysis of variance table for WGHT, using Adjusted SS for tests

Source	DF	Seq SS	Adj SS	Adj MS	F	P
RLEG	1	3627.7	83.3	83.3	2.93	0.090
LLEG	1	66.0	66.0	66.0	2.32	0.131
Error	97	27601.1	2761.1	28.5		
Total	99	6454.8				

The **sequential sum of squares** is the amount of variation explained by a variable when the preceding terms in the model have been statistically eliminated. Hence in the output above, the sequential sum of squares for RLEG is the same in the first and second analysis, as in the first it is the only explanatory variable, while in the second analysis it is the first explanatory variable. The sequential sum of squares for RLEG is quite high, as it is a good predictor of WGHT (as shown in Box 4.4a).

The **adjusted sum of squares** is the amount of variation explained by a variable when all other explanatory variables in the model have been statistically eliminated. So the adjusted sum of squares for RLEG is the amount of variation explained by RLEG once any variation explained by LLEG has been taken into account. Consequently there is a dramatic drop in the sum of squares for RLEG between the first and second column of the ANOVA table in Box 4.4(b).

Left leg length (LLEG) is the second explanatory variable in the word equation —so its sequential sum of squares indicates the variability explained after the inclusion of all preceding terms. As it is also the last term in the model, its sequential and adjusted sums of squares are equal.

So the reason for both *p*-values being insignificant in this analysis, is that a different question is being posed: that is, 'Does the length of the RLEG or the LLEG provide additional predictive power if the other explanatory variable is already known?' Box 4.4(b) illustrates that neither variable provides additional information if the other variable is already known. These two explanatory variables are essentially providing the same information. The ANOVA table not only provides answers via the *p*-values, but also gives additional clues through a comparison of the sequential and adjusted sums of squares. For example, in this case, the drop in value of the sequential to the adjusted sum of squares for RLEG would alert us to the possibility that if LLEG was dropped from the model, then RLEG would become significant—it would not be necessary to actually conduct the two separate analyses to deduce this.

A visual analogy of this analysis is to represent a sum of squares as a line, the length of the line being indicative of the size of the sum of squares. In the RLEG–LLEG example, the sum of squares for the whole model, is only slightly larger than either of the solo sum of squares (Fig. 4.4).

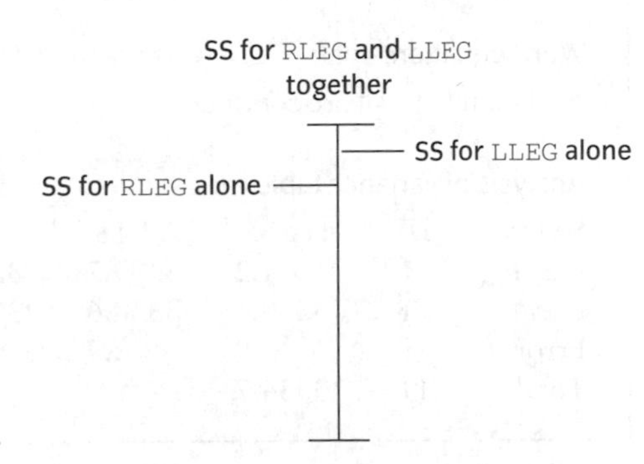

Fig. 4.4 The whole sum of squares of two variables which share information is only slightly larger than the solo sums of squares.

In summary, a sequential sum of squares is the sum of squares explained by a variable when *all preceding terms* in the model have been statistically eliminated. The adjusted sum of squares is the sum of squares explained by a variable *when all other terms* in the model have been statistically eliminated. For this reason, the sequential and adjusted sums of squares of the last term in a model will always be the same.

Eliminating a third variable makes the second more informative

The aim of the analysis in this example is to predict the age at death of a number of distinguished poets. The first piece of information you are given is their birth dates (contained in variable BYEAR). This explains very little of the variation you observe in age at death. With this one piece of information, the predictive power of your model is virtually nil ($p = 0.954$ for BYEAR in Box 4.5(a)). The second piece of information is the year of their death (contained in DYEAR—full

BOX 4.5(a) Age of poets cannot be predicted from birth date alone

General Linear Model

Word equation: POETSAGE = BYEAR

BYEAR is continuous

Analysis of variance table for POETSAGE, using Adjusted SS for tests

Source	DF	Seq SS	Adj SS	Adj MS	F	P
BYEAR	1	1.2	1.2	1.2	0.00	0.954
Error	10	3333.5	3333.5	333.4		
Total	11	3334.7				

BOX 4.5(b) Age of poets can be accurately predicted from birth and death dates

General Linear Model

Word equation: POETSAGE = BYEAR + DYEAR

BYEAR and DYEAR are continuous

Analysis of variance table for POETSAGE, using Adjusted SS for tests

Source	DF	Seq SS	Adj SS	Adj MS	F	P
BYEAR	1	1.2	3299.7	3299.7	1.0E+04	0.000
DYEAR	1	3330.6	3330.6	3330.6	1.0E+04	0.000
Error	9	2.9	2.9	0.3		
Total	11	3334.7				

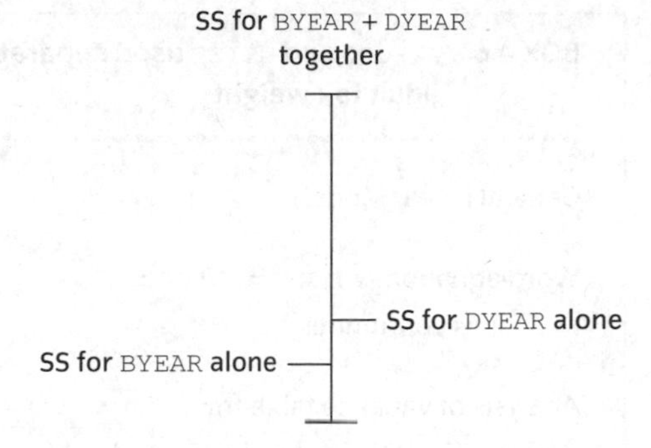

Fig. 4.5 The sum of squares of BYEAR and DYEAR together are much greater than the sum of the solo sums of squares.

dataset in *poets*). Inclusion of this second variable causes the predictive power of the model to increase greatly, as now the age at death can be predicted almost exactly (not always correctly, as the poet could have died before or after her birthday in her death year). Now $p < 0.0005$ for BYEAR in Box 4.5(b).

Examining the output of the model including two explanatory variables, it can be seen that the adjusted sum of squares for BYEAR is greater than the sequential sum of squares. On its own, BYEAR has very little explanatory power, but in conjunction with DYEAR its explanatory power is very much increased. The low sequential sum of squares for BYEAR in Box 4.5(b) indicates that this variable would be insignificant if DYEAR were removed from the model.

Returning to the visual analogy of the length of a line demonstrating the magnitude of a sum of squares, in this example the full model is much more than the sum of the solo sums of squares (Fig. 4.5).

We will adopt the convention of printing out the sequential and adjusted sums of squares side by side in most of the ANOVA tables we discuss. This is not the convention adopted automatically by most statistical packages, but the comparison of the two provides useful information for learning purposes, and is either supplied directly (Minitab) or can be constructed from the output given (see package specific supplements).

4.4 Urban Foxes—an example of statistical elimination

In this section the comparison of sequential and adjusted sums of squares is used to assist in the interpretation of an analysis. If the reader is only interested in the basic concepts, then skip this section. If the reader is interested in seeing how the concepts work in practice, then this example illustrates both increased and decreased informativeness from the presence of another variable.

These data are drawn from a study on urban foxes, in which the investigator was concerned about the factors which may influence their overwinter survival. They live in social groups that have territories, within which they scavenge for food. Sources of food include household waste and earthworms. Thirty

BOX 4.6 AVFOOD and GSIZE used separately as predictors of mean adult fox weight

General Linear Model

Word equation: WEIGHT = AVFOOD

AVFOOD is continuous

Analysis of variance table for WEIGHT

Source	DF	Seq SS	Adj SS	Adj MS	F	P
AVFOOD	1	0.0631	0.0631	0.0631	0.14	0.716
Error	28	13.0948	13.0948	0.4677		
Total	29	13.1579				

General Linear Model

Word equation: WEIGHT = GSIZE

GSIZE is continuous

Analysis of variance table for WEIGHT

Source	DF	Seq SS	Adj SS	Adj MS	F	P
GSIZE	1	0.7972	0.7972	0.7972	1.81	0.190
Error	28	12.3607	12.3607	0.4415		
Total	29	13.1579				

groups were studied over the course of three winters, and the following data were recorded:

GSIZE the number of foxes in a social group

WEIGHT the mean weight of adult foxes within the group

AVFOOD an estimate of the amount of food available in a territory

AREA the area of the territory.

Certain features of the data were noted at the outset. GSIZE and AVFOOD are correlated, because if there is more food available, then the territory can sustain more adults. Also AREA and AVFOOD are correlated, because larger areas tend to have more food. In the first instance, the investigator uses AVFOOD and GSIZE to explain variation in the mean weight of adult foxes, but he does this in two separate analyses (Box 4.6).

In neither case are the explanatory variables significant ($p = 0.716$ for AVFOOD and $p = 0.190$ for GSIZE). In isolation, these analyses would suggest that mean adult fox weight cannot be predicted using either of these variables. However, the analysis is then repeated using both variables in one model (Box 4.7). Now both of these variables are significant ($p < 0.0005$ in both cases)

BOX 4.7 Food availability and group size as predictors of mean adult fox weight

General Linear Model

Word equation: WEIGHT = AVFOOD + GSIZE

AVFOOD and GSIZE are continuous

Analysis of variance table for WEIGHT

Source	DF	Seq SS	Adj SS	Adj MS	F	P
AVFOOD	1	0.0631	4.7039	4.7039	16.59	0.000
GSIZE	1	5.4380	5.4380	5.4380	19.18	0.000
Error	27	7.6568	7.6568	0.2836		
Total	29	13.1579				

—what is the statistical and biological explanation for this? Examination of the sequential and adjusted sums of squares for AVFOOD in Box 4.7 illustrates that the addition of GSIZE to the model has greatly increased AVFOOD's explanatory power. Both of these variables are required to predict weight. The reason for this is that mean fox weight will largely be determined by the amount of food available *per fox present in that territory*. Both variables are required to accurately predict weight (this is akin to the age of poets example).

Finally, the investigator includes the third explanatory variable in his model, the territory area. As a consequence of this, the significance of AVFOOD decreases to a p-value of 0.026 (though this is still significant). Why has this happened? In this case, a comparison of the adjusted sum of squares for AVFOOD in the third and fourth models (Boxes 4.7 and 4.8) indicates a drop. In the first instance,

BOX 4.8 Food availability, group size and territory area as predictors of mean adult fox weight

General Linear Model

Word equation: WEIGHT = AVFOOD + GSIZE + AREA

AVFOOD, GSIZE and AREA are continuous

Analysis of variance table for WEIGHT

Source	DF	Seq SS	Adj SS	Adj MS	F	P
AVFOOD	1	0.0631	1.4938	1.4938	5.57	0.026
GSIZE	1	5.4380	5.8434	5.8434	21.79	0.000
AREA	1	0.6841	0.6841	0.6841	2.55	0.122
Error	26	6.9728	6.9728	0.2682		
Total	29	13.1579				

the adjusted sum of squares is the variability explained by AVFOOD when GSIZE is already known, while in the second, it is the variability explained by AVFOOD when GSIZE *and* AREA are already known. The inclusion of AREA has made AVFOOD less informative—this is because the two variables share information. Given that AREA itself is not significant with AVFOOD already present in the model, the simplest solution would be to leave it out. The model of Box 4.7 would then become our final choice as the clearest representation of our data.

This example illustrates the insights that may be gained from a comparison of the sequential and adjusted sums of squares—both within one ANOVA table, and between ANOVA tables. It has also touched upon another subject area—that of model choice. Given the choice between two models, the one in which insignificant unnecessary terms are not included is preferable. This is discussed in much greater detail in Chapters 10 and 11.

4.5 Statistical elimination by geometrical analogy

Partitioning and more partitioning

In the first example (Chapter 1), the total sum of squares was partitioned into the variation explained by fitting the model, and the variation which remained unexplained. This chapter has illustrated that the model may consist of more than one variable, so the variability has been partitioned into finer components. This principle can be illustrated for two variables. For example, the original *MFY* triangle can be thought of as lying on the vertical diagonal of a cube (Fig. 4.6). The vector *MF* represents the variability explained by fitting the model as before (Fig. 1.11). Now, however, this vector has also been resolved into two components, *MD* and *DF*, corresponding to the two explanatory variables of the model.

Relating this to a specific example, the volume of wood contained within a tree can be explained by its diameter and height (Box 4.9—we first used this

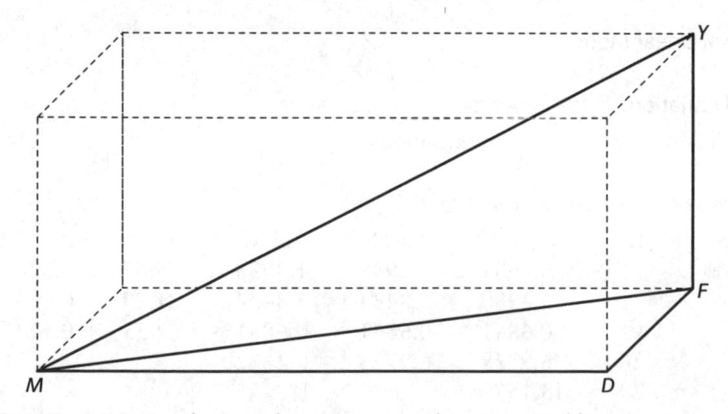

Fig. 4.6 Partitioning a sum of squares between two explanatory variables.

BOX 4.9 Explaining the volume of trees by their height and diameter

General Linear Model

Word equation: LVOL = LDIAM + LHGT

LDIAM and LHGT are continuous

Analysis of variance table for LVOL, using Adjusted SS for tests

Source	DF	Seq SS	Adj SS	Adj MS	F	P
LDIAM	1	7.9289	4.6234	4.6234	701.33	0.000
LHGT	1	0.1987	0.1987	0.1987	30.14	0.000
Error	28	0.1846	0.1846	0.0066		
Total	30	8.3122				

dataset in Chapter 2—*trees*). The whole regression is represented by the *MFY* triangle, in which *MF* represents the regression sum of squares (the sum of squares explained by the two variable model) and *FY* represents the error sum of squares (0.1846 in this instance). Vector *MF* is then further partitioned into *MD*, the sum of squares explained by LDIAM (7.9289), and *DF*, the sum of squares explained by LHGT when LDIAM is already known (0.1987). Note that it is the sequential sum of squares that is being partitioned, as it is the sequential SS that add up to the total SS (just as by Pythagoras, $MF^2 = MD^2 + DF^2$).

Just as with one explanatory variable, there are three parallel partitions: the vectors, the degrees of freedom and the sums of squares. These are illustrated for the partition between regression and error, and then the partitioning of regression into the components explained by the two variables LDIAM and LHGT, in Table 4.1.

Table 4.1 Two sets of three parallel partitions between two explanatory variables.

Vector	Df	Squared length
Set 1		
MF	2	SS for LDIAM and LHGHT together = Regression SS
+ FY	+ 28	+ Error SS
= MY	= 30	= Total SS
Set 2		
MD	1	SS explained by LDIAM alone
+ DF	+ 1	+ SS explained by LHGHT, controlling for LDIAM
= MF	= 2	= SS for LDIAM and LHGHT together = Regression SS

BOX 4.10 Changing the order of the explanatory variables

General Linear Model

Word equation: LVOL = LHGT + LDIAM

LDIAM and LHGT are continuous

Analysis of variance table for LVOL, using Adjusted SS for tests

Source	DF	Seq SS	Adj SS	Adj MS	F	P
LHGT	1	3.5042	0.1987	0.1987	30.14	0.000
LDIAM	1	4.6234	4.6234	4.6234	701.33	0.000
Error	28	0.1846	0.1846	0.0066		
Total	30	8.3122				

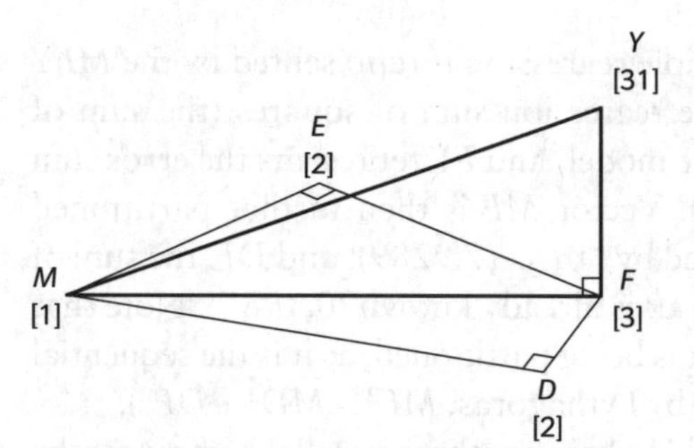

Fig. 4.7 Partitioning the sum of squares by two different routes.

It is also worth considering what would happen if the two variables had been fitted in a different order (as in Box 4.10). Certain parts of the ANOVA table (as we present it) will change, and certain parts will remain the same. Which and why? The total, regression and error sums of squares will remain the same—so the original *MFY* triangle is unchanged. The adjusted sum of squares for an explanatory variable will also be unchanged, provided that all the same explanatory variables are included. However, the model (and therefore the point *F*) has been reached by a different route (Fig. 4.7).

The vector *ME* represents the sum of squares explained by LHGT (3.5042) and *EF* the sum of squares explained by LDIAM when LHGT is already known (4.6234). So the partitioning of *MF* is rather more equal in this instance (the triangle *MEF* is closer to an isosceles). The regression sum of squares however is the same in both cases (8.3122)—indicative of the fact that both routes end up at the same point *F*.

The degrees of freedom also add up in the same manner regardless of route. At each corner, the number in the square brackets represent the number of parameters estimated (Fig. 4.7). For *M*, only the grand mean has been estimated, for both *D* and *E*, a slope and an intercept has been estimated, and for *F*, two slopes and an intercept have been estimated. The difference between these values represent the degrees of freedom for the line (and so the sum of squares) between those two points. Whichever route you take between *M* (the mean) and *F* (the model), the degrees of freedom will add up to 2 (the regression degrees of freedom).

So in summary, changing the order of the variables will change the sequential sums of squares—but all other aspects of the ANOVA table (the total, error, regression, and adjusted sums of squares, and the degrees of freedom associated with these) will stay the same.

Picturing sequential and adjusted sums of squares

The two examples discussed in Section 4.3 (RLEG and LLEG sharing information, and BYEAR and DYEAR increasing each other's informativeness), can be represented in pictorial equivalents of Fig. 4.7. If the *MFY* triangle on the vertical is ignored, then the relative positions of *M*, *D*, *E* and *F* can be drawn on one plane. The lines *MD* and *DF* then represent the route followed by fitting the variables in one order, while *ME* and *EF* represent the route followed by fitting the same variables in reverse order. The same end point is reached (*F*) because the model fitted is exactly the same.

Considering first the two variables which share information (the RLEG–LLEG example), the first variable fitted will always have a high sequential sum of squares, while the second will have a low SS. This will be true regardless of the order in which they are fitted, and is illustrated in Fig. 4.8. In both cases, the fitting of the first variable is equivalent to travelling along *MD* or *ME*—both *D* and *E* are close to the final fitted values at point *F*, and some distance from *M*, the grand mean. The fitting of the second variable moves the model along the short remaining distance (*DF* or *EF*) to the final point *F* in which

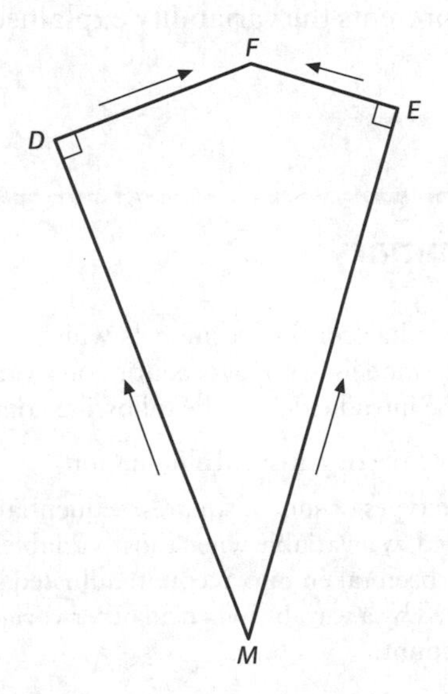

Fig. 4.8 The geometrical analogy of the RLEG–LLEG example.

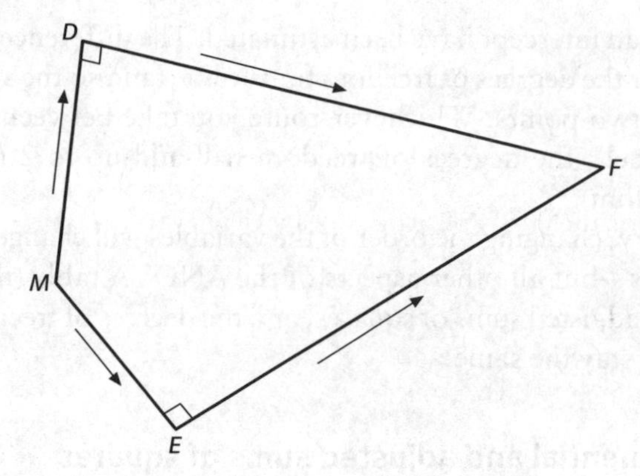

Fig. 4.9 The geometrical analogy of predicting the poets' age from birth and death years.

both explanatory variables are included. So *MD* is equal to the variability in weight explained by fitting RLEG alone, while *DF* is the variability explained by fitting LLEG when RLEG is already known. Similarly, *ME* is the variability in weight explained by fitting LLEG alone, with *EF* being the variability explained by fitting RLEG when LLEG is already known.

In contrast, predicting the age of the famous poets at death from their birth years and death years is illustrated by the points *E* and *D* both being some distance from *F* (and close to *M*, the grand mean)—with the lines *EF* and *DF* being much greater than *ME* and *MD* (Fig. 4.9). Here, the line *ME* represents the sum of squares explained by fitting BYEAR alone as an explanatory variable. The resulting model does not have much explanatory power (*E* is not far from *M*)—but on fitting DYEAR in addition to the first variable, the point *F* is reached at some considerable distance from *M*. Vector *EF*, the variability explained by DYEAR when BYEAR is already known, is much greater than *MD*, which represents the variability explained by DYEAR alone.

4.6 Summary

This chapter has considered models which contain two explanatory variables. At this point, only models with two continuous variables have been considered, which has involved the introduction of the following concepts:

- The principle of statistical elimination.
- The two types of sum of squares: sequential sum of squares = the amount of variation explained by a variable when those variables that precede it in the word equation have already been taken into account; adjusted sum of squares = the amount of variation explained by a variable when all other variables in the word equation have been taken into account.

- Statistical packages will test one or other or both types of sums of squares. You should always be aware of the type you are looking at. Make sure you know how to get both types out of the package you are using. See the package specific supplements.
- Comparison of the sequential and adjusted sums of squares for a variable can provide information on the way in which two explanatory variables may be connected: they may share information (be explaining the same thing) or they may increase each others' informativeness.

4.7 Exercises

The cost of reproduction

Life history theory assumes that there is a trade off between survival and reproduction. Data were collected to test this assumption using the fruit fly *Drosophila subobscura*. Twenty-six female flies laid eggs over more than one day. Reproductive effort was measured as the average number of eggs laid per day over the lifetime of the fly. Survival was recorded as the number of days the fly survived after the first egg laying day. Their size was measured as the length of the prepupa at the beginning of the experiment, before emergence and egg laying began. Three variables were created: LSIZE, LLONGVTY and LEGGRATE, in which these data were logged. These variables are stored in the *Drosophila* dataset.

In Box 4.11 the researcher asked the question 'How does reproductive effort affect survival?'

BOX 4.11 GLM of survival against reproductive rate for *Drosophila subobscura*

General Linear Model

Word equation: LLONGVTY = LEGGRATE

LEGGRATE is continuous

Analysis of variance table for LLONGVTY, using Adjusted SS for tests

Source	DF	Seq SS	Adj SS	Adj MS	F	P
LEGGRATE	1	7.738	7.738	7.738	5.83	0.024
Error	23	30.507	30.507	1.326		
Total	24	38.245				

Coefficients table

Term	Coef	SECoef	T	P
Constant	1.7693	0.2313	7.65	0.000
LEGGRATE	0.2813	0.1165	2.42	0.024

BOX 4.12 GLM of survival against size and reproductive rate for *Drosophila subobscura*

General Linear Model

Word equation: LLONGVTY = LSIZE + LEGGRATE

LSIZE and LEGGRATE are continuous

Analysis of variance table for LLONGVTY, using Adjusted SS for tests

Source	DF	Seq SS	Adj SS	Adj MS	F	P
LSIZE	1	26.240	21.842	21.842	55.46	0.000
LEGGRATE	1	3.340	3.340	3.340	8.48	0.008
Error	22	8.665	8.665	0.394		
Total	24	38.245				

Coefficients table

Term	Coef	SECoef	T	P
Constant	1.6819	0.1266	13.28	0.000
LSIZE	1.4719	0.1976	7.45	0.000
LEGGRATE	−0.28993	0.09956	−2.91	0.008

A second analysis was then conducted in Box 4.12, which included the size of the flies.

(1) Calculate a confidence interval for the slope of LLONGVTY on LEGGRATE based on the analysis in Box 4.11.

(2) Calculate a confidence interval for the slope of LLONGVTY on LEGGRATE based on the analysis in Box 4.12, in which LSIZE has been eliminated.

(3) The graph of Fig. 4.10 is a plot of LLONGVTY against LEGGRATE with each point being allocated to one of six groups depending upon size (group 1 being the smallest up to

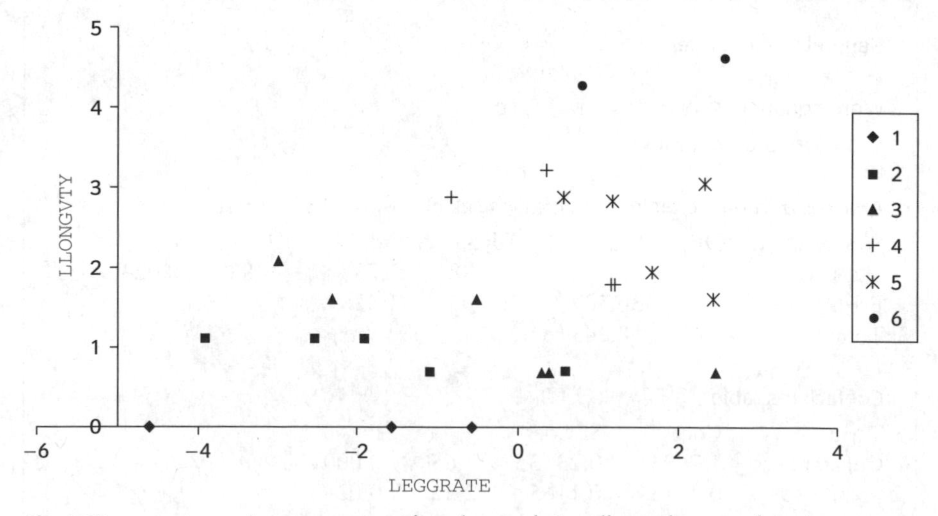

Fig. 4.10 LLONGVTY against LEGGRATE with each point being allocated to one of six size groups.

group 6 being the largest). Why is there such a discrepancy between the two slopes at the centre of the confidence intervals calculated in 1 and 2?

Investigating obesity

As part of an investigation into obesity, three measurements were taken from a sample of 39 men. These were: FOREARM, the thickness of a skin fold on the forearm, which is taken as an indicator of obesity; height (HT) and weight (WT). These data are found in the dataset *obesity*.

(1) Taking FOREARM as the response variable, which of the two explanatory variables HT and WT is the best predictor of obesity when used alone in a GLM?
(2) If the two explanatory variables are used together to predict FOREARM in a GLM, do they increase or detract from each other's informativeness and why?
(3) How could you predict the patterns found in the analyses you conducted in question 1 from the analysis you conducted in question 2?

5 Designing experiments—keeping it simple

In the previous chapter, the principle of statistical elimination was introduced. This allows the influence of one variable to be eliminated, so clarifying the influence of a third variable. This technique is very useful when designing experiments, and this is one of the themes explored here. This chapter introduces some fundamental ideas in the design of experiments which are best made in relation to field experiments. The same ideas apply, but in more sophisticated ways, to other areas of biology.

In designing an experiment, the aim is to manipulate explanatory variables which we suspect may influence the response variable (the data variable). However, there will inevitably be other influences on our response variable which will not be part of the experimental design, which will cause variation which appears to be random. So two fundamental aims of experimental design are:

- reduce or take account of the impact of extraneous variables
- reduce error variation.

In doing this, we will be increasing the power of our statistical tests (that is, the probability of detecting a difference if it is there), and increasing the precision of all our parameter estimates.

5.1 Three fundamental principles of experimental design

The first three principles of experimental design are replication, randomisation and blocking.

Replication

The whole idea of statistics is to look for differences by accumulating independent pieces of evidence. The greater the number of pieces of evidence, the more certain are your conclusions. A set of **replicates** are a series of measurements of exactly the same thing—for example, a number of plots that have received the same treatment. In the fertiliser example of Chapter 1, each of the three fertilisers were applied to 10 plots, so the level of replication here is 10. While the concept is straightforward, obtaining true replicates in practice is not quite so easy. This is because in many aspects of life sciences, particularly ecology, everything varies. In the enthusiasm of maximising the level of replication, it is easy to fall into the trap of **pseudoreplication**. Considering the fertiliser experiment of Chapter 1, it may be that an overconscientious researcher recorded the yield of each plant in the plot individually, so producing a huge dataset. Is it possible to consider each plant as an independent estimate for that fertiliser? Unfortunately the answer is no. Each plant is, in fact, a pseudoreplicate. Anything that happens to that plot (for example the impact of herbivores, pathogens or water level) is likely to have a similar impact on all plants within the plot, giving them all lower or higher yields. The plants are not providing individual independent estimates of how that fertiliser may influence yield, as they are in such close proximity, that they will influence each other, and be influenced by the same extraneous factors. Only the plot as a whole will provide one independent estimate of the impact of fertiliser on yield. Other examples of pseudoreplication include taking repeated measurements from the same place, or the same individual. Spotting and avoiding this problem is one of the skills of experimental design, and mistakes can occur at the design stage or during the analysis. If an analysis of variance has a very large number of degrees of freedom, then it is likely that this mistake has been made, especially if this is combined with surprising significant results. (Pretending that there are 100 independent pieces of evidence for increased yields with a particular fertiliser, rather than 10, will inflate the mean square for 'Fertiliser', and so inflate *F*-ratios). It is usually possible to calculate the correct degrees of freedom from a description of the experimental design. This problem is returned to in Chapter 12.

So in designing an experiment, one aim is to maximise the true number of replicates, as in doing so, we are maximising the number of independent pieces of evidence which will answer our question. The level of replication is reflected in both tables of the GLM output.

1. The ANOVA table partitions the variation into that explained by the model, and that which remains unexplained. The level of replication is reflected in the error degrees of freedom. The greater the error degrees of freedom, the better is our estimate of the unexplained variance. (In terms of population and sample parameters discussed in Chapter 3, s is likely to be closer to σ).
2. The unexplained variance then contributes to the calculation of the standard errors in the coefficients table. The number of replicates for each treatment combination also contributes to this calculation. For example, in Chapter 1, the standard error of a mean was found to be $\frac{s}{\sqrt{n}}$. The n referred to in this instance is the number of replicates that contribute to the calculation of that

mean (e.g. the mean of fertiliser A), and therefore not the total number of data points. So the lower the level of replication, the higher the standard error of the mean, as we would be less certain about our estimate of this mean. In general, the formula for the standard error of the coefficient would be more complicated than $\frac{s}{\sqrt{n}}$, but it would still be a function of s and n, and the same principle would apply. In general, the level of replication affects the precision of all our parameter estimates.

When designing an experiment, the level of replication has to be decided. If you have a priori information on the expected magnitude of the error variance, then the desired level of replication can be calculated. However, this is rarely the case. One useful rule of thumb is to aim to have at least 10 degrees of freedom for error, when the full model has been fitted (see Mead (1990), *The design of experiments*, Cambridge University Press). In reality however one is often forced to limit the degree of replication owing to resource constraints.

Randomisation

For a statistically analysed experiment, treatments should be allocated to experimental plots or units in a **random** manner. There is actually a link between randomisation and replication. The unit to which you randomly allocate each treatment is the true replicate. This can be useful in distinguishing true from pseudo replicates. If you do not randomise, you run the risk of committing one of three cardinal sins:

1. *Systematic designs*

In planning an experiment to compare four fertilisers, the farmer may wish to lay out the plots as in Fig. 5.1, for the sake of ease of application.

Plots adjacent to each other however are likely to be very similar—more so than plots on either end of the field. There may be a river along one side of the field, raising soil moisture content. Even if the field appears to be uniform, subtle local similarities between plots are likely to be present. What would be the

A	A	C	C
A	A	C	C
B	B	D	D
B	B	D	D

Fig. 5.1 A systematic design.

consequences when the data are analysed? If local similarities and differences coincide with the different fertiliser treatments (for example, the top left hand corner of the field has higher nutrient levels), then this may cause differences in yield, which you would think, erroneously, were due to fertiliser treatment. Also, the error variation, which is the variation between plots given the same treatment, would be underestimated. Again, this would contribute to artificially high *F*-ratios, and so you may conclude that there are significant differences between fertilisers, when no such differences exist.

2. *Unconscious biases*

It is wise to scrutinise exactly how you intend to execute an experiment, as unconscious bias can enter at any stage. For example, in a laboratory experiment to test the susceptibility of some larvae to a series of pathogens, it would be a mistake to take the first thirty larvae from the holding box, and allocate them to one treatment, followed by the second set for treatment two etc. This is because the first larvae drawn from the box may be the active ones (in which case they may be more susceptible), or the large ones (and so less susceptible). It may be difficult to predict the direction the bias may take, or indeed whether it is present at all. Without randomisation however there will always be this possibility.

3. *Undesirable subjectivity*

It is often tempting to take a short cut, and instead of rigorously randomising the allocation of treatments to replicates, we simply try to be haphazard. Humans are notoriously bad at being genuinely random. In allocating treatments to plots in a field, for example, we tend to be extremely reluctant to place two treatments together in adjacent plots, as a reaction against systematic designs. In fact, we actually do this less frequently than would be expected by chance, with the end result that plots containing the same treatment are more widely dispersed than they would be through random allocation. It is impossible in such situations to actually quantify the effect on the error variance.

The bottom line is that the best way in which treatments can be allocated to replicates is to randomise, and to be rigorous in the way this is done. If randomisation is not part of the experimental design, then there needs to be a well thought out reason why this is not the case. Even then, you will suffer from the statistical disadvantages of lack of randomisation.

To work out a randomised allocation, in a completely randomised experiment first list all the units to which a treatment is going to be applied. Then make a separate list of the same length containing all the treatment combinations, each combination present as many times as it will be replicated. Use a program to randomise the order of this second list, and then match it with the first. Details are given in the package specific supplements.

Blocking

In the analysis of experiments, the variation in the data is partitioned between the explanatory variables manipulated in the experiment, and error variation. The smaller the error variation, the more powerful is the experiment. **Blocking** is a design tool which can be used to minimise the error variation, and it makes use of the principle of statistical elimination described in the previous chapter.

Returning to the fertiliser example, it may be that the experiment was conducted on a steep gradient. At the top of the field, soil conditions are very dry, while at the bottom of the field soil moisture content is much higher. Water level is likely to affect yield, and so if all the plots of fertiliser A were located at the top of the field, with all of fertiliser D at the bottom of the field, fertiliser effects and water level effects would be confounded—there would be no way of distinguishing which of these two possible factors were responsible for any differences in yield (Figure 5.2).

One solution would be to fully randomise the allocation of treatments to plots. The variation due to water level would then contribute to the error variation, and would not cause any bias between the fertiliser comparisons. However, there is an even better solution. We can take advantage of the fact that we actually know something about the way the field varies, and so incorporate the variation in water level as part of our experimental design. Variation due to water level can then be statistically eliminated, and the impact of fertiliser can be examined, once water level has been taken into account. The way to do this is by blocking. The field is divided into sections—either two or four sections in this instance. The blocks should be internally homogenous, but as different as possible from each other. Within each block, a 'mini experiment' is conducted, with each of the four fertilisers represented. The allocation of fertiliser treatments within blocks should be randomised, but each block should contain each treatment combination (Fig. 5.3).

When the data are analysed, the results from all four mini-experiments are combined, to get as much information as possible about our treatment comparisons. For each plot, both its block and its treatment are recorded, and then

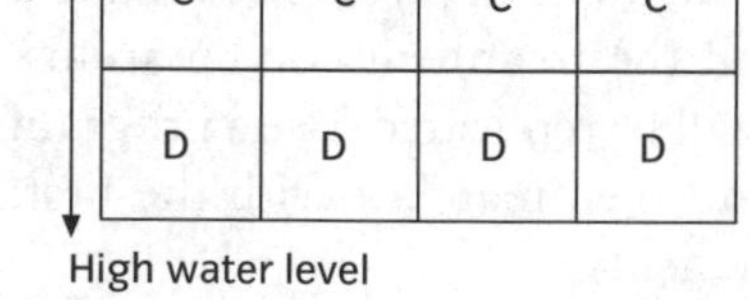

Low water level

Gradient

A	A	A	A
B	B	B	B
C	C	C	C
D	D	D	D

High water level

Fig. 5.2 Fertiliser treatments (A to D) are confounded with water level.

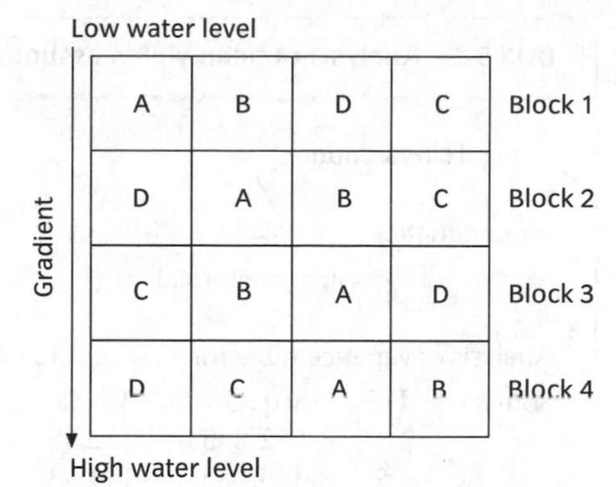

Fig. 5.3 Blocking for water level.

BOX 5.1 Analysis of bean yields assuming a fully randomised design

General Linear Model

Word equation: YIELD = BEAN

BEAN is categorical

Analysis of variance table for YIELD, using Adjusted SS for tests

Source	DF	Seq SS	Adj SS	Adj MS	F	P
BEAN	5	444.435	444.435	88.887	14.59	0.000
Error	18	109.690	109.690	6.094		
Total	23	554.125				

the effect of block is statistically eliminated by including it as a categorical variable in the word equation. The advantage of blocking in this manner is that we increase the precision of our analysis, and increase our chances of finding a significant result.

This is illustrated by an experiment comparing six varieties of bean. The researcher wished to know which variety was most suited to growing in the soil type found on his farm. He had 24 plots in total, distributed between four blocks (stored in the *beans* dataset). Within each block, there was a plot of each of the six varieties. The data were then analysed in two ways. In Box 5.1, he analysed the experiment ignoring the blocks. In Box 5.2, BLOCK has been included as a second explanatory variable. In the first analysis, the question posed is 'Are there significant differences between the bean varieties?'. In the second analysis, the question is 'Are there significant differences between the bean varieties, once any differences between the blocks have been taken into account?' The *F*-ratio for BEAN is significant in both cases, but it is much higher in the blocked analysis (23.48 compared to 14.59). The error mean square is

BOX 5.2 Analysis of bean yields assuming a randomised block design

General Linear Model

Word equation: YIELD = BLOCK + BEAN

BLOCK and BEAN are categorical

Analysis of variance table for YIELD, using Adjusted SS for tests

Source	DF	Seq SS	Adj SS	Adj MS	F	P
BLOCK	3	52.895	52.895	17.632	4.66	0.017
BEAN	5	444.435	444.435	88.887	23.48	0.000
Error	15	56.795	56.795	3.786		
Total	23	554.125				

also lower in the second analysis (3.79 compared to 6.09), and thus the precision of our model has been increased.

Blocking is therefore a means of eliminating variation which would otherwise be assigned to error. It therefore carries with it all the advantages discussed earlier of reducing the error variation.

In summary, there are two rules for creating blocks:

- Blocks should be used to represent a factor believed to influence the response variable.
- Blocks should be internally as homogenous as possible (and therefore as different from each other as possible).

If possible, the interpretation of experiments are easier if all treatments are represented in all blocks—and the experiments reviewed here conform to this.

Just as in Chapter 4, the sums of squares and degrees of freedom have been partitioned three ways instead of two, by partitioning between and within blocks, and then a further partition between treatment and error (Fig. 5.4).

The term 'blocking' originated from agricultural experiments, where there were clear reasons for dividing up the experimental area into blocks of land—just as discussed here. However, this technique has much wider applicability. It is often true that in gathering data, there are other reasons why sets of data are likely to have similarities, for example:

1. Many things change over time. When conducting experiments in the laboratory, the temperature and humidity may vary on a daily basis. If it is not possible to carry out the entire experiment within one day, it would be wise to block according to day, and carry out some replicates of all treatment combinations on each day.

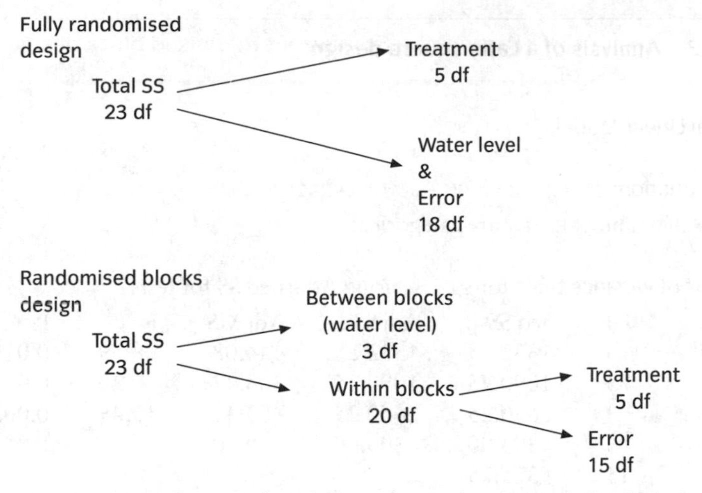

Fig. 5.4 Partitioning sums of squares for a fully randomised and a randomised block design.

Column blocks

A	B	D	C
D	A	C	B
C	D	B	A
B	C	A	D

Row blocks

Fig. 5.5 Latin square design.

2. Animal behaviour experiments may involve animals from more than one litter. Within a litter, animals are likely to be more genetically homogenous than between litters. This could affect them in unpredictable ways, so it would be wise to block for litter.

Returning to the field experiment, it may not be obvious whether the blocks should run across the field, or down the field. However, it is actually possible to do both, by using a **Latin square design**. This is blocking by row and column simultaneously (Fig. 5.5). In a four treatment design, this would need at least a four by four arrangement, with each treatment appearing once in every row and every column. To randomly allocate treatments to rows and columns, while still fulfilling the requirement that each treatment appears once in every row and every column, one useful trick is to write the design out systematically initially, then randomly shuffle the rows and columns. In analysing such a design, there are effectively two blocking factors (each with four blocking levels). For example, in Box 5.3 is an analysis of a Latin square experiment, in which the mean number of seeds per plant for an oilseed rape field trial was recorded for each of sixteen plots (laid out as in Fig. 5.5, data stored in the *oilseed rape*

BOX 5.3 Analysis of a Latin square design

General Linear Model

Word equation: SEEDS = COLUMN + ROW + TREATMT

COLUMN, ROW and TREATMT are categorical

Analysis of variance table for SEEDS, using Adjusted SS for tests

Source	DF	Seq SS	Adj SS	Adj MS	F	P
COLUMN	3	1332.25	1332.25	444.08	8.79	0.013
ROW	3	1090.25	1090.25	363.42	7.20	0.021
TREATMT	3	2650.25	2650.25	883.42	17.49	0.002
Error	6	303.00	303.00	50.50		
Total	15	5375.75				

dataset). The two blocks are significant (rows and columns are categorical variables, just as blocks are). The ANOVA is then effectively asking how the mean number of seeds per plant is influenced by treatment once the horizontal and vertical heterogeneity in the field has been taken into account.

Just as with all other blocks, this design is not only applicable to the layout of plots on land. Any two factors can be considered as blocking factors: for example, as mentioned earlier, time is often an important blocking factor. If, over a period of four days, four balances were being used to weigh the nestlings of sixteen different nests, then DAY and BALANCE would be the two blocking factors (with FEEDING as treatment). Alternatively, both OPERATOR and DAY may be important factors influencing the results in a laboratory experiment, in which case these would be the two factors in a Latin square design.

There are, however, drawbacks to this apparently neat design. The Latin square relies on the experimental design having the same number of treatments as blocks. If applied to a field experiment, it also relies on the variation being horizontal and/or vertical (rather than diagonal!). If neither of these conditions are the case, then it is not necessarily worth shoe-horning the design to fit a Latin square—some other solution might well be more appropriate. A third problem is that if the design fits neatly into one square, then it is not possible to look for interactions (see later in this chapter). So while this appears to be a convenient solution, it is rarely the best design in practice.

Before leaving the topic of blocking, it is perhaps worth noting that although it is a basic component of experimental design, there are situations in which it is not deemed to be appropriate. In designing a greenhouse experiment, it would appear that there is considerable internal heterogeneity in sunlight, temperature and humidity, all of which are likely to affect the growth of plants, and that blocking, perhaps from top to bottom, could allow us to eliminate some of this variation. In fact, within a greenhouse there is so much internal variation that there is a view that it is too heterogeneous to solve by blocking. Instead, one recommendation is that all plant pots are randomly shuffled around every day. This is an attempt to expose all plants within an experiment to the different conditions within the greenhouse, and illustrates that you might have to think of imaginative solutions to reduce error variation, or minimise the impact of extraneous variables, effectively and without bias.

5.2 The geometrical analogy for blocking

Partitioning two categorical variables

Blocking is another form of statistical elimination, which was first discussed in the previous chapter. The fundamental difference is that the blocks are categorical variables, so the analysis involves two categorical variables, rather than two continuous ones. The principle of partitioning is the same: with the first partition being between (BLOCK + TREATMENT) and error, that is *MF* and *FY* respectively, with the second being the partitioning of the first term into BLOCK and TREATMENT, represented by *ME* and *EF* (Fig. 5.6).

The partitioning of the degrees of freedom can be calculated by considering the number of parameters fitted at each point, as before. The difference in the number of parameters fitted when moving from one point to the next, represents the degrees of freedom associated with that partition. Point *Y* represents *N* data points, and *M*, 1 grand mean (or the constant). In moving from *M* to *E*, this represents the variability explained by blocks (52.895 in the example described in Box 5.2). If the number of blocks is *b*, then *b* means have been fitted to the data, to calculate the between block variability. Thus in moving from *M* to *E*, $b - 1$ degrees of freedom have been used. Vector *EY* then represents the within block variability, which we now go on to partition further. Vector *EF* represents the variability explained by treatment, when block has already been taken into account. If the number of treatments is *t*, this partition will involve another $t - 1$ degrees of freedom.

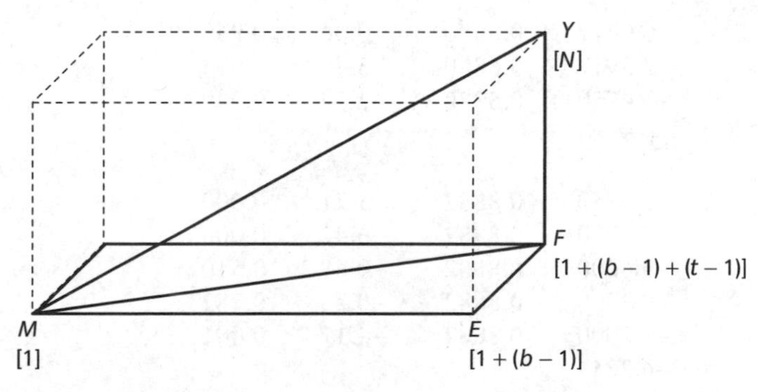

Fig. 5.6 The geometrical analogy of statistical elimination with categorical variables.

Calculating the fitted model for two categorical variables

Returning to the blocked bean experiment, the output is reproduced in Box 5.4, with the coefficients table in addition to the ANOVA table. While the first table tells us which of the explanatory variables are important in determining yield, the second tells us how we can predict yield from the variables block and bean. This is an extension of the process by which YIELD was predicted from FERTIL in Chapter 1. Following the convention we adopted earlier, the coefficient given for 'constant' is the grand mean, that is, an unweighted average of the group means. This is our chosen reference group, but as before this will vary depending upon your package. The figures given for the blocks are the deviations of the block means from the constant—and so the fourth deviation can be calculated, as all deviations must sum to zero. For the six different bean varieties, five deviations have been given, and the sixth can be calculated. So the other parameters are the difference of the other groups from the reference group. How

BOX 5.4 ANOVA and Coefficient tables for a randomised blocked design

General Linear Model

Word equation: YIELD = BLOCK + BEAN

BLOCK and BEAN are categorical

Analysis of variance table for YIELD, using Adjusted SS for tests

Source	DF	Seq SS	Adj SS	Adj MS	F	P
BLOCK	3	52.895	52.895	17.632	4.66	0.017
BEAN	5	444.435	444.435	88.887	23.48	0.000
Error	15	56.795	56.795	3.786		
Total	23	554.125				

Coefficients table

Term	Coef	SECoef	T	P
Constant	16.6750	0.3972	41.98	0.000
BLOCK				
1	0.0417	0.6880	0.06	0.953
2	2.3917	0.6880	3.48	0.003
3	−1.4750	0.6880	−2.14	0.049
4	*−0.9584*			
BEAN				
1	5.0750	0.8882	5.71	0.000
2	5.7000	0.8882	6.42	0.000
3	−0.6000	0.8882	−0.68	0.510
4	−0.2500	0.8882	−0.28	0.782
5	−3.7000	0.8882	−4.17	0.001
6	*−6.225*			

Table 5.1 The parameters corresponding to the coefficients for a randomised blocked design.

Term		Coef
Constant		μ
Block	1	α_1
	2	α_2
	3	α_3
	4	$-\alpha_1 - \alpha_2 - \alpha_3$
Treatment	1	β_1
	2	β_2
	3	β_3
	4	β_4
	5	β_5
	6	$-\beta_1 - \beta_2 - \beta_3 - \beta_4 - \beta_5$

these parameters would be presented in other packages is illustrated in the package specific supplements.

So the model for two categorical variables follows the general pattern:

$$y = \mu + \begin{pmatrix} \text{Block} & \text{Difference} \\ 1 & \alpha_1 \\ 2 & \alpha_2 \\ 3 & \alpha_3 \\ 4 & -\alpha_1 - \alpha_2 - \alpha_3 \end{pmatrix} + \begin{pmatrix} \text{Treatment} & \text{Difference} \\ 1 & \beta_1 \\ 2 & \beta_2 \\ 3 & \beta_3 \\ 4 & \beta_4 \\ 5 & \beta_5 \\ 6 & -\beta_1 - \beta_2 - \beta_3 - \beta_4 - \beta_5 \end{pmatrix} + \varepsilon$$

with estimates of the parameters being found in the coefficient table (Table 5.1).

It can now be seen that the unitalicised numbers in the square brackets of the last equation correspond to the numbers of parameters fitted at that point and given in the coefficient table, the italicised ones being fixed once the others are known. The final symbol, ε, represents the scatter around the model—in other words, it is a random variable, drawn independently for each data point, from a Normal distribution with mean zero and variance equal to the error variance and estimated as the error mean square in the ANOVA table. The coefficients table provides us with the best estimates of the parameters, and these are used to calculate the fitted values (taking the mean of ε, zero). The fitted values estimate the signal in the data. The error variance estimates the noise in the data. The aim of the analysis is to separate the signal and the noise.

Drawing together the two previous sections, the geometrical analogy reflects the model fitting process. There are three steps to the partitioning—fitting the grand mean, then the block means, and then the treatment means. The number of extra parameters calculated at each step account for the degrees of freedom. The three steps of the model fitting process are also reflected in the way in which we have chosen to represent the model parameters (grand mean, block deviations and treatment deviations). This is why we have adopted this convention, similar to that followed by Minitab. Other packages present the fitted values differently, and this is covered in the package specific supplements. Which form of aliassing is used, (i.e. how the parameters are presented), is just a matter of internal bookkeeping. The number of parameters estimated and the essential outcome of the analysis will always be the same.

5.3 The concept of orthogonality

The perfect design

In the previous chapter, the concept of adjusted and sequential sums of squares was introduced, in the context of analysing two continuous variables. How can this same concept be applied to two categorical variables? Inspecting the adjusted and sequential sums of squares for BLOCK in Box 5.4, it transpires that they are exactly the same. This means that the variability explained by block is the same, regardless of whether bean variety has been taken into account or not, so BLOCK is **orthogonal** to BEAN. Two variables are said to be orthogonal when knowledge of one variable provides no information about another variable. In the case of two continuous variables, this is unlikely to be the case, as the two variables would have to be *perfectly* uncorrelated (with a correlation coefficient of zero). When two continuous variables are measured, then they are likely to be slightly correlated by chance alone, and so would influence each other's informativeness. How can we recognise two categorical variables as orthogonal?

If a table is drawn up of the number of replicates in each block–treatment combination, then this can be used to illustrate orthogonal designs. In Plan A of Fig. 5.7, all block–treatment combinations have exactly the same number of replicates. Within each treatment, each block is represented exactly equally, therefore even if there is a block effect (for example, Block 2 is very nutrient poor), this would have no influence on the treatment comparisons (because all treatment comparisons have the same proportion of plots in Block 2—it will affect all treatments equally). The same is true of treatment within block. This idea can also be illustrated with probabilities. By choosing a plot at random, even if the block it originates from is known, this does not provide any information about which treatment was applied (the probability that Treatment 1 was applied remains at $^1/_4$, regardless of knowledge of block). So the variation explained by treatment remains the same, regardless of whether block has been controlled for or not (and vice versa), and the sequential and adjusted sums of squares for each of these two variables will be equal.

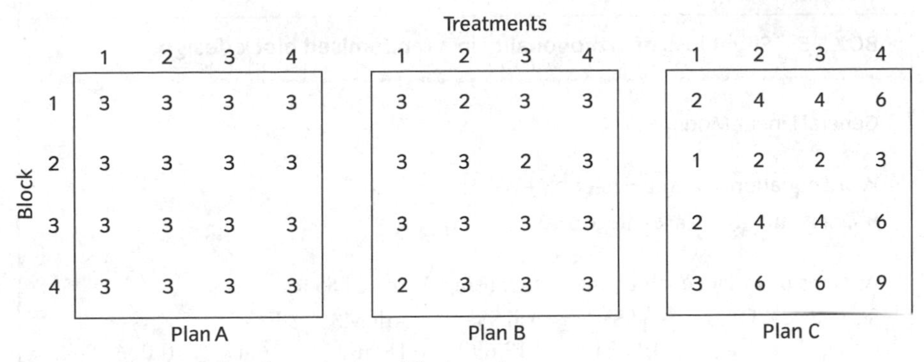

Fig. 5.7 Allocation of replicates between blocks and treatments.

Plan C illustrates a rather unusual design which is orthogonal, yet not balanced. This is achieved by allocating replicates to treatments and blocks proportionately. For example, Treatment 1 has 2, 1, 2 and 3 replicates in Blocks 1 to 4 respectively. The other treatments have some multiple of this pattern. If one plot is chosen at random, then knowledge of which block it is allocated to will still not provide any information about which treatment has been applied to it. Similarly, the same proportion of plots are allocated to each block within one treatment (always 1/8th of the plots are in Block 2 for example). Thus again any block effects will influence all treatments equally—making this design orthogonal also. However, the imbalance in the allocation of replicates to treatments and blocks results in the experimenter acquiring very much more information about Treatment 4 (24 replicates) than Treatment 1 (8 replicates). So these designs are *not* balanced. In practice it is usual to be equally interested in obtaining information on all treatments, and so most common designs are both orthogonal and balanced.

Why is it desirable to design our experiments to be orthogonal? The answer to this is that it makes the interpretation of output so much easier. Unlike the examples considered in the last chapter, it is not necessary to carefully consider what is going on by comparing the sequential and adjusted sums of squares, because they are equal. Thus all the considerations of shared information or increased informativeness which complicated our interpretations with two continuous variables are not relevant to two orthogonal variables. This advantage of orthogonality will become most useful when the models fitted are rather more complicated (for example see Chapter 10).

In real life, however, it is not uncommon for an experiment which was designed to be orthogonal to suffer from the occasional mishap. For example, Plan B of Fig. 5.7 has had three plots destroyed during the course of the trial. This will result in slight differences between the sequential and adjusted sums of squares in the resulting ANOVA table.

Returning to the bean varieties dataset, the analysis is repeated in Box 5.5 —but this time, two plots are omitted. In comparing this with the original orthogonal analysis of Box 5.4, it just so happens that in this case the F-ratios

BOX 5.5 Slight loss of orthogonality in a randomised block design

General Linear Model

Word equation: MYIELD = MBLOCK + MBEAN

MBLOCK and MBEAN are categorical

Analysis of variance table for MYIELD, using Adjusted SS for tests

Source	DF	Seq SS	Adj SS	Adj MS	F	P
MBLOCK	3	49.291	49.690	16.563	6.47	0.006
MBEAN	5	449.413	449.413	89.883	35.09	0.000
Error	13	33.296	33.296	2.561		
Total	21	532.000				

Coefficients table

Term	Coef	SECoef	T	P
Constant	16.7007	0.3529	47.33	0.000
MBLOCK				
1	0.0160	0.5813	0.03	0.978
2	2.3660	0.5813	4.07	0.001
3	−1.5007	0.5813	−2.58	0.023
4	*−0.8813*			
MBEAN				
1	5.0493	0.7425	6.80	0.000
2	6.7389	0.8435	7.99	0.000
3	−0.6257	0.7425	−0.84	0.415
4	−0.2757	0.7425	−0.37	0.716
5	−3.7257	0.7425	−5.02	0.000
6	*−7.1611*			

for both explanatory variables are higher in this reduced dataset, and the conclusions do not alter. As a consequence of the loss of orthogonality however the adjusted and sequential sums of squares for block do differ a little (those for bean variety are the same, as this is the last term in the model). A second consequence is that the standard errors in the coefficient table also differ slightly. The standard error of a mean is calculated by $\frac{s}{\sqrt{n}}$, where n is the number of data points which contribute to that mean. The standard error for bean Variety 2 is higher than for other varieties, as one of the missing plots was of that variety. So any confidence interval calculated for that mean will be wider than for Variety 1 (for example). So while our conclusions have not altered, we have lost some of the precision in those means affected.

Had the loss of orthogonality been more extensive, there would have been greater differences between sequential and adjusted sums of squares. In losing the advantages of orthogonality, we move back towards experiencing the problems of interpretation discussed with two continuous variables in the last chapter, and should consider the possibilities of shared information or increased informativeness between the two explanatory variables.

Three pictures of orthogonality

Returning to the original definition of orthogonality, it was stated that two factors are orthogonal to each other if the variability explained by each factor is the same regardless of whether the other factor has been taken into account or not. So if we represent the amount of variability explained by a factor as a line, the length of this line will be the same in the solo and the additive models. The line representing the additive model is simply the sum of the lines for the two solo models (Fig. 5.8).

In Fig. 5.6, the partitioning of variation into that explained by the model and that explained by error is represented by the vertical triangle. The further partitioning of variation into the two explanatory variables (block and treatment for example), is represented by the triangle in the horizontal plane. This is reproduced below (Fig. 5.9). In the case of orthogonal variables (and only then), if the order of the variables in the word equation is changed, this is simply represented by going around the rectangle in the other direction. In other words, $ME_1 = E_2F$ and $ME_2 = E_1F$. This is just another expression of the definition of orthogonality.

Finally, taking the horizontal plane, the orthogonal equivalent of Figs 4.8 and 4.9 can be drawn. The result is a rectangle (Fig. 5.10). The word 'orthogonal' actually means 'at right angles'—so in fact the statistical use of this word derives from the geometrical analogy. The right angle in question is $(E_1)M(E_2)$ —so the two variables are at right angles. But $M(E_1)F$ must also be at right angles (from our earlier discussions of the geometrical analogy, with the fitted values being as close as possible to the data). So the end result is that $F(E_1)M(E_2)$ must be a rectangle. This forces the SS of each of the two variables to be the same, regardless of whether the other variable has been eliminated or not.

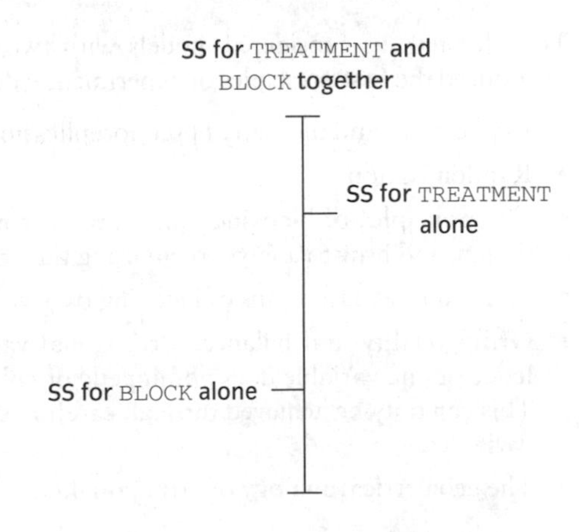

Fig. 5.8 The sum of squares of two orthogonal variables together equal the sum of squares for the model containing both variables.

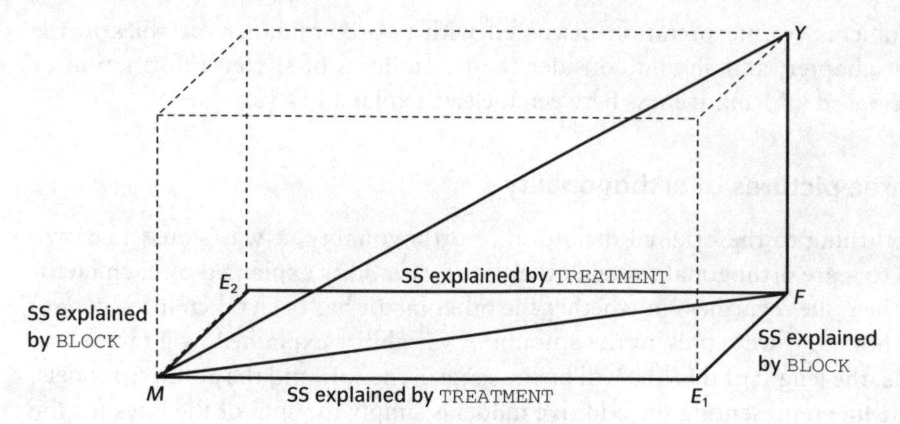

Fig. 5.9 Partitioning the sums of squares by two different routes.

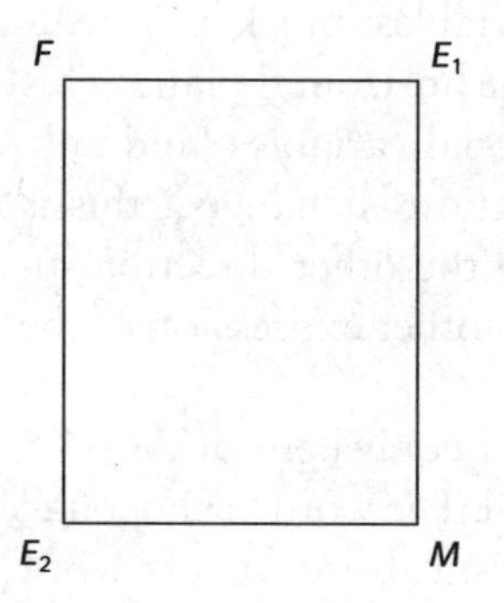

Fig. 5.10 The geometrical analogy of orthogonality.

5.4 Summary

This chapter has considered models with two categorical explanatory variables, and introduced the first principles of experimental design. Specifically:

- Replication, and the trap of pseudoreplication.
- Randomisation.
- The principles of blocking: both how to construct blocks effectively as part of the design, and how to analyse them using the principle of statistical elimination.
- Latin squares as a means of blocking two ways.
- Orthogonality and balance. Orthogonal variables are those for which the knowledge of one variable does not impede or enhance the predictive power of the other. This can only be achieved through careful experimental design involving categorical variables.
- The geometrical analogy of orthogonality.

5.5 Exercises

Growing carnations

A flower grower decided to investigate the effects of watering and the amount of shade on the number of saleable carnation blooms produced in his nursery. He designed his experiment to have three levels of watering (once, twice or three times a week) and four levels of shade (none, 1/4, 1/2 and fully shaded). To conduct this experiment he needed to grow the carnation plots in three different beds. In case these beds differed in fertility or other important features, he decided to use these beds as blocks. After four weeks, he analysed the data by counting the number of blooms, and using the square root as the response variable, SQBLOOMS. He then fitted a GLM with three categorical explanatory variables: BED, WATER and SHADE (the data are stored in the *blooms* dataset). The output is shown in Box 5.6.

BOX 5.6 Analysis of the number of carnation blooms with bed, water and shade

General Linear Model

Word equation: SQBLOOMS = BED + WATER + SHADE

BED, WATER and SHADE are categorical variables

Analysis of variance table for SQBLOOMS, using Adjusted SS for tests

Source	DF	Seq SS	Adj SS	Adj MS	F	P
BED	2	4.1323	4.1323	2.0661	9.46	0.001
WATER	2	3.7153	3.7153	1.8577	8.50	0.001
SHADE	3	1.6465	1.6465	0.5488	2.51	0.079
Error	28	6.1173	6.1173	0.2185		
Total	35	15.6114				

Term	Coef	SECoef	T	P
Constant	4.02903	0.07790	51.72	0.000
BED				
1	0.0620	0.1102	0.56	0.578
2	0.3805	0.1102	3.45	0.002
3	*−0.4425*			
WATER				
1	−0.4110	0.1102	−3.73	0.001
2	0.3731	0.1102	3.39	0.002
3	*0.0379*			
SHADE				
1	0.0965	0.1349	0.72	0.480
2	0.2934	0.1349	2.17	0.038
3	−0.1191	0.1349	−0.88	0.385
4	*−0.2708*			

BOX 5.7 The carnation bloom analysis without bed used as a block

General Linear Model

Word equation: SQBLOOMS = WATER + SHADE

WATER and SHADE are categorical variables

Analysis of variance table for SQBLOOMS, using Adjusted SS for tests

Source	DF	Seq SS	Adj SS	Adj MS	F	P
WATER	2	3.7153	3.7153	1.8577	5.44	0.010
SHADE	3	1.6465	1.6465	0.5488	1.61	0.209
Error	30	10.2496	10.2496	0.3417		
Total	35	15.6114				

Term	Coef	SECoef	T	P
Constant	4.02903	0.09742	41.36	0.000
WATER				
1	−0.4110	0.1378	−2.98	0.006
2	0.3731	0.1378	2.71	0.011
3	*0.0379*			
SHADE				
1	0.0965	0.1687	0.57	0.572
2	0.2934	0.1687	1.74	0.092
3	−0.1191	0.1687	−0.71	0.486
4	*−0.2708*			

1. Is the analysis in Box 5.6 orthogonal?

He then wondered whether it had been worth treating the beds as blocks, or whether future experiments could be fully randomised across beds. So he repeated the analysis without using bed as a blocking factor. This is shown in Box 5.7.

2. Was it worthwhile blocking for bed? If so, why?

He then discovered that a visitor had picked carnations from three of his experimental plots. He decided that because the final bloom numbers for these plots were inaccurate, he would exclude them from his analysis. So he produced a new, shorter data variable SQ2 for the square root of blooms, and explanatory variables B2, W2 and S2 for bed, water and shade levels respectively. This third analysis is presented in Box 5.8.

3. Which parts of the output differ in Box 5.8 but are the same in Box 5.6 and why? Does this fundamentally alter our conclusions?
4. Using the coefficient table given in Box 5.6, draw histograms illustrating how the number of blooms vary with level of water and level of shade.

BOX 5.8 Analysis of the carnation blooms with three plot values removed

General Linear Model

Word equation: SQ2 = B2 + W2 + S2

B2, W2 and S2 are categorical

Analysis of variance table for SQ2, using Adjusted SS for tests

Source	DF	Seq SS	Adj SS	Adj MS	F	P
B2	2	2.7626	2.6490	1.3245	8.03	0.002
W2	2	5.0793	4.6764	2.3382	14.18	0.000
S2	3	0.8072	0.8072	0.2691	1.63	0.207
Error	25	4.1213	4.1213	0.1649		
Total	32	12.7704				

The dorsal crest of the male smooth newt

Male smooth newts (*Triturus vulgaris*) develop a dorsal crest during the breeding season. During courtship the male releases pheromones and waggles his tail. The crest is thought to help the male waft the pheromones past the female's snout. A student conducted a survey to investigate variation in the size of the dorsal crest. She visited 10 local ponds, and measured a total of 87 male newts over a period of two weeks. The following data are recorded in the *newt* dataset:

LSVL: Logarithm of the snout-vent length in mm—a measure of skeletal size.

LCREST: Logarithm of the height of the dorsal crest in mm.

POND: A code from 1 to 10 for the pond at which the male was captured.

DATE: The day of the study on which the male was measured.

1. Taking LCREST as the response variable, analyse the data to investigate if the size of the dorsal crest reflects the body size of the newt.
2. Why is it probably a good idea to include POND in a model of this sort? Does it seem to matter in this case?
3. What circumstances might make it desirable to include DATE? How would you detect these circumstances?

6 Combining continuous and categorical variables

6.1 Reprise of models fitted so far

In the first five chapters, several models have been fitted to data. All of these models have a number of things in common. In each case, they are aiming to predict one variable (the *Y* variable, or data variable) from one or two explanatory variables. The *Y* variables considered so far are always continuous, but the data variables are either categorical or continuous. The models fitted are specified by word equations. Some of the models that have been used up to this point are listed below:

YIELD = FERTIL
YIELDM = VARIETY
VOLUME = HEIGHT
MATHS = ESSAYS
SPECIES2 = SPECIES1

AMA = YEARS + HGHT
FINALHT = INITHT + WATER
WGHT = RLEG + LLEG
POETSAGE = BYEAR + DYEAR
LVOLUME = LDIAM + LHGHT

YIELD = BLOCK + BEAN
SEEDS = COLUMN + ROW + TREATMT.

Analyses with categorical explanatory variables are looking for differences between the means, while those with continuous variables are looking for linear relationships between the *Y* and the *X* variables. With general linear models (GLMs), both ANOVA and regression can be conducted under the same umbrella. The results may then be summarised as two tables of output:

1. The ANOVA table. This table sums up the evidence bearing on *whether* the explanatory variables are related to the data variable. The answer to this is contained within the *p*-values—with low *p*-values indicating that the null hypothesis should be rejected. The *p*-values may be based on the

sequential or adjusted sums of squares. When based on the latter, the question answered is 'Does this variable have any predictive power when the other explanatory variables are already known?'

2. The coefficients table. This tells us *how* the variables are related. From this table an equation can be drawn up, which will enable the data variable to be predicted from the explanatory variables. The different levels of a categorical variable are represented as deviations, and the continuous variables are related to the data variable through a slope. Both the deviations and the slopes are the best guess the dataset can provide of the true population parameters. The standard deviations of those parameters (or standard errors) are also given in the coefficient table, and show how reliably the relationship has been estimated by the data.

The fitted value equation can be viewed as estimating the signal in the data. The differences between the data and the model are the residuals, the variance of which is the error variance (that is, the error mean square of the ANOVA table). So, in summary, a GLM analysis can be seen as separating the signal from the noise, and informs us whether the signal is sufficiently strong to be included in representations of the data.

The models that have been fitted so far have consisted of either continuous or categorical variables, with up to three explanatory variables in one model (as in the *urban foxes* data analysis). It is possible to add many explanatory variables (though as will be discussed in Chapter 11, this can lead to difficulties in interpretation). The only limit to the number of variables is the number of degrees of freedom—though the most useful models aim to represent the relationship using as few variables as possible.

It is also possible to combine continuous and categorical variables—this is often referred to as analysis of covariance.

6.2 Combining continuous and categorical variables

To combine continuous and categorical variables within the same model, the word equation is specified as before, including any commands to distinguish categorical and continuous variables. In fact, in introducing the principle of statistical elimination, one example has been included in Chapter 4—the case of measuring growth rates of tree seedlings in different watering regimes. A second example is considered here in more detail.

Looking for a treatment for leprosy

Leprosy is caused by a bacillus. Three treatments are being compared, to see which proves the most effective in reducing the bacillus score. There are 10 patients in each treatment group—so the simplest analysis would be to

BOX 6.1 A treatment for leprosy

General Linear Model

Word equation: BACAFTER = BACBEF + TREATMT

BACBEF is continuous, TREATMT is categorical

Analysis of variance table for BACAFTER, using Adjusted SS for tests

Source	DF	Seq SS	Adj SS	Adj MS	F	P
BACBEF	1	587.48	515.01	515.01	35.22	0.000
TREATMT	2	83.35	83.35	41.67	2.85	0.076
Error	26	380.15	380.15	14.62		
Total	29	1050.98				

Coefficients table

Term	Coef	SECoef	T	P
Constant	−0.013	1.806	−0.01	0.994
BACBEF	0.8831	0.1488	5.93	0.000
TREATMT				
1	−1.590	1.012	−1.57	0.128
2	−0.726	1.002	−0.72	0.475
3	*−2.316*			

compare bacillus score after treatment between the three groups (a straightforward ANOVA). However, the score will depend to a certain extent on how sick the patients were in the first place—so it would be a more powerful test if these initial differences could be taken into account. The question to be asked becomes 'How does the final bacillus score vary between treatments, once the initial differences in bacillus levels have been taken into account?' The variables BACAFTER, TREATMT and BACBEF are stored in the *leprosy* dataset and analysed in Box 6.1.

What is the best fit equation derived from this model? If the model fitted had just been a regression between BACBEF and BACAFTER, then the model would be a straight line. If BACBEF was not being used as a covariate, and the model only involved the three treatments, then the fitted values would be three means. Combining the two explanatory variables allows the mean level of bacillus after treatment to vary depending upon the level before treatment. So the model is now three parallel lines, one for each treatment.

The model derived from the coefficients table is:

$$\text{BACAFTER} = -0.013 + 0.8831 \times \text{BACBEF} + \begin{pmatrix} \text{TREATMT} & \\ 1 & -1.590 \\ 2 & -0.726 \\ \mathit{3} & \mathit{2.316} \end{pmatrix}.$$

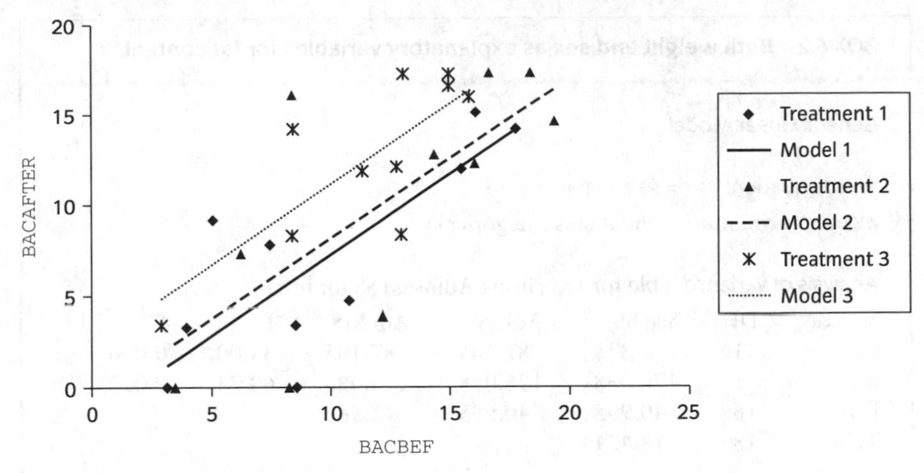

Fig. 6.1 The best fit lines plotted with the data for the leprosy experiment.

Note that the coefficient for treatment 3 (in italics) is calculated from the other two, as the deviations must sum to zero, as before (using our chosen method of aliassing). The main effect of treatment alters the intercept, and so after simplification, the equations for the three lines are:

BACAFTER =	$-1.603 + 0.8831 \times$ BACBEF	TREATMT 1
BACAFTER =	$-0.739 + 0.8831 \times$ BACBEF	TREATMT 2
BACAFTER =	$2.303 + 0.8831 \times$ BACBEF	TREATMT 3

These three lines are drawn with the data in Fig. 6.1. So why are the three lines parallel? At this point, they have actually been constrained to be parallel, by the word equation that has been fitted. In reality, the data may suggest something different, e.g. that the third treatment does work, but only on very sick patients. Then a better fit between model and data would be provided by the line for Treatment 3 having a shallower slope (or maybe a curve would fit better than a straight line). To ask these sorts of questions, more complex models need to be fitted, and these are explored in Chapters 7 and 11.

Sex differences in the weight—fat relationship

Total body fat was estimated as a percentage of body weight from skinfold measurements of 19 students in a fitness program. In Section 2.10, this dataset (*fats*) was examined to look at the relationship between weight and fat, with the aim being to predict fat content from the weight of a person. No significant relationship could be found. This same dataset is reanalysed here with additional information on the sex of each participant.

The data have been plotted in Fig. 6.2, with females represented as diamonds, and males as squares.

BOX 6.2 Both weight and sex as explanatory variables for fat content

General Linear Model

Word equation: FAT = WEIGHT + SEX

WEIGHT is continuous and SEX is categorical

Analysis of variance table for FAT, using Adjusted SS for tests

Source	DF	Seq SS	Adj SS	Adj MS	F	P
WEIGHT	1	1.328	87.105	87.105	34.00	0.000
SEX	1	176.098	176.098	176.098	68.73	0.000
Error	16	40.995	40.995	2.562		
Total	18	218.421				

Coefficients table

Term	Coef	SECoef	T	P
Constant	13.010	2.679	4.86	0.000
WEIGHT	0.21715	0.03724	5.83	0.000
SEX				
1	3.9519	0.4767	8.29	0.000
2	*−3.9519*			

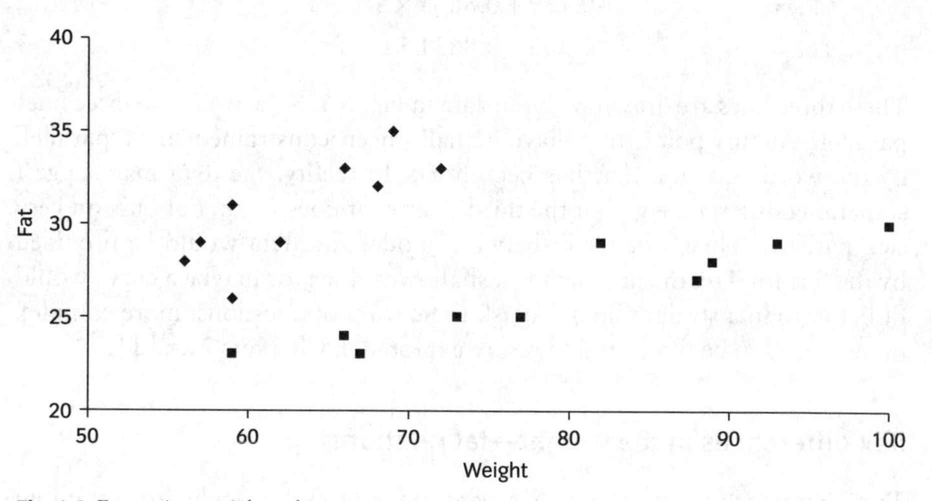

Fig. 6.2 Fat against weight, taking sex into account.

In the plot of Fig. 6.3, WEIGHT is ignored, and the variable FAT is plotted simply by SEX.

From the plots, it can be seen that females have higher fat content than males, and that it looks likely that there is a relationship between FAT and WEIGHT for both sexes, but that this relationship may be different. The GLM

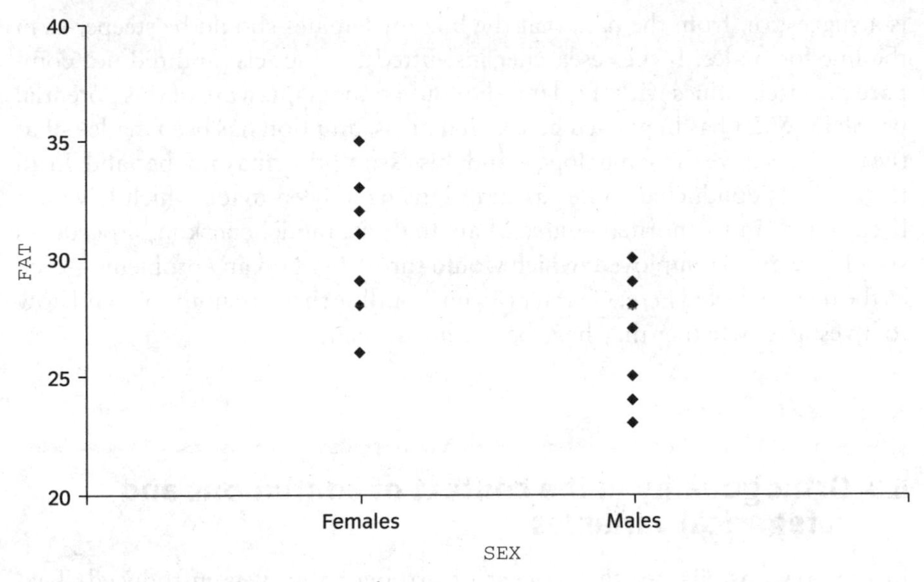

Fig. 6.3 Fat against sex, ignoring weight.

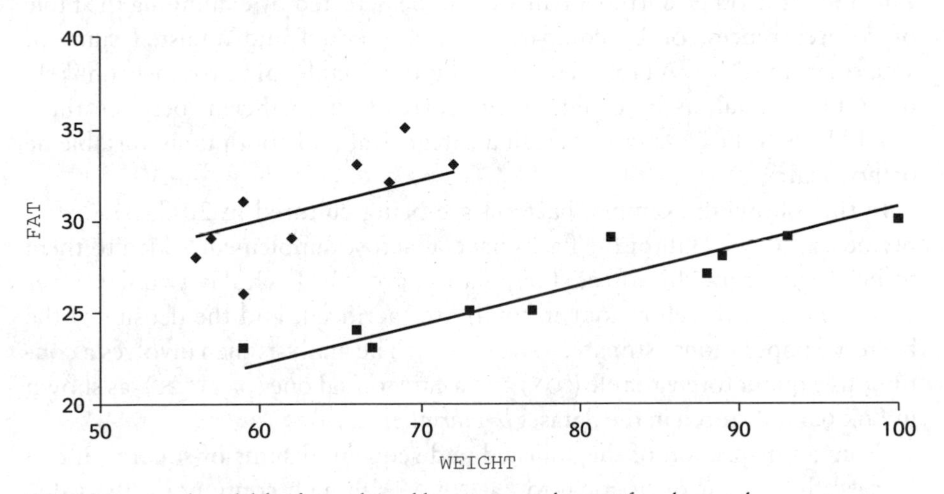

Fig. 6.4 Data and model for fat explained by WEIGHT and SEX, ◆ female, ■ male.

in Box 6.2 illustrates that both WEIGHT and SEX are significant predictors of FAT, giving a fitted value equation of:

$$\text{FAT} = 13.01 + 0.21717 \times \text{WEIGHT} + \begin{bmatrix} \text{SEX} & \\ 1 & 3.9519 \\ 2 & -3.9519 \end{bmatrix}.$$

The model and data are compared in Fig. 6.4. As in the previous example, the model fitted has constrained the two lines to be parallel. However, there is a suggestion from the data that the line for females should be steeper than the line for males. If the researcher just fitted the models, and did not compare the fitted values with the data, she could remain unaware of this potential problem. What has happened here is that an assumption has been made—that the two lines have the same slope—and this assumption may not be valid. In all the analyses conducted so far, assumptions have been made which have not been tested. In the normal course of an analysis, model checking procedures should always be employed, which would throw light on any problems—such as the one suspected here. Chapters 8 and 9 outline these assumptions and how to investigate whether they have been contravened.

6.3 Orthogonality in the context of continuous and categorical variables

In the previous chapter the concept of orthogonality was introduced. Two categorical variables are orthogonal if knowledge of one provided no information about the other. Orthogonality could be detected by examining the table of co-occurrences, or by comparing the sequential and adjusted sums of squares in an ANOVA table. Two continuous variables are extremely unlikely to be orthogonal, as by definition the correlation coefficient between them would have to be exactly zero. Can a categorical and continuous variable be orthogonal?

In the following example, bacteria are being cultured in 20 flasks over a period of 5 days. Half of the flasks have a lactose supplement added to them to investigate how this affected population growth. Each day two flasks per treatment (and therefore four in total) are sacrificed, and the density of the bacterial populations estimated (BACTERIA). The analysis then involves a continuous explanatory variable (DAY) and a categorical one (LACTOSE), as shown in Box 6.3 (all stored in the dataset *bacterial growth*).

From an inspection of the adjusted and sequential sums of squares, it can be seen that DAY is orthogonal to LACTOSE. Although continuous, the value of DAY is set within the experimental design, and therefore exactly the same values are recorded for the two levels of LACTOSE (namely 1, 2, 3, 4 and 5). A table of co-occurrences would illustrate orthogonality in the same way as between two categorical variables.

Orthogonality between a continuous and categorical variable can be illustrated in another way. In Box 6.4, DAY has been taken as the response variable, and LACTOSE as the explanatory variable—asking 'To what extent can DAY be explained by the level of LACTOSE?' The definition of two orthogonal variables is that one provides no information about the other. No explanatory power results in an F-ratio of zero, as shown in Box 6.4.

BOX 6.3 Bacterial growth at two levels of lactose

General Linear Model

Word equation: BACTERIA = DAY + LACTOSE

DAY is continuous and LACTOSE is categorical

Analysis of variance table for BACTERIA, using Adjusted SS for tests

Source	DF	Seq SS	Adj SS	Adj MS	F	P
DAY	1	297.97	297.97	297.97	130.56	0.000
LACTOSE	1	397.07	397.07	397.07	173.00	0.000
Error	17	38.80	38.80	2.28		
Total	19	733.85				

Coefficients table

Term	Coef	SECoef	T	P
Constant	3.0939	0.7922	3.91	0.001
DAY	2.7293	0.2389	11.43	0.000
LACTOSE				
1	−4.4557	0.3378	−13.19	0.000
2	*4.4557*			

BOX 6.4 Illustrating orthogonality between continuous and categorical variables

General Linear Model

Word equation: DAY = LACTOSE

LACTOSE is categorical

Analysis of variance table for DAY, using Adjusted SS for tests

Source	DF	Seq SS	Adj SS	Adj MS	F	P
LACTOSE	1	0.000	0.000	0.000	0.00	1.000
Error	18	40.000	40.000	2.222		
Total	19	40.000				

Coefficients table

Term	Coef	SECoef	T	P
Constant	3.0000	0.3333	9.00	0.000
LACTOSE				
1	0.0000	0.3333	−0.00	1.000
2	*0.0000*			

So just as the correlation coefficient, r, must equal 0 for two continuous variables to be orthogonal, so the F-ratio must equal 0 for a continuous and categorical variable to be orthogonal. The average value of DAY must be exactly the same at each level of LACTOSE.

6.4 Treating variables as continuous or categorical

Some variables may be legitimately treated as either continuous or categorical. The explanatory variable DAY in the previous dataset provides an example of this—it is treated as continuous, but could equally well be categorical with five levels. Is there any particular advantage in treating it as one or the other?

In Box 6.5 the analysis of Box 6.3 is repeated, but DAY is fitted as a categorical variable. With exactly the same dataset DAY is still significant, but the F-ratio has dropped from 130.6 to 27.3. Why is this?

BOX 6.5 Analysing bacterial growth with DAY as a categorical variable

General Linear Model

Word equation: BACTERIA = DAY + LACTOSE

DAY and LACTOSE as categorical

Analysis of variance table for BACTERIA, using Adjusted SS for tests

Source	DF	Seq SS	Adj SS	Adj MS	F	P
DAY	4	298.51	298.51	74.63	27.30	0.000
LACTOSE	1	397.07	397.07	397.07	145.28	0.000
Error	14	38.26	38.26	2.73		
Total	19	733.85				

Coefficients table

Term	Coef	SECoef	T	P
Constant	11.2820	0.3697	30.52	0.000
DAY				
1	−5.2730	0.7393	−7.13	0.000
2	−2.8054	0.7393	−3.79	0.002
3	−0.2044	0.7393	−0.28	0.786
4	2.6235	0.7393	3.55	0.003
5	*−5.6593*			
LACTOSE				
1	−4.4557	0.3697	−12.05	0.000
2	*4.4557*			

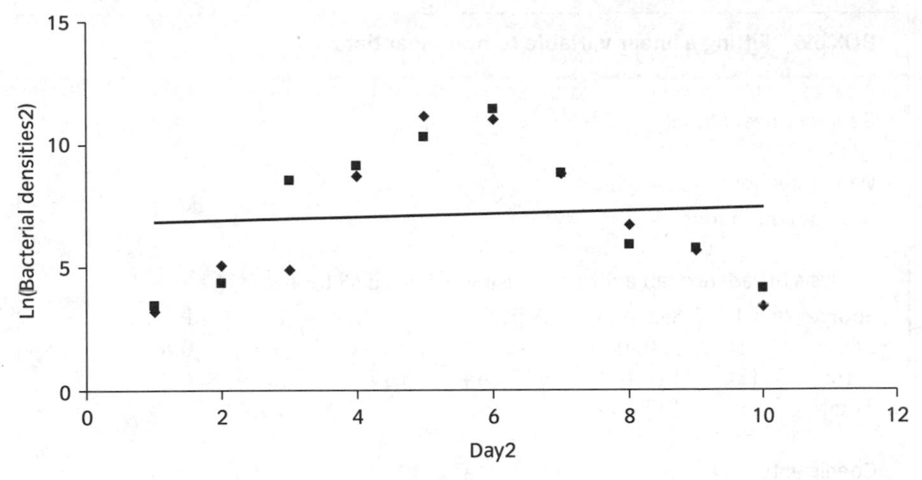

Fig. 6.5 Fitting a line through nonlinear population data.

The adjusted sum of squares explained by DAY is fractionally larger. This is not suprising, as fitting five means is likely to produce a final model that is closer to the data than fitting a straight line. In producing the mean square for DAY however the sum of squares is divided by the degrees of freedom. A categorical variable with 5 levels has 4 degree of freedom, whereas a line has only 1 degree of freedom. It is this which causes the substantial difference in the mean square, and therefore in the *F*-ratio. So if DAY is analysed as a continuous variable, the *F*-ratio is likely to be larger and therefore significant. So the analysis of Box 6.3 is more **powerful** than the analysis of Box 6.5 (more likely to reject the null hypothesis when it is not true).

One other point to note is that in spite of the fact that the analysis in which DAY is continuous is the more powerful, the error SS is slightly larger. This is because the line fits less well than 5 means, and the nonlinear part of this fit will remain unexplained in error. We investigate ways in which this nonlinear part of the fit may also be analysed in Chapter 7.

Does this mean that variables which fall into this ambiguous area of being categorical and continuous should always be treated as continuous? In the example illustrated here, the continuous analysis is superior, but this is partly because the factor DAY has a linear effect on bacteria numbers. In another dataset (*bacterial growth2*) monitoring bacterial growth under different conditions, the bacteria numbers rose to a peak on day 5, and then from days 6 to 10 dropped. In this case there may be significant differences between the days, yet a straight line through these points may not have a slope significantly different from zero (Fig. 6.5 and Box 6.6). If your analysis is combined with a graphical inspection of the data you would be aware of these possibilities.

BOX 6.6 Fitting a linear variable to nonlinear data

General Linear Model

Word equation: BAC2 = DAY2

DAY2 as continuous

Analysis of variance table for BAC2, using Adjusted SS for tests

Source	DF	Seq SS	Adj SS	Adj MS	F	P
DAY2	1	0.494	0.494	0.494	0.06	0.809
Error	18	147.144	147.144	8.175		
Total	19	147.638				

Coefficients table

Term	Coef	SECoef	T	P
Constant	6.666	1.381	4.83	0.000
DAY2	0.0547	0.2226	0.25	0.809

How best to analyse such nonlinear data is covered in more detail in Chapter 7, and the pros and cons of analysing variables as continuous where possible is discussed in more detail in Chapter 10. In summary, data which follow a trend are best analysed with a continuous variable, given a choice, as this proves to be the more powerful method of analysis.

6.5 The general nature of General Linear Models

General Linear Models are a comprehensive set of techniques that cover a wide range of analyses. For historical reasons, many of these tests have different names, and are still frequently referred to by these names in the scientific literature. In fact, the General Linear Model principle covers such a range of analyses that this includes some that do not have separate names. Table 6.1 summarises the GLMs covered so far, and their traditional names.

In the final two rows of this table, models are mentioned which are yet to be covered. For multiple regression, see Chapter 11, and for factorial experiments, see Chapter 7. Now not only the power and flexibility of general linear modelling is becoming apparent, but also the simplicity. One umbrella suits many situations. The key to success is learning to phrase your hypothesis as a word equation. It is also true that the same assumptions apply to all GLMs, and the same techniques can be used to test your model, to see if these assumptions have been contravened. Up to this point, it has been assumed that the models

Table 6.1 Comparing word equations with traditional tests

Example	Traditional test	GLM word equation
Comparing the yield between two fertilisers	Two sample *t*-test	YIELD = FERTIL
Comparing the yield between three or more fertilisers	One way analysis of variance	YIELD = FERTIL
Comparing the yield between fertilisers in a blocked experiment	One way blocked analysis of variance	YIELD = BLOCK + FERTIL
Investigating the relationship between fat content and weight	Regression	FAT = WEIGHT
Investigating the relationship between fat content and sex, controlling for weight differences	Analysis of covariance	FAT = WEIGHT + SEX
Investigating which factors may influence the likelihood of spotting whales on a boat trip	Multiple regression	LGWHALES = CLOUD + RAIN + VIS
Investigating the factors which affect the number of blooms on prize roses	Two way analysis of variance	SQBLOOMS = SHADE | WATER

fitted have been appropriate and valid—however, all GLMs should be subject to model criticism, and this is the subject of Chapters 8 and 9.

6.6 Summary

In summary

- Analyses have been introduced in which continuous and categorical variables have been mixed in an additive model.
- Orthogonality between a categorical and continuous variable has been defined.
- Some variables may be treated as categorical or continuous. Continuous explanatory variables provide a more powerful analysis for data following a trend.
- The generality of general linear models has been summarised, in which many traditional tests have been represented by word equations.

6.7 Exercises

Conservation and its influence on biomass

An ecological study was conducted into the effect of conservation on the biomass of vegetation supported by an area of land. Fifty plots of land, each one hectare, were sampled at random from a ten thousand hectare area in Northern England. For each plot, the following variables were recorded:

BIOMASS: An estimate of the biomass of vegetation in kg per square metre.

ALT: The mean altitude of the plot in metres above sea level.

CONS: A categorical variable, which was coded as 1 if the plot was part of a conservation area, and 2 otherwise.

SOIL: A categorical variable crudely classifying soil type as 1 for chalk, 2 for clay and 3 for loam.

These data are stored in the *conservation* dataset. The output in Box 6.7 analyses BIOMASS as explained by the other three variables.

BOX 6.7 Conservation and biomass analysis

General Linear Model

Word equation: BIOMASS = CONS + ALT + SOIL

ALT is continuous, CONS and SOIL are categorical

Analysis of variance table for BIOMASS, using Adjusted SS for tests

Source	DF	Seq SS	Adj SS	Adj MS	F	P
CONS	1	0.7176	0.0249	0.0249	2.80	0.101
ALT	1	5.8793	4.4273	4.4273	498.10	0.000
SOIL	2	0.3953	0.3953	0.1977	22.24	0.000
Error	45	0.4000	0.4000	0.0089		
Total	49	7.3922				

Term	Coef	SECoef	T	P
Constant	2.21156	0.02486	88.97	0.000
CONS				
1	−0.02443	0.01460	−1.67	0.101
2	*0.02443*			
ALT	−0.002907	0.000130	−22.32	0.000
SOIL				
1	0.10574	0.02057	5.14	0.000
2	0.01952	0.01889	1.03	0.307
3	*−0.12526*			

1. On the basis of this analysis, what biomass would you predict for a plot with a mean altitude of 200m in a conservation area with chalk soil?
2. What biomass would you predict for a plot with mean altitude of 300m, with loam soil and not in a conservation area?
3. How strong is the evidence that the biomass of vegetation depends upon being in a conservation area? In what direction is the effect?
4. How strong is the evidence that SOIL affects BIOMASS? Which soil types are associated with the highest and lowest biomass values?
5. Give a 95% confidence interval for the effect of an additional metre of altitude on biomass.
6. Comment on the discrepancy between the sequential and adjusted sums of squares for CONS.
7. Given that it is impractical to conduct a randomised experiment in a study of this kind, what kind of uncertainties must remain in the conclusions that can be drawn?

Determinants of the Grade Point Average

The academic performance of some students in the USA is evaluated as a Grade Point Average (GPA) each year. Faculty are concerned to admit good students, and assess students via tests that are broken down into verbal skills (VERBAL) and mathematical skills (MATH). A hundred students from each of two years (YEAR) had their marks analysed, to investigate whether verbal or mathematical skills were more important in determining a student's GPA. The variables GPA, YEAR, VERBAL and MATH are recorded in the *grades* dataset.

1. How good is the evidence that MATH, VERBAL or YEAR predicts GPA? In what direction is the effect for each of these variables?
2. What GPA would you expect from a student in the first year whose verbal score was 700 and mathematical score was 600? What about a student in the second year whose verbal score was 600 and mathematical score was 700?

7 Interactions—getting more complex

The chapter covers one major new concept—the concept of interactions.

7.1 The factorial principle

Many experiments require a considerable investment in terms of time and resources, so it is always in our interests to design them as efficiently as possible. The designed experiments of Chapter 5 involved the manipulation of one variable (for example, BEAN), with a second variable, BLOCK, being added to eliminate some of the unavoidable variation. However, frequently we will be interested in more than one explanatory variable, as in the following example.

There are two varieties of wheat on the market, and farmers wish to know which variety will give the greater yields, and which of four sowing densities will produce maximum returns. Two possible sets of experiments to answer this question are compared:

Experimental Plan 1

To compare the four sowing rates, only variety 1 is used. Sixteen plots are assigned at random to have four each of the four sowing rates. In a separate experiment on varieties, only density 1 is used. Eight plots are assigned at random to have four each of the two varieties. In total, 24 plots have been used, and effectively two separate experiments have been carried out; one to compare sowing rate and one to compare variety. The treatment combinations are outlined in Fig. 7.1 as a table of co-occurrences.

Suppose the first experiment shows that sowing density 4 gives the highest yields for Variety 1, and from the second experiment, that Variety 2 yields more grain than Variety 1. However, the two varieties have only been compared at Sowing Density 2.. Would it be safe to extrapolate our conclusions, and say that for the greatest yields, we should use Variety 2 at Sowing Density 4? This particular treatment combination has not actually been grown in either experiment. It is possible that the two varieties respond differently to changes in sowing density, and from this experimental design we have no way of knowing.

	S1	S2	S3	S4
V1	4	4	4	4
V2				

PLUS

	S1	S2	S3	S4
V1	8			
V2	8			

Fig. 7.1 Comparing sowing densities and comparing varieties: S1 = sowing rate 1 etc; V1 = variety 1 etc.

Experimental Plan 2

An alternative plan is to sow every density for both varieties, giving every possible combination of sowing rate and variety. With three replicates of each treatment combination, again 24 plots of land will have been used. The treatment combinations are illustrated in Fig. 7.2 as a table of co-occurrences.

Now it is possible to find the variety–sowing rate combination that gives the greatest yield. This second experimental design actually asks three questions: how variety and sowing rate affect yield, and does variety have the same effect on yield at different sow rates? In statistical terms, this third question is asking 'Is there an **interaction** between sow rate and variety in the way they affect yield?' It is only by including all eight combinations of sow rate and variety that this question can be answered.

The treatment combinations outlined in Fig. 7.2 are of a **factorial** experiment. Such designs that include all possible treatment combinations are ideal to test for interactions between our explanatory variables. However, even if there is no interaction between the two variables, there is still a distinct advantage to

	S1	S2	S3	S4
V1	3	3	3	3
V2	3	3	3	3

Fig. 7.2 Comparing sowing densities and varieties in a factorial design: codes as for Fig. 7.1.

the factorial design. In Plan 1, there were four replicates of sowing rate and four of variety in the two separate experiments. In Plan 2, there are effectively six replicates of each sowing rate, and twelve replicates of each variety. Considering first the four sow rates, each level of SOWRATE has three plots of Variety 1 and three of Variety 2 (so the design is orthogonal). Sowing rates can therefore be compared, without reference to variety. Similarly, each variety has three replicates of each sowing rate, and because they are equally represented, the two varieties can be compared without reference to sowing rate.

So our second experimental plan has used the same number of plots, yet has an effectively greater level of replication. This is the second advantage of factorial designs, namely **hidden replication**. Even if there is no interaction, each plot is taking part in two comparisons—a sowing rate comparison and a variety comparison.

In summary, the factorial design has two advantages over two one-factor experiments: (i) it allows us to test for interactions; (ii) in the absence of significant interactions it still provides hidden replication. It is therefore a very efficient way to design experiments. Is it necessary for all treatment combinations to be present to test for an interaction? There may be some experiments for which it is hard, if not impossible to get all combinations. Under these constraints, it is possible to test for an interaction, but with reduced degrees of freedom. However, while accepting that sometimes these constraints are unavoidable, the most usual and best way to test for interactions is to include all treatment combinations.

7.2 Analysis of factorial experiments

The principle behind analysing factorial experiments is once again the partitioning of sums of squares, and their associated degrees of freedom. This can be demonstrated by considering the factorial yield experiment in detail. This was carried out in three blocks—so each block contained one replicate of each of the eight treatment combinations. The simplest way to analyse these data (stored in the data set *wheat*) would be to do a blocked one-way ANOVA, similar to the analyses carried out in Chapter 5. The eight treatment combinations could be coded as 1 to 8, and the word equation would be 'YIELD = BLOCK + TRTMT'. This analysis is carried out in Box 7.1.

This analysis is perfectly valid, and our conclusions would be that there is a significant difference between the treatments (but not between the blocks). However, some of the detail of the experimental design has been lost in this analysis. The treatment sums of squares can be partitioned further, to answer the three questions posed by our factorial experiment: namely (i) does sow rate affect yield; (ii) does variety affect yield; (iii) does the yield of both varieties respond to sow rate in the same way (i.e. is there an interaction)? To answer

BOX 7.1 A blocked one-way ANOVA of the yield data

General Linear Model

Word equation: YIELD = BLOCK + TRTMT
BLOCK and TRTMT are categorical

Analysis of variance table for YIELD, using Adjusted SS for tests

Source	DF	Seq SS	Adj SS	Adj MS	F	P
BLOCK	2	0.3937	0.3937	0.1968	0.52	0.606
TRTMT	7	8.0776	8.0776	1.1539	3.04	0.036
Error	14	5.3069	5.3069	0.3791		
Total	23	13.7782				

Table 7.1 Recoding the data set to look for interactions

TRTMT	1	2	3	4	5	6	7	8
SOWRATE	1	2	3	4	1	2	3	4
VARIETY	1	1	1	1	2	2	2	2

BOX 7.2 A factorial ANOVA of the yield data

General Linear Model

Word equation: YIELD = BLOCK + VARIETY + SOWRATE + VARIETY * SOWRATE
BLOCK, VARIETY and SOWRATE are categorical

Analysis of variance table for YIELD, using Adjusted SS for tests

Source	DF	Seq SS	Adj SS	Adj MS	F	P
BLOCK	2	0.3937	0.3937	0.1968	0.52	0.606
VARIETY	1	2.1474	2.1474	2.1474	5.67	0.032
SOWRATE	3	5.8736	5.8736	1.9579	5.16	0.013
VARIETY * SOWRATE	3	0.0566	0.0566	0.0189	0.05	0.985
Error	14	5.3069	5.3069	0.3791		
Total	23	13.7782				

these questions, SOWRATE and VARIETY are coded for as separate explanatory variables (Table 7.1), and they are included in the analysis as shown in Box 7.2.

The sum of squares for treatment has now been partitioned three ways: into the **main effect** of sowing rate, the **main effect** of variety, and the **interaction** between the two. So instead of asking the one question 'Are there significant

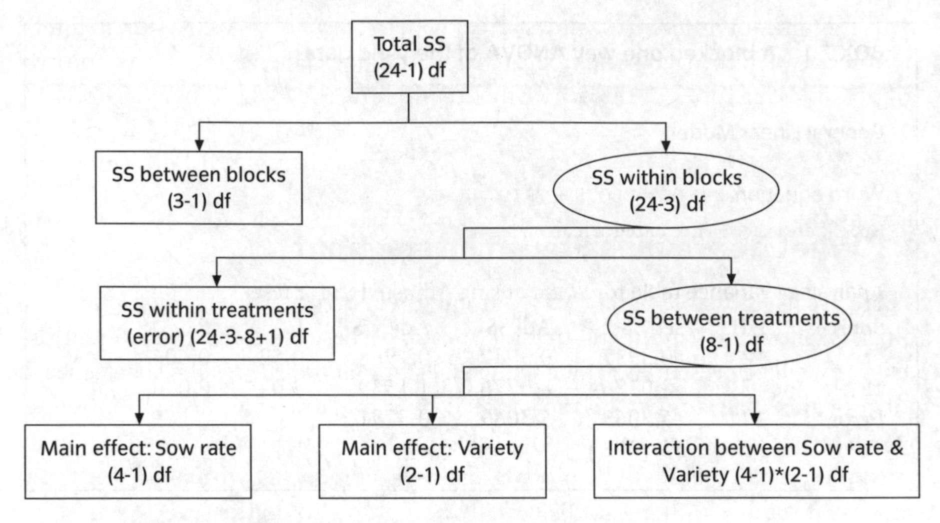

Fig. 7.3 Further partitioning of sums of squares.

differences between the treatments?', three questions are posed, one for each source in the ANOVA table.

In comparing Boxes 7.1 and 7.2, notice that the sums of squares and degrees of freedom for the three new terms (SOWRATE, VARIETY and SOWRATE * VARIETY) add up to those for TRTMT. So a further level of partitioning has taken place. Second, notice how this was done. The new model formula is SOWRATE + VARIETY + SOWRATE * VARIETY. The model fitted in Box 7.2 has partitioned the treatment sums of squares into the three components of SOWRATE + VARIETY + SOWRATE * VARIETY, as illustrated in Fig. 7.3. These three terms correspond to the three questions we wished to ask:

1. Does the way in which SOWRATE affects YIELD differ between the two VARIETYs? The answer is no ($p = 0.985$). This is actually precisely the same question as 'Does the way in which VARIETY affects YIELD differ between the SOWRATEs?'
2. Does SOWRATE affect YIELD, when differences in VARIETIES have been taken into account? The answer is yes ($p = 0.013$).
3. Does VARIETY affect YIELD when differences in SOWRATES have been taken into account? The answer is yes ($p = 0.032$).

In interpreting factorial analyses, it is good practice to start with the most complicated question, so the first question to be considered involves the interaction. Quite why this is the case is discussed in more detail in Chapter 10.

So it would be concluded at this point that there is no significant interaction between these two factors, but that both main effects are significant. In the next section we learn what that means! It can also be seen that in breaking the differences between treatments into three components, we have been able to detect significant differences, and ascribe them to SOWRATE, VARIETY or an interaction between the two, when before we could not. The next section explores all possible outcomes of this experiment, and how they would be interpreted.

7.3 What do we mean by an interaction?

There are five possible outcomes from the sowing rate—variety experiment (if block is ignored to simplify matters). For each of these outcomes, an interaction diagram will be presented, which is a visual summary of the results. The general form of this diagram can be represented by a line drawing, just by knowing which terms are significant, without referring to the coefficient table. For these examples, mean yield will be plotted against sowing rate, with a line for each variety (if required). It would be equally acceptable to do the reverse—that is, plot yield against variety, with four lines for the four sowing rates, if required.

Neither sow rate nor variety affect yield

If all terms are insignificant, the most concise way to describe the data is by the grand mean (Fig. 7.4), averaged over the three blocks.

The corresponding model formula would be: YIELD = BLOCK.

Variety of wheat affects yield, but sowing rate does not

If the two varieties are significantly different, then a mean value for each variety would appear as two parallel lines. These lines are horizontal, because sow rate has no affect on yield (see Fig. 7.5).

In this case the corresponding model formula would be YIELD = BLOCK + VARIETY.

Sowing rate affects yield but variety does not

In this case, one line can represent both varieties because they are not significantly different, but it is no longer parallel to the sow rate axis (Fig. 7.6).

This corresponds to the model formula YIELD = BLOCK + SOWRATE.

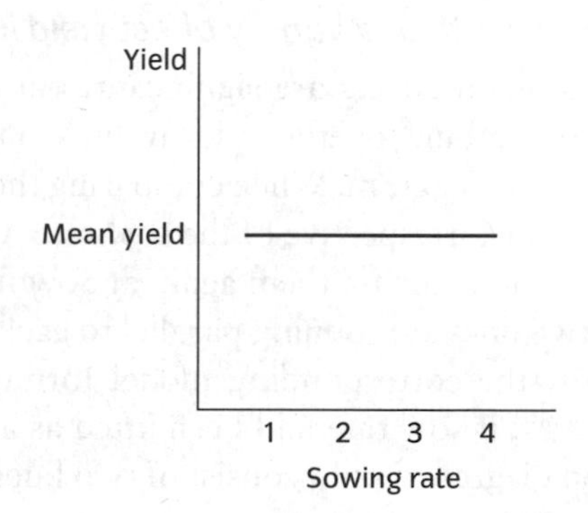

Fig. 7.4 Interaction diagram when neither varieties nor sowing rates are different.

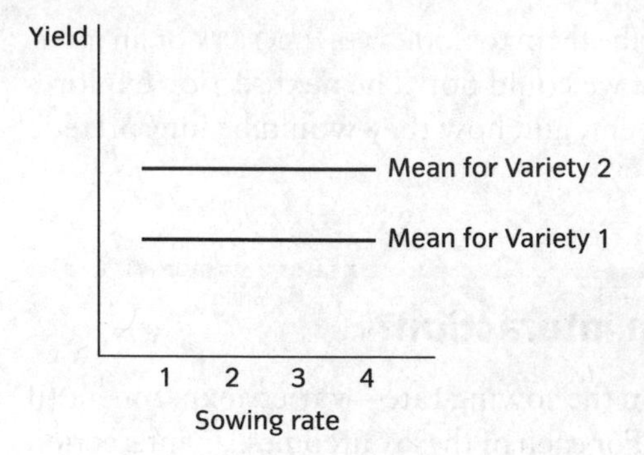

Fig. 7.5 Varieties are different, but sowing rates are not.

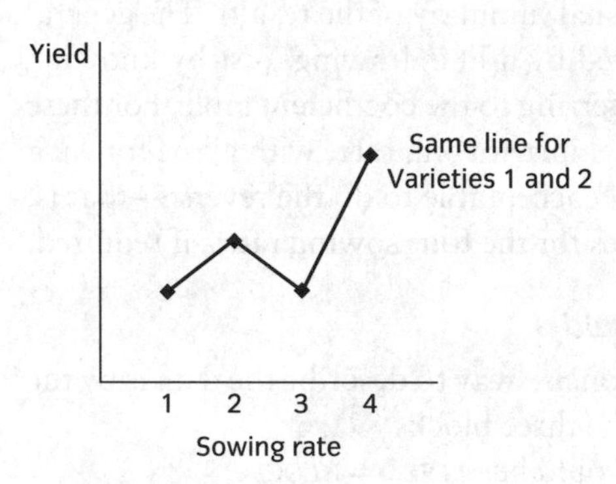

Fig. 7.6 Sowing rates are different but varieties are not.

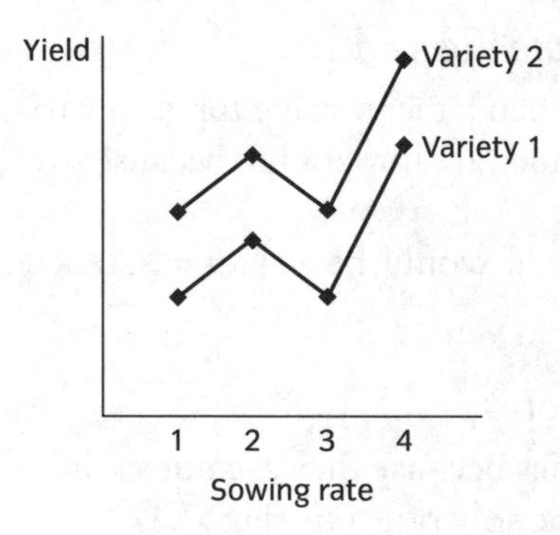

Fig. 7.7 Variety and sowing rate combine together additively to predict yield.

Both sow rate and variety affect yield in an additive manner

If both main effects are significant, but the interaction is not, then the two factors combine together additively. What this means is well illustrated by the interaction diagram. When comparing the two varieties, the difference in yield is the same, irrespective of the sow rate. Compare the two varieties at Sowing rate 1, and compare them again at Sowing rate 4—the difference is the same. The two lines are running parallel to each other (Fig. 7.7).

Now the corresponding model formula is `YIELD = BLOCK + VARIETY + SOWRATE`. If sow rate had been fitted as a continuous variable, then the interaction diagram would consist of two linear parallel lines (see Section 7.5).

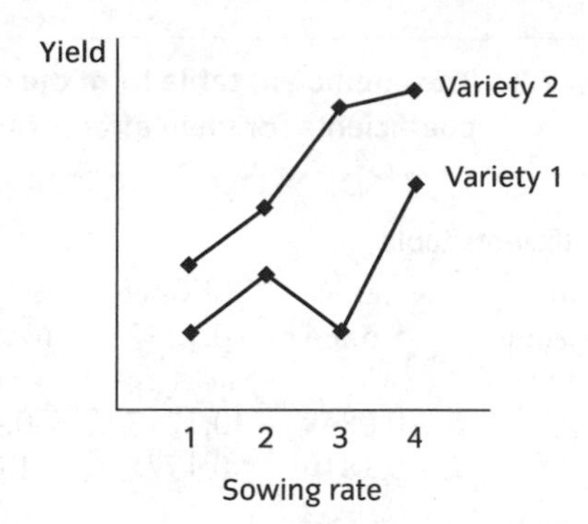

Fig. 7.8 Variety and sowing rate combine together non-additively to affect yield.

There is an interaction between sow rate and variety

Finally, if the interaction is significant, then sowing rate and variety combine in a non-additive way to affect yield. Now the difference in yield between the two varieties differs, depending upon the sow rate. At Sowing rate 3, for example, the two varieties differ greatly in yield, whereas at Sowing rates 1 and 2, they are more similar. For some experiments, the strength of the interaction may be such that the two lines cross over. It is not necessary for the two lines to cross for there to be an interaction. All that is required is that the two lines are significantly different from parallel, as in Fig. 7.8.

The corresponding model formula is now `YIELD = BLOCK + VARIETY + SOWRATE + SOWRATE * VARIETY`.

It is now possible to return to the ANOVA table presented in Box 7.2, to decide which of these best represents the results. The aim is to present the message contained in the analysis as clearly and simply as possible. So the most complicated term should be considered first, which in this case is the interaction between sow rate and variety. It is not significant ($p = 0.985$), so it is not necessary. However, both main effects are significant, so both should be retained. Hence the interaction diagram of the form shown in Fig. 7.7 would be the most suitable.

These interaction diagrams have been schematic, indicating different possible underlying truths about the influence of `SOWRATE` and `VARIETY` on `YIELD`. Interaction diagrams are also used as an effective way of presenting the data itself, as described in the next section.

7.4 Presenting the results

Factorial experiments with insignificant interactions

To obtain the fitted value equation for the relevant interaction diagram, the appropriate terms should be used from the coefficient table. With designed experiments, it is usual to fit the full model, which contains any blocks, and all the main effects and their interactions (just as was done in Box 7.2). Interpretation

BOX 7.3 The coefficient table from the factorial analysis of Box 7.2 with coefficients for main effects only

Coefficients table

Term	Coef	SECoef	T	P
Constant	8.5364	0.1257	67.92	0.000
BLOCK				
1	−0.0848	0.1777	−0.48	0.641
2	0.1810	0.1777	1.02	0.326
3	*−0.0962*			
VARIETY				
1	0.2991	0.1257	2.38	0.032
2	*−0.2991*			
SOWRATE				
1	−0.2804	0.2177	−1.29	0.219
2	−0.4469	0.2177	−2.05	0.059
3	−0.1034	0.2177	−0.47	0.643
4	*0.8307*			

then should always proceed from the bottom of the ANOVA table upwards—in other words starting with the interaction. In this case, it was not significant, but the main effects both are. To obtain the fitted value equation for the interaction diagram which does not contain a significant interaction, the coefficients for the main effects, but not the interaction, should be used (Box 7.3).

The fitted value equation may be constructed following the same principles as for the blocked one-way ANOVA of Chapter 5, with the addition of an extra term

$$\texttt{YIELD} = 8.5364 + \begin{bmatrix} & \texttt{BLOCK} \\ 1 & -0.0848 \\ 2 & 0.181 \\ \mathit{3} & \mathit{0.0848 - 0.181} \end{bmatrix} + \begin{bmatrix} & \texttt{SOWRATE} \\ 1 & -0.2804 \\ 2 & -0.4469 \\ 3 & -0.1034 \\ \mathit{4} & \mathit{0.2804 + 0.4469 + 0.1034} \end{bmatrix}$$

$$+ \begin{bmatrix} & \texttt{VARIETY} \\ 1 & 0.2991 \\ \mathit{2} & \mathit{-0.2991} \end{bmatrix}.$$

As before, the first term (the constant of the coefficient table) is the grand mean, and all other terms are deviations from that mean, so the values within a conditional bracket must sum to zero. The terms in italics are those that can be calculated, given the unitalicised terms. In calculating `YIELD` for our interaction diagram the final bracket (the interaction bracket) for the interaction term has been ignored because in this data set, the interaction is not significant. The interaction bracket would include additional deviations in yield due to the interaction, so by ignoring this bracket we are assuming that the average

BOX 7.4 Summarising the results when the interaction is not significant

Least squares means for YIELD

SOWRATE	Mean	St Error
1	8.256	0.2514
2	8.089	0.2514
3	8.433	0.2514
4	9.367	0.2514
VARIETY		
1	8.835	0.1777
2	8.237	0.1777

deviation due to the interaction term is zero. How to obtain this fitted value equation from output will depend upon your statistical package, and is explained more fully in the package specific supplements.

So how should we present the results? Because the interaction is not significant, the clearest way would be to present the means for the four sow rates and the means for the two varieties. This could be in a tabular form, giving the means and standard errors of the means as in Box 7.4 or as two histograms.

This data set is both orthogonal and balanced. Dropping terms is more complicated in other cases. However, packages will produce 'least square means' which will allow the interaction diagram to be plotted, and indeed, the packages will, if asked nicely, produce the actual plots. See the package specific supplements for further details.

The standard deviations given in this table are the standard errors of the means, as described in Chapter 1 $\left(\frac{s}{\sqrt{n}}\right)$. The error variance, s^2, is taken from the ANOVA table of Box 7.2 as 0.3791, and n for each sowing rate is 6 (the number of data points that contribute to that mean). So $\sqrt{\frac{0.3791}{6}} = 0.2514$ as given in Box 7.4. This explains why the standard errors are constant within sowing rate—because the experiment had equal numbers of plots allocated to each level of sowing rate, and the pooled estimate of error variance is always used to calculate the standard errors. For variety, s will be the same, but n is 12. Hence the standard errors are $\sqrt{\frac{0.3791}{12}} = 0.1777$.

So for this example, in which the interaction is not significant, the results can be presented as two simple stories—one describing how YIELD changes with SOWRATE, and one describing the difference between the two VARIETYs (Fig. 7.9).

Finally, on analysing a factorial experiment such as this, and on discovering that the interaction is not significant, it may be tempting to remove the interaction term and refit the simpler model just containing the main effects. If the

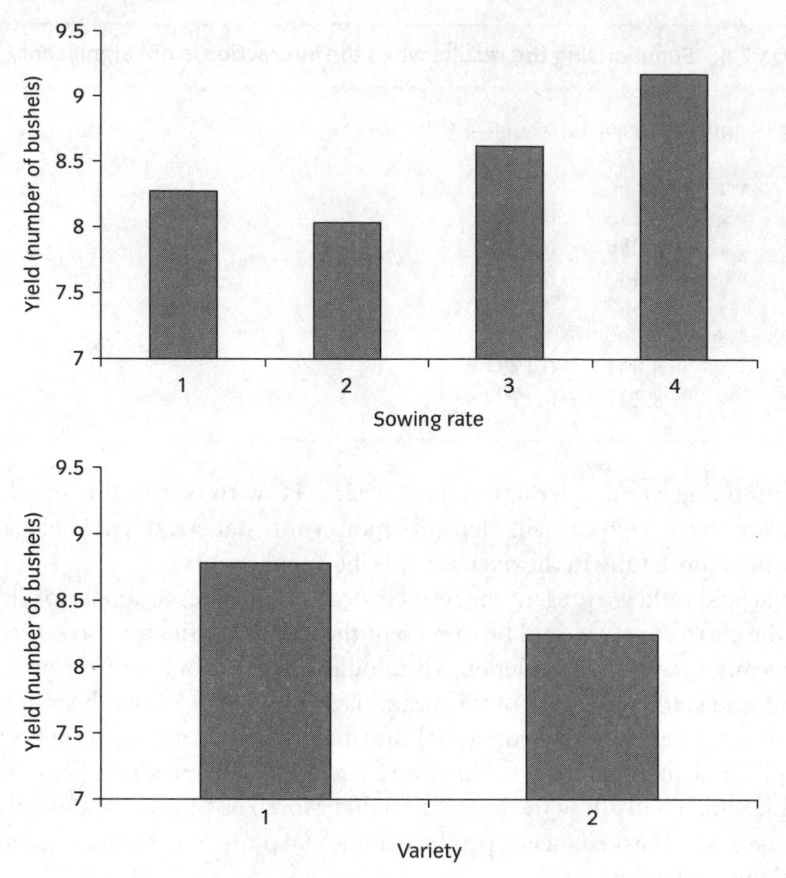

Fig. 7.9 The two simple stories from the wheat experiment.

'true model' in the population does not contain an interaction, then the error mean square of this model (s^2) would be a fair estimate of the unexplained population variance (σ^2). However, it may be that these two variables do interact, but that your experiment has failed to detect it (a Type I error). Under these circumstances, if the interaction term is omitted, we are selecting a special subset of cases (those experiments in which a Type I error has been made). Now s^2 will not on average reflect σ^2, so all the standard errors will be biased. Thus it is good practice to fit the model which reflects the experimental design, and to use that estimate of s^2 to calculate the standard errors.

Factorial experiments with significant interactions

We will now consider the analysis and presentation of experimental results when the interaction is significant. An experiment was carried out to determine the optimum conditions for growing tulips. The number of flower heads was taken as the response variable, with the treatments being three levels of shade and three different watering regimes. The experiment was carried out in three flower beds, each of which was divided into nine plots. The plots in a block receive one each of the nine treatment combinations, making the design orthogonal. Data are stored in the *tulips* dataset.

An analysis of this experiment is shown in Box 7.5 which includes the interaction term. Looking first at the interaction between shade and water, it

BOX 7.5 Analysing a factorial experiment with a significant interaction

General Linear Model

Word equation: BLOOMS = BED + WATER + SHADE + WATER * SHADE
BED, WATER and SHADE are categorical variables

Analysis of variance table for BLOOMS, using Adjusted SS for tests

Source	DF	Seq SS	Adj SS	Adj MS	F	P
BED	2	13 811	13 811	6 906	3.88	0.042
WATER	2	103 626	103 626	51 813	29.11	0.000
SHADE	2	36 376	36 376	18 188	10.22	0.001
WATER * SHADE	4	41 058	41 058	10 265	5.77	0.005
Error	16	28 477	28 477	1 780		
Total	26	223 348				

Coefficients table

Term		Coef	SECoef	T	P
Constant		128.994	8.119	15.89	0.000
BED					
1		−31.87	11.48	−2.78	0.014
2		13.59	11.48	1.18	0.254
3		*18.28*			
WATER					
1		−77.72	11.48	−6.77	0.000
2		3.85	11.48	0.33	0.742
3		*73.87*			
SHADE					
1		51.44	11.48	4.48	0.000
2		−19.67	11.48	−1.71	0.106
3		*−31.77*			
WATER *	SHADE				
1	1	−72.67	16.24	−4.47	0.000
1	2	12.94	16.24	0.80	0.437
1	*3*	*59.73*			
2	1	29.92	16.24	1.84	0.084
2	2	−6.48	16.24	−0.40	0.695
2	*3*	*−23.44*			
3	*1*	*42.75*			
3	*2*	*−6.46*			
3	*3*	*−36.29*			

can be concluded that this is significant ($p = 0.005$). The number of blooms is therefore sensitive to water regime and shade, and the response to water level depends upon which of the two shade treatments has been applied. Instead of two simple stories, the results indicate one more complicated picture, with the two factors no longer being additive. The fitted value equation will have an extra term, which corresponds to the interaction. Because there is a significant interaction, each combination of water and shade requires an additional term to predict the expected number of blooms. However, in the coefficient table, there are only 4 unitalicised deviations given for the interaction, because the remaining 5 can be calculated from the first 3. Packages will differ in how many and which coefficients will be provided in the output, and these details are provided in the package specific supplements. The interaction term is best thought of as a 3 × 3 table, as shown in Table 7.2.

The deviations across rows and down columns must sum to zero. This illustrates why the interaction term for this model has 4 degrees of freedom—because only four of the nine interaction coefficients are free to vary. The degrees of freedom for an interaction are obtained by multiplying together the corresponding degrees of freedom for the main effects (if all treatment combinations are included).

Now we are in a position to write out the fitted value equation for the full factorial model:

$$\texttt{BLOOMS} = 128.994 + \begin{bmatrix} \texttt{BED} & \\ 1 & -31.87 \\ 2 & 13.59 \\ \mathit{3} & \mathit{18.28} \end{bmatrix} + \begin{bmatrix} \texttt{WATER} & \\ 1 & -77.72 \\ 2 & 3.85 \\ \mathit{3} & \mathit{73.87} \end{bmatrix}$$

$$+ \begin{bmatrix} \texttt{SHADE} & \\ 1 & 51.44 \\ 2 & -19.67 \\ \mathit{3} & \mathit{-31.77} \end{bmatrix} + \begin{bmatrix} \texttt{WATER * SHADE} & & \\ 1 & 1 & -72.67 \\ 1 & 2 & 12.94 \\ 1 & 3 & \mathit{59.73} \\ 2 & 1 & 29.92 \\ 2 & 2 & -6.48 \\ 2 & 3 & \mathit{-23.44} \\ 3 & 1 & \mathit{42.72} \\ 3 & 2 & \mathit{-6.46} \\ 3 & 3 & \mathit{-36.29} \end{bmatrix}.$$

When the model contains a significant interaction, it is no longer appropriate to represent the data as two sets of means (in this case for water and for shade), because the response of the flowers to water depends upon the level of shade and vice versa.

The means displayed in Box 7.6 can be obtained directly from the fitted value equation, by adding the appropriate terms for the main effects and the

Table 7.2 Calculating the coefficients for the interaction term

		Water level			Total
		1	2	3	
Shade level	1	−72.67	29.92	42.75	0
	2	12.94	−6.48	−6.46	0
	3	59.73	−23.44	−36.29	0
Total		0	0	0	0

BOX 7.6 Summarising the results with a significant interaction

Least squares means for BLOOMS

WATER *	SHADE	Mean	SEMean
1	1	30.04	24.36
1	2	44.55	24.36
1	3	79.22	24.36
2	1	214.20	24.36
2	2	106.69	24.36
2	3	77.63	24.36
3	1	297.05	24.36
3	2	176.74	24.36
3	3	134.82	24.36

interaction. For example, the number of blooms predicted per plant when grown in Shade level 1 and Water level 3 is 128.994 + 51.44 + 73.87 + 42.75 = 297.05 (to 2 decimal places). (It is usual to average across the blocks when presenting the means). The standard errors are equal across all shade and water levels, because each combination of water and shade has the same number of replicates.

So the results of this experiment require one complicated picture, in contrast to the two simple pictures of the previous section, as shown in Fig. 7.10. As water level increases, so does BLOOMS, and this increase is greatest when shade is lowest.

Error bars

When presenting the results of an experiment in a figure, it is usual to give a representation of the amount of variation in the data. This is most frequently done in one of three ways: (1) standard errors; (2) standard deviations; (3) standard error of the difference.

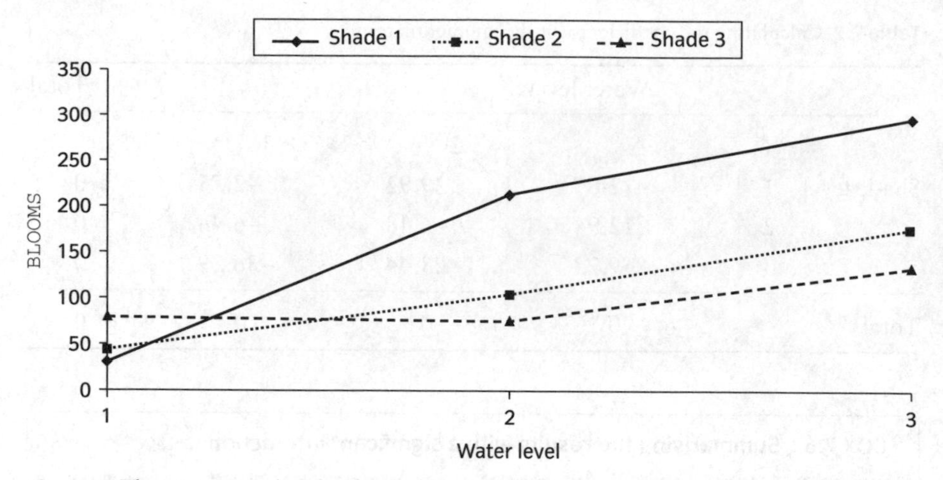

Fig. 7.10 The interaction diagram for the blooms experiment.

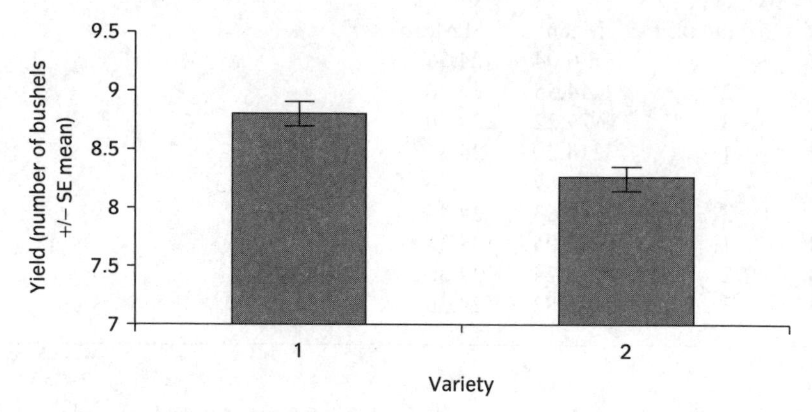

Fig. 7.11 Using standard errors as error bars.

Standard errors (of the means)

The first example is illustrated in Fig. 7.11, where the means have been plotted in a histogram, plus and minus one standard error (the same information may be given in a table). Because the level of replication is the same for both treatment groups, the error bars are the same size. This gives an indication of the accuracy of the means you have estimated.

Standard deviations

The second example is illustrated in Fig. 7.12, where the same histogram is plotted, but using plus or minus one standard deviation as the error bar. The standard deviations are estimated separately from each group of data, which is why they vary in size (in this case only slightly). This indicates the variability in your data.

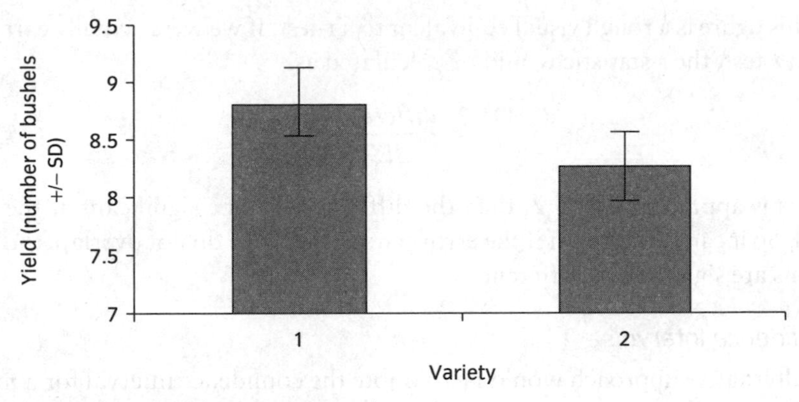

Fig. 7.12 Using standard deviations as error bars.

Standard error (of the difference)

If you are particularly interested in comparing two means, one option is to use error bars which represent the standard error of the difference. For orthogonal designs this is calculated as:

$$SE_{\text{diff}} = \sqrt{SE_1^2 + SE_2^2}$$

This could be used in presenting the results for the two varieties from the yield experiment, in which the standard error of the difference would be calculated as:

$$SE_{\text{diff}} = \sqrt{0.1777^2 + 0.1777^2} = 0.2514.$$

(It is co-incidental that this equals the standard error of the sowing rate means). The error bars of Fig. 7.13 are given as standard errors of the difference.

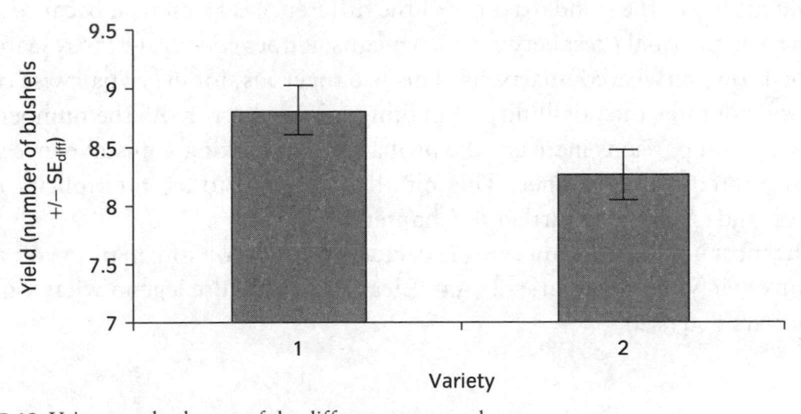

Fig. 7.13 Using standard error of the difference as error bars.

This figure is a rough visual equivalent to a t-test. If we were actually carrying out a t-test, the t-statistic would be calculated as:

$$t_s = \frac{difference}{SE_{diff}}.$$

If this is approximately > 2, then the difference will be significant at the 5% level. So it can be seen that if the error bars in Fig. 7.13 do not overlap, the two means are significantly different.

Confidence intervals

An alternative approach would be to quote the confidence interval for a mean (as discussed in Revision Section 1.3). For the yield experiment, the error degrees of freedom would be taken from the full model (Box 7.2), and so are 14. For a 95% confidence interval, this gives a critical t-value of 2.1448. Using the general formula for a confidence interval of:

$$\text{estimate} \pm \text{standard error of the estimate} \times \text{critical } t\text{-value}$$

and substituting in the appropriate values for, say, the mean for Sowing rate 2, we obtain

$$8.089 \pm 0.2514 \times 2.1448$$

which becomes

$$(7.550, 8.628).$$

In summary, when drawing error bars on histograms the two most commonly used options are: (1) the standard deviation of the data represented by that particular histogram bar; (2) the standard error of the mean. The first option will tell you how variable the data are within that histogram bar. The second option tells you something about the accuracy of your model estimate. The usual approach is to use the standard error of the mean, and to take the standard error estimates from the full model, for the reasons discussed earlier. While the use of the standard error of the difference is attractive, because it allows a rough visual t-test between two means, it does encourage the researcher to focus on pairwise comparisons. This is dangerous, for every pairwise comparison contains the possibility of making a Type I error. As the numbers of pairwise comparisons increase, the probability of making a mistake increases in an unpredictable manner. This pitfall is referred to as 'multiplicity of p values' and is discussed further in Chapter 11.

The most important points are (1) decide what kind of information you wish to convey in your error bars; (2) state clearly in the figure legend what kind of error bars you used.

7.5 Extending the concept of interactions to continuous variables

Mixing continuous and categorical variables

The explanatory variable SOWRATE in the last two sections was treated as a categorical variable with four levels. However, it is one of those variables that could be treated as continuous, if levels 1 to 4 correspond to increasing sowing densities. If we imagine it as continuous, what would an interaction mean in this context? An interaction means that the effect of one variable depends upon the value of another variable. In this case, the slope of the continuous variable depends upon the level of the categorical variable.

Returning to the leprosy data set from Section 6.2, the graph of the fitted values drawn in Fig. 6.1 was of three parallel lines. The lines were constrained to be parallel, because an additive model had been fitted. Fitting the interaction will be investigating whether the best fit lines are significantly different from parallel (Box 7.7).

BOX 7.7 Analysing interactions with categorical and continuous variables: the leprosy dataset

General Linear Model

Word equation: BACAFTER = TREATMT + BACBEF + TREATMT * BACBEF

TREATMT is categorical and BACBEF is continuous

Analysis of variance table for BACAFTER, using Adjusted SS for tests

Source	DF	Seq SS	Adj SS	Adj MS	F	P
BACBEF	1	587.48	482.63	482.63	30.57	0.000
TREATMT	2	83.35	5.83	2.91	0.18	0.833
TREATMT * BACBEF	2	1.25	1.25	0.62	0.04	0.961
Error	24	378.90	378.90	15.79		
Total	29	1050.98				

Coefficient table

Term	Coef	SECoef	T	P
Constant	−0.126	1.955	−0.06	0.949
BACBEF	0.8894	0.1609	5.53	0.000
TREATMT				
1	−0.946	2.520	−0.38	0.711
2	−0.896	2.701	−0.33	0.743
3	*1.842*			
BACBEF * TREATMT				
1	−0.0611	0.2174	−0.28	0.781
2	0.0167	0.2128	0.08	0.938
3	*0.0444*			

Inspection of the ANOVA table reveals that the interaction is not significant ($p = 0.961$). So we would conclude that the more appropriate model was that given by Box 6.1 previously. This output however will be used to demonstrate how the fitted values equation would be constructed. The first three terms follow the same model as the additive model, with the interaction adding a fourth term:

$$\text{BACAFTER} = -0.126 + \begin{bmatrix} & \text{TREATMT} \\ 1 & -0.946 \\ 2 & -0.896 \\ \mathit{3} & \mathit{0.946 + 0.896} \end{bmatrix} + 0.8894 \times \text{BACBEF}$$

$$+ \begin{bmatrix} & \text{TREATMT} \\ 1 & -0.0611 \\ 2 & 0.0167 \\ \mathit{3} & \mathit{0.0611 - 0.0167} \end{bmatrix} \times \text{BACBEF}$$

So the interaction term allows the slope to vary, depending upon the level of treatment. This expression can be simplified to three equations with different slopes and intercepts, see Table 7.3.

Converting this to an interaction diagram gives a graph as in Fig. 7.14.

Table 7.3 Calculating equations with interactions between continuous and categorical variables

Treatment	Equation
1	BACAFTER = −1.072 + 0.8283 × BACBEF
2	BACAFTER = −1.022 + 0.9061 × BACBEF
3	BACAFTER = 1.716 + 0.9338 × BACBEF

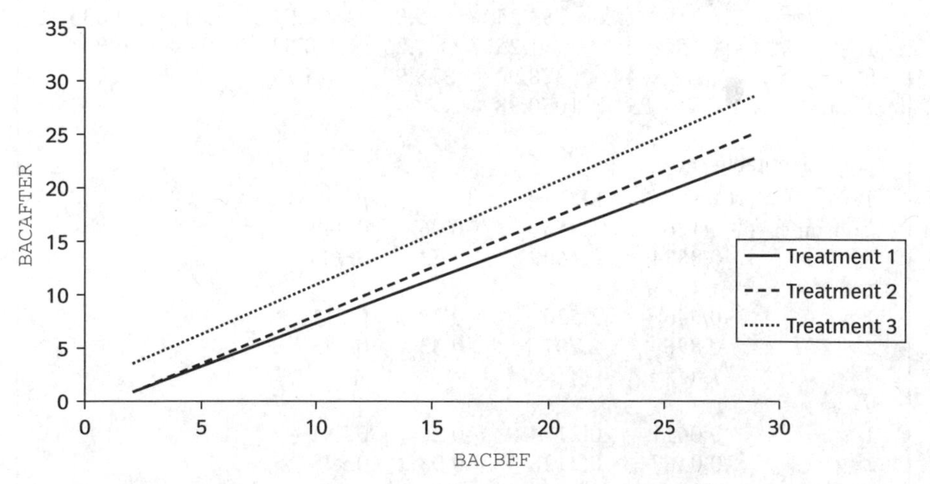

Fig. 7.14 An interaction diagram for the leprosy data set fitted with an interaction between continuous and categorical variables.

While we have fitted three different lines, we know from the ANOVA table that these three lines are not significantly different in slope.

Adjusted means (or least square means in models with continuous variables)

When presenting the results from a model which mixes continuous and categorical variables, we can no longer simply use the means to compare the treatments. Taking the leprosy data set as an example, it is no longer appropriate to take just the means in BACAFTER for the three treatments. The actual means may be different because of treatment differences, but also because of differences in the bacillus score before treatment. So to get a fair comparison between the three treatments, we pretend all the patients started off with the same bacillus score. This is taken to be the average value of BACBEF. Means for the three treatments are then adjusted for this initial bacillus score, by simply substituting this value into the fitted value equation for treatment.

The example given in Box 7.8 is from the model in which the interaction was fitted (Box 7.7). Therefore the adjusted means are obtained by substituting 11.19 for BACBEF into the fitted value equations for the three treatments given in Table 7.3. Thus the adjusted mean for Treatment 3, for example, is: $1.716 + 0.934 \times 11.19 = 12.17$.

We can draw an analogy between adjusted sums of squares and adjusted means. A set of three adjusted means are the means of BACAFTER for the three treatments, when the influence of BACBEF has been eliminated (by substituting the same value for BACBEF in each of the three equations). Similarly, the adjusted sums of squares for TREATMT, is the variation explained by treatment, when the influence of BACBEF has been statistically eliminated (i.e. BACBEF has been included in the model as a continuous variable).

BOX 7.8 Obtaining adjusted means to compare treatments, when the model contains a covariate

Means for continuous variables

	Mean	SEMean
BACBEF	11.19	4.904

Are then substituted into the fitted value equation to obtain the least squares adjusted means for BACAFTER

TREATMT	Mean	SEMean
1	8.201	1.313
2	9.121	1.286
3	12.171	1.262

It is also worth noting at this point that if we lose orthogonality, then the least squares means will also differ from simple means. However, in all cases the least squares mean may be requested from your package (see package specific supplements for details).

Finally, note that this is the best way to illustrate the results when the interaction is not significant. If the lines in Fig. 7.14 were significantly different from parallel, then the comparisons between the treatments would depend upon which value of BACBEF was substituted into the equation. For significant interactions, a full interaction diagram best illustrates the results.

Confidence intervals for interactions

Just as confidence intervals for slopes may be calculated, so can a confidence interval be calculated for the difference between two slopes. This will be illustrated with an example.

Sixty great tits (*Parus major*) were captured briefly during May in a woodland in Northern Scotland. Their weight in grams (WGHT) and tarsus length in millimetres (TARSUS) were measured. A track ran through the woods, and management had been different in the two parts. The side of the track on which the bird had been captured was also recorded (LOCATION, a categorical variable with two levels). The data are stored in the *great tits* dataset. The aim of the study was to investigate the relationship between WGHT and TARSUS, and to see if it differed between the two sides of the track (i.e. with LOCATION).

The full model was fitted (Box 7.9), from which it can be seen that the interaction is not significant: that is, the slope of the line describing the relationship between tarsus and weight is not significantly different between the two sides of the road.

The information in the coefficient table can be used to construct a confidence interval for the interaction. As revised earlier in this chapter, the general formula for a confidence interval is:

$$\text{estimate} \pm SE_{\text{estimate}} \times t_{\text{crit}}.$$

We can construct the confidence interval for any parameter following this formula. The slope between WGHT and TARSUS for Location 1 is (3.9929 – 0.0354) and for Location 2 is (3.9929 + 0.0354). The estimate of –0.0354 given in the coefficient table is half the difference between the two slopes, and the standard error of half the difference is 0.1458. Both of these values need to be doubled to give us the difference between the slopes, and the standard error of that difference. Once again, the critical *t*-value has its degrees of freedom defined by the error (56). Putting all this information together, we obtain an expression for the confidence interval of the difference:

$$(0.0354 \times 2) \pm (0.1458 \times 2) \times 2.004$$

which comes to:

$$(-0.514, 0.655).$$

BOX 7.9 Predicting the weight of great tits—the full model

General Linear Model

Word equation: WGHT = LOCATION + TARSUS + LOCATION * TARSUS
LOCATION is categorical and TARSUS is continuous

Analysis of variance table for WGHT, using Adjusted SS for tests

Source	DF	Seq SS	Adj SS	Adj MS	F	P
LOCATION	1	11.0	1.5	1.5	0.19	0.661
TARSUS	1	5682.0	5599.8	5599.8	750.41	0.000
LOCATION * TARSUS	1	0.4	0.4	0.4	0.06	0.809
Error	56	417.9	417.9	7.5		
Total	59	6111.4				

Coefficients table

Term	Coef	SECoef	T	P
Constant	−41.993	2.408	−17.44	0.000
LOCATION				
1	1.062	2.408	0.44	0.661
2	*−1.062*			
TARSUS	3.9929	0.1458	27.39	0.000
TARSUS * LOCATION				
1	−0.0354	0.1458	−0.24	0.809
2	*0.0354*			

One point to note about this confidence interval is that it contains zero. This is because it is not significant—the two properties follow one from the other.

In this particular example, it has been possible to construct a confidence interval for the interaction because the categorical variable involved had only two levels. In other words, a pairwise comparison is necessary to be able to construct a confidence interval for the difference using this approach.

Interactions between continuous variables

In many ways, the interaction between two continuous variables is the easiest kind of interaction to understand, because the interaction sign in the model formula (* or . depending upon your package) now really does mean multiplication. For example, consider the model $Y = X + Z$, where X and Z are continuous. If we wish to fit an interaction, one way of doing it would be to multiply the two columns together to create a third continuous variable, XTIMESZ and fit it as an additional term

$$Y = X + Z + \texttt{XTIMESZ}.$$

BOX 7.10 Testing for interactions between two continuous variables

General Linear Model

Word equation: VOLUME = DIAMETER + HEIGHT + DIAMETER * HEIGHT

DIAMETER and HEIGHT are continuous variables

Analysis of variance table for VOLUME, using Adjusted SS for tests

Source	DF	Seq SS	Adj SS	Adj MS	F	P
DIAMETER	1	7581.8	68.1	68.1	9.29	0.005
HEIGHT	1	102.4	128.6	128.6	17.52	0.000
DIAMETER * HEIGHT	1	223.8	223.8	223.8	30.51	0.000
Error	27	198.1	198.1	7.3		
Total	30	8106.1				

Coefficients table

Term	Coef	SECoef	T	P
Constant	69.40	23.84	2.91	0.007
DIAMETER	−5.856	1.921	−3.05	0.005
HEIGHT	−1.2971	0.3098	−4.19	0.000
DIAMETER * HEIGHT	0.13465	0.02438	5.52	0.000

Alternatively, we could simply fit the interaction between the two variables as follows:

$$Y = X \mid Z = X + Z + X * Z$$

If the slope of XTIMESZ (or equivalently the slope of the term $X * Z$) is significantly different from zero, then there is a significant interaction.

This can be illustrated by an example. Returning to the data set *trees*, we can attempt to predict the volume by using the height and diameter.

In this example (Box 7.10), the interaction term is very significant ($p < 0.0005$). What exactly does this mean? In this instance, it means that the effect of an extra foot of height on volume is greater, the greater the diameter of the tree (as would be expected).

7.6 Uses of interactions

Interactions have two main functions. This chapter has focused on the first, and touched on the second (which will be expanded in Chapter 9).

Is the story simple or complicated?

For experiments that have a factorial design, fitting an interaction is an essential part of the analysis. It is also the first *p*-value whose significance must be considered. If the model involves two variables, and the interaction between them is not significant, then we can describe the effects of those two variables on the response variable quite separately. This was the case with the experiment in which yield of wheat was predicted from sowing rate and variety. Because the interaction was not significant, the effects of sowing rate on yield could be summarised in one figure, while the effects of variety on yield could be summarised in another figure. If the interaction had been significant, then we could not answer the question 'How does sowing rate affect yield?', without first needing to know which variety was being used. In the *tulips* example, there was a significant interaction between water and shade. So we need to describe the impact of water level on blooms twice—once for each level of shade. So the significance of the interaction tells us if we are dealing with two simple stories, or one complicated one.

Is the best model additive?

The second function of an interaction is to test for additivity. If the model we fit does not include any interaction terms, then we are assuming that the variables combine together additively. This is true of both categorical and continuous explanatory variables. This is an assumption we have made without much questioning until now. Adding an interaction allows us to test if this assumption is actually justified.

The example analysed in Box 7.10 has a highly significant interaction. This illustrates that it is not appropriate to assume that height and diameter have additive effects on volume. If a tree trunk is anything like a cylinder, the volume will follow an equation similar to:

$$Volume = \pi\left(\frac{Diameter}{2}\right)^2 Height.$$

So it should be expected that diameter and height combine together multiplicatively. An alternative is to transform all three variables into logarithms, so converting a multiplicative relationship into an additive one, in other words fitting a model along the lines of:

$$\log(Volume) = \log(Height) + \log(Diameter).$$

This raises the question of which of these two alternatives is the best one? Testing your alternative models to see which is best is the theme of Chapter 9. In fact, a significant interaction can be an indication that it is more appropriate to take logs.

The theme of testing additivity and using interactions to solve problems of non-additivity is greatly expanded in Chapter 9.

7.7 Summary

- Factorial experimental designs have two advantages: they allowing testing for interactions; and in the absence of significant interactions, it provides the great advantage of hidden replication.
- An interaction was defined to mean that the additive effect of one *X*-variable on *Y* is different at the different levels of another *X* variable.
- Model formulae were introduced which allow the partitioning of sums of squares between main effects and interactions.
- Interaction diagrams are introduced. Plotting the data in a diagram allows a clear visual summary of the results. Ideal versions of the diagrams help us understand null hypotheses and the meaning of significance tests.
- When interactions are not significant, there are often two simple stories. If interactions are significant, there will be one more complicated story.
- Standard errors and confidence intervals for both means and differences were introduced as alternative ways of presenting the results.
- Interactions between categorical and continuous variables, and also between two continuous variables, were introduced.
- Interactions have two basic functions: (1) is the story simple or complicated?; (2) is the best model additive?

7.8 Exercises

Antidotes

An experiment was conducted into the effectiveness of two antidotes to four different doses of toxin. The antidote was given five minutes after the toxin, and twenty-five minutes later the response was measured as the concentration of related products in the blood. There were three subjects at each combination of the antidote and dose level. The data are stored in the dataset *antidotes*. The results of the factorial ANOVA are given in Box 7.11.

(1) Draw the full interaction diagram.
(2) What are the conclusions from the ANOVA table?
(3) What is the most useful way to summarise the results of this experiment?

BOX 7.11 Interaction in the *antidotes* data

General Linear Model

Word equation: BLOOD = ANTIDOTE + DOSE + ANTIDOTE * DOSE

ANTIDOTE and DOSE are categorical

Analysis of variance table for BLOOD, using Adjusted SS for tests

Source	DF	Seq SS	Adj SS	Adj MS	F	P
ANTIDOTE	1	1396.90	1396.90	1396.90	23.68	0.000
DOSE	3	1070.09	1070.09	356.70	6.05	0.006
ANTIDOTE * DOSE	3	835.88	835.88	278.63	4.72	0.015
Error	16	943.68	943.68	58.98		
Total	23	4246.55				

Coefficients table

Term		Coef	SE Coef	T	P
Constant		8.697	1.568	5.55	0.000
ANTIDOTE					
1		7.629	1.568	4.87	0.000
2		*−7.629*			
DOSE					
5		−8.186	2.715	−3.01	0.008
10		−4.119	2.715	−1.52	0.149
15		3.097	2.715	1.14	0.271
20		*9.208*			
ANTIDOTE * DOSE					
1	5	−7.244	2.715	−2.67	0.017
1	10	−3.551	2.715	−1.31	0.209
1	15	2.573	2.715	0.95	0.358
1	*20*	*8.222*			
2	*5*	*7.244*			
2	*10*	*3.551*			
2	*15*	*−2.573*			
2	*20*	*−8.222*			

Weight, fat and sex

Returning to the *fats* data set, we are now in the position to do a more detailed analysis taking the sex of the participants into account. Analyse the data to answer the following:

(1) What is the best fitting line through the male data?

(2) What is the best fitting line through the female data?

(3) How strong is the evidence that the slopes differ?

8 Checking the models I: independence

All statistical tests rely on assumptions. General Linear Models are a group of **parametric tests**, which make the four assumptions of: **independence**; **homogeneity of variance**; **normality of error**; and **linearity/additivity**. The first of these, independence, is fundamental to all statistics. It is also the single most important cause of serious statistical problems.

Why is it so important that these assumptions are upheld? The aim of hypothesis testing in a GLM analysis is to ask if certain factors are important in explaining the variability in our data. This is summarised in a p value, which weighs up the strength of the evidence that our null hypothesis is false. This p value is accurate only if the assumptions are true. If they are contravened, not only is the p value misleading, it is rarely possible to tell how inaccurate it is. Furthermore, the parameter estimates as presented in the coefficients table will not be the best estimates possible, and the standard errors of these coefficients will not be correct. The worth of the whole analysis is cast into doubt. Clearly the assumptions of a GLM are potentially very important.

So checking the assumptions is a vital part of any GLM analysis. To explore exactly what is meant by each of the four assumptions in turn, it is useful to think through an analysis in reverse. In other words, imagine Nature creating data sets by writing a program. It is then our aim to guess how the program was written by analysing the resulting data. Such a program would need to contain the following elements:

- the noise (for example, draw 30 points at random from a Standard Normal Distribution, and multiply each by the square root of the 'true error variance')
- the signal (create the data variable Y by specifying the same word equation that would be used in the analysis of the data, but also include the 'true parameter estimates').

This could be represented as $Y = 2 * X1 + 4.5 * X2 + Noise$.

Once created, the data set can be analysed using the correct word equation (which in this case would be $Y = X1 + X2$). The aim of the analysis is to distinguish the noise from the signal: except that you now have the a priori knowledge of what the signal is. This artificially constructed data set conforms to the assumptions of a GLM because it has been designed to do so. It is a useful tool in illustrating exactly what is meant by these assumptions.

In this chapter, the first of the four assumptions, independence, is examined in detail. The other three assumptions are discussed in Chapter 9. A formal definition of independence is that *datapoints are independent if knowing the error of one or a subset of datapoints provides no knowledge of the error of any others*. So how is this illustrated by the artificial data set? The residuals represent the scatter around the model, and have been produced by the numbers drawn at random from the Standard Normal Distribution. Given that they have been drawn at random, they must be independent of each other. How does this relate to real data sets? The easiest way to envisage this is to consider a situation in which independence would be contravened.

8.1 Heterogeneous data

An ecologist hypothesised that a certain species of caterpillar was reaching such high densities in the field that there was competition for food. He collected data from three habitats, measuring the weight gain of marked individuals over 5 days (WGTGAIN), and the mean number of caterpillars per shoot on 10 randomly chosen plants (POPDEN). These data are stored in the dataset *caterpillars*. He then looked for a relationship between these two variables as shown in Box 8.1.

This analysis appears to confirm the hypothesis that at higher population densities, the caterpillar experiences lower weight gains—the relationship between these two variables being negative. However, the fact that these data

BOX 8.1 Weight gain and population density, ignoring habitat heterogeneity

General Linear Model

Word equation WGTGAIN = POPDEN

POPDEN is continuous

Analysis of variance table for WGTGAIN, using Adjusted SS for tests

Source	DF	Seq SS	Adj SS	Adj MS	F	P
POPDEN	1	354.12	354.12	354.12	36.82	0.000
Error	28	269.32	269.32	9.62		
Total	29	623.44				

Coefficients table

Term	Coef	SECoef	T	P
Constant	14.708	1.082	13.59	0.000
POPDEN	−3.2338	0.5330	−6.07	0.000

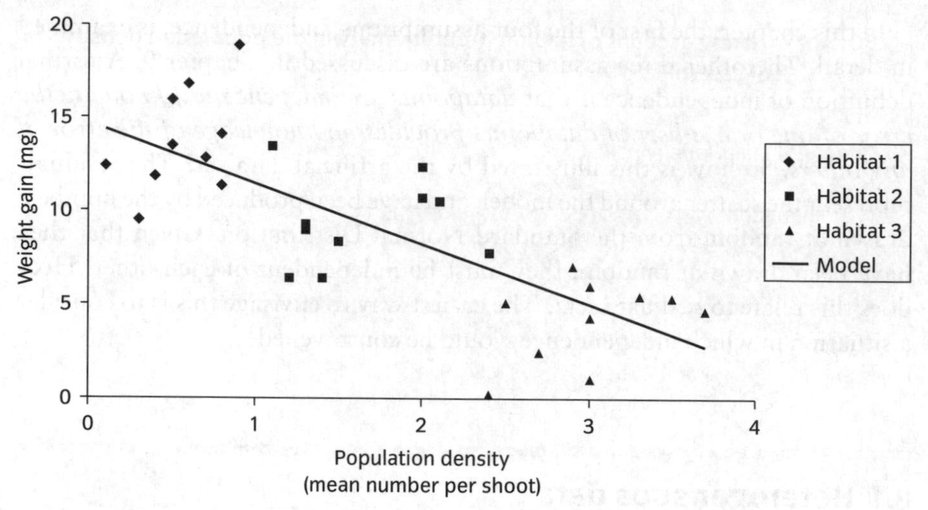

Fig. 8.1 The relationship between weight gain and population density.

have been gathered from three different habitats has been ignored—is this important?

The data, together with a line representing the fitted model, are illustrated in Fig. 8.1. The three different habitats are represented by diamonds, squares and triangles respectively. It appears that the population densities are significantly different in Habitats 1 and 3—so what happens if HABITAT is included in the model? This is done in Box 8.2 below.

Now it can be seen that the relationship between POPDEN and WGTGAIN has disappeared—the slope between these two variables is not significantly different from zero once the differences between the habitats have been statistically eliminated.

So the best representation of the data would be by three horizontal lines as depicted in Fig. 8.2. While the mean POPDEN and the mean WGTGAIN is different between the habitats, there is no significant relationship between them within habitats. By ignoring the fact that the data were collected in three different habitats, an artefactual association was created between the two continuous variables, or at least, the evidence was exaggerated.

So how does this relate to the assumption of independence? The original data set was not drawn at random from one homogenous pool of data, but consisted of three subsets. The model of Fig. 8.1 implies that the data vary at random around the line, whereas a better representation is that they vary around three different means. All the data points drawn from one habitat have something in common—knowing the value of some of these residuals would help you predict the likely values of other residuals from the same habitat. However, with the correct model of Fig. 8.2, the data points are now grouped around the means for each habitat—so knowing the residual of one point will not help you predict the likely residuals of other data points. If the grouping

BOX 8.2 Taking account of habitat heterogeneity in the population density of caterpillars

General Linear Model

Word equation: WGTGAIN = HABITAT + POPDEN

HABITAT is categorical and POPDEN is continuous

Analysis of variance table for WGTGAIN, using Adjusted SS for tests

Source	DF	Seq SS	Adj SS	Adj MS	F	P
HABITAT	2	458.36	118.85	59.43	10.27	0.001
POPDEN	1	14.62	14.62	14.62	2.53	0.124
Error	26	150.47	150.47	5.79		
Total	29	623.44				

Coefficients table

Term	Coef	SECoef	T	P
Constant	5.784	2.141	2.70	0.012
HABITAT				
1	6.853	1.547	4.43	0.000
2	0.5455	0.6328	0.86	0.397
3	*−7.3985*			
POPDEN	1.924	1.211	1.59	0.124

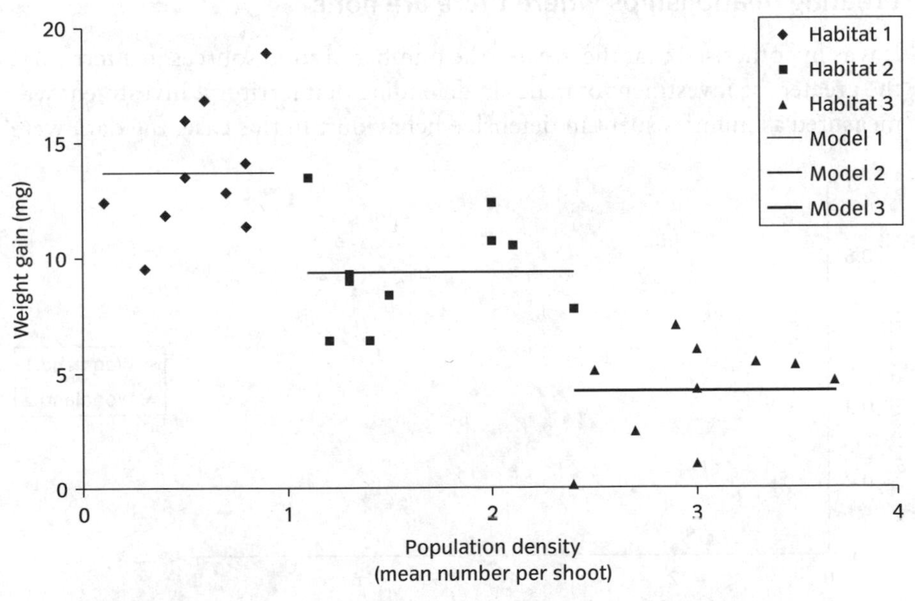

Fig. 8.2 The relationship between weight gain, population density and habitat.

within the data is ignored, the data are **heterogeneous**, and so contravene the assumption of independence.

It is possible to imagine all kinds of scenarios in which, ignoring the natural subsets within a data set, the wrong conclusion would be drawn. Biologists used to ignore heterogeneity because they lacked the statistical tools needed to include extra variables in whatever model they wanted to fit. Now there is no excuse.

Same conclusion within and between subsets

The number of caterpillars of the Orange Tip butterfly *Anthocharis cardamines* were counted on ten plants in each of six heavily infested patches in two woodlands. Caterpillar density (CATDEN) was recorded as the mean number of caterpillars per plant for each patch. They were then monitored until pupation, and the number that were parasitised by an ichneumonid recorded (PARA). These data are stored in the *Orange Tips* dataset, and shown graphically in Fig. 8.3.

If the data are analysed ignoring the two woodlands, then we would conclude that parasitism is strongly density dependent. If WOOD is included as a factor in the model, then the same conclusion would be reached, but in addition to this, it would become apparent that densities of caterpillars are higher in one woodland than the other. So in this example, we would reach the same basic conclusion, but if WOOD is ignored, extra information on the distribution of the butterfly would be lost.

Creating relationships where there are none

It was hypothesised that the greater the number of food sources in a territory, the greater the investment of males in defending that territory. Investment was measured as minutes spent in defensive behaviour. In this case, the data were

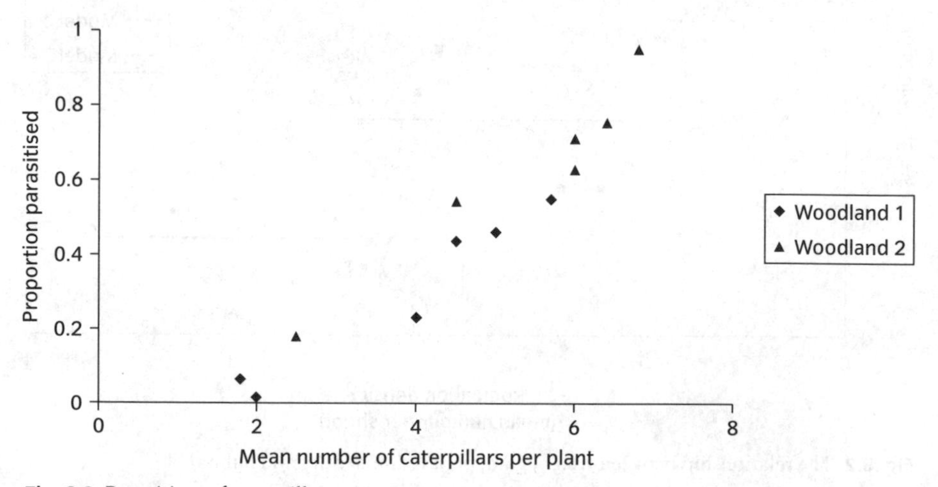

Fig. 8.3 Parasitism of caterpillars.

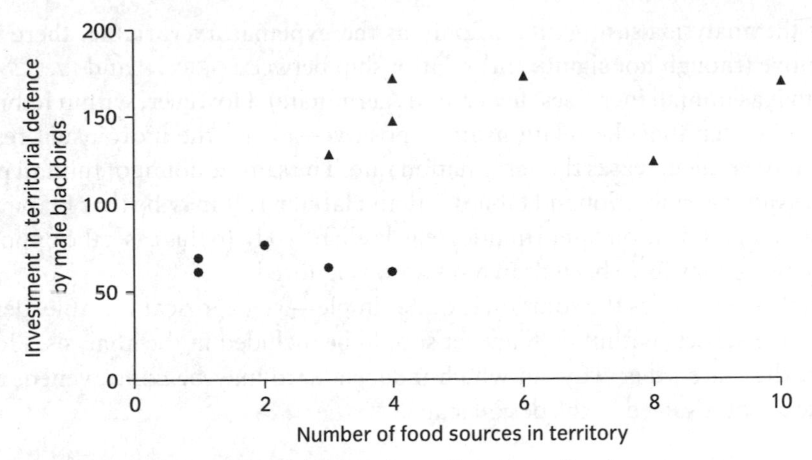

Fig. 8.4 Territorial defence by male blackbirds: ● = Year 1; ▲ = Year 2.

collected over two breeding seasons, and the variables INVEST, YEAR and FSOURCE are stored in the *blackbirds* dataset. See Fig. 8.4.

It can be seen that if YEAR is ignored, the data would support the hypothesis. If YEAR is included as a factor, then within years there is no such relationship (there is no trend with the dots and there is also no trend with the triangles). It may be that in Year 2 there was greater food availability, increasing the number of food sources, and allowing the males to spend more time on defending their territories. So in this case the wrong conclusion would be drawn from omitting YEAR.

Concluding the opposite

The germination rate of tree seedlings was recorded at a number of sites (GERMIN), and regressed against rainfall at those sites (RAINFALL). The sites were drawn from two very different habitat types (HAB), see Fig. 8.5. The data are recorded in the *rain and germination* dataset.

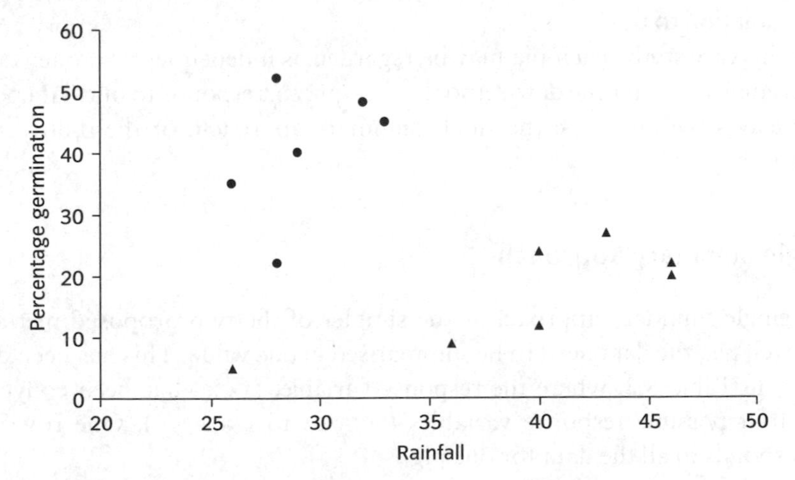

Fig. 8.5 Germination rates with rainfall: ● = Habitat 1; ▲ = Habitat 2.

If the analysis uses RAINFALL only as the explanatory variable, there is a negative (though not significant) relationship between GERMIN and RAINFALL (so that as rainfall increases, fewer seeds germinate). However, within habitat, it can be seen that the relationship is positive—giving the more usual result that more rain increases the germination rate. The same amount of rainfall produces more germination in Habitat 2 than Habitat 1. It may be that Habitat 1 has fewer germination opportunities (e.g. less bare soil). In this case, the opposite conclusion may have been drawn if HAB was ignored.

In these examples the solution is quite simple—a categorical variable identifying the subsets within the data set should be included in the analysis. However, there are other ways in which independence may be contravened, and these are best solved at the design stage.

8.2 Repeated measures

When an individual is measured more than once, and these measures are treated as independent, then the fundamental assumption of independence is contravened. This is illustrated by the following example.

A farmer wishes to find out which diet fattens his pigs up best. He has 10 pigs with which to compare two diets, so 5 are placed on each diet. He then weighs them on four separate dates: 3, 8, 20 and 60 weeks, and presents the data in a table (Table 8.1).

These data are stored in the *pigs* dataset. Ostensibly, this data set has 40 data points. However, each pig has been measured four times, and these four data points are not independent. If a pig is relatively large on day 1, then it is likely to remain relatively large throughout the experiment. If this large pig happened to be on Diet 1, then it would provide four pieces of evidence that Diet 1 is superior (when it should only provide one piece of evidence). So what is the solution to this?

In this case study, each pig may be regarded as independent. So one way of proceeding is to alter the data set so that one pig corresponds to one data point. This leaves two avenues: the single summary approach, or the multivariate approach.

Single summary approach

The single summary approach is the simpler of the two proposed methods. For each pig, the data need to be summarised in one value. This has been done below in Table 8.2, where the response variable, LOGWT has been converted into four possible response variables (LOGWT3 to LOGWT60). One row now corresponds to all the data for one pig.

Table 8.1 The pig dataset

DIET	PIG	SAMPLE	LGWT	DIET	PIG	SAMPLE	LGWT
1	1	1	0.78846	2	6	1	0.74194
1	1	2	1.70475	2	6	2	1.66771
1	1	3	3.72810	2	6	3	3.71357
1	1	4	4.68767	2	6	4	4.52504
1	2	1	0.69315	2	7	1	0.58779
1	2	2	1.58924	2	7	2	1.45862
1	2	3	3.83298	2	7	3	3.58074
1	2	4	4.53903	2	7	4	4.37450
1	3	1	0.69315	2	8	1	0.64185
1	3	2	1.64866	2	8	2	1.52606
1	3	3	3.73050	2	8	3	3.62700
1	3	4	4.60517	2	8	4	4.35927
1	4	1	0.78846	2	9	1	0.53063
1	4	2	1.60944	2	9	2	1.45862
1	4	3	3.63495	2	9	3	3.46574
1	4	4	4.45783	2	9	4	4.26690
1	5	1	0.83291	2	10	1	0.91629
1	5	2	1.72277	2	10	2	1.68640
1	5	3	3.87743	2	10	3	3.79098
1	5	4	4.64150	2	10	4	4.62301

Table 8.2 The 'one-pig-one-datapoint' dataset. As each pig is a separate datapoint, there is no need for the variable PIG

LOGWT3	LOGWT8	LOGWT20	LOGWT60	DIET
0.7885	1.7048	3.7281	4.6877	1
0.6931	1.5892	3.8330	4.5390	1
0.6931	1.6487	3.7301	4.6052	1
0.7885	1.6090	3.6350	4.4578	1
0.8329	1.7228	3.8774	4.6415	1
0.7419	1.6677	3.7136	4.5250	2
0.5878	1.4586	3.5807	4.3745	2
0.6419	1.5261	3.6270	4.3593	2
0.5306	1.4586	3.4657	4.2669	2
0.9163	1.6864	3.7910	4.6230	2

The choice of summary value will depend upon the aim of the experiment. For example, if the farmer was interested in which diet produced the greatest weight after 60 weeks, then it would be most appropriate to use the final weight (LOGWT60), and discard the rest of the data. However, if he was more interested in the fastest growth rate over the first 20 weeks, then it might be more appropriate to use (LOGWT20 – LOGWT3) and discard the final data point. The key points to remember are:

- Any summary variable will do, but it is important that it encapsulates the information you are really interested in.
- For many summary variables, you may end up discarding data. While this is not ideal, it is vastly preferable to contravening the assumption of independence. Sometimes extra measurements are taken in a futile attempt to provide more datapoints. Understanding about independence will help to avoid this wasted effort.
- It is possible to choose more than one summary statistic. However, care must be taken here. If you analyse the data in different ways, you are having a number of bites at the cherry. The probability of finding a significant result by chance alone will increase. (This problem is called 'multiplicity of *p* values', and is discussed in more detail in Chapter 11).

The farmer decided that he was most interested in their final weight, so he conducted the analysis summarised in Box 8.3.

BOX 8.3 Analysing the pigs' final weight

General Linear Model

Word equation: LOGWT60 = DIET
DIET is categorcial

Analysis of variance table for LOGWT60, using Adjusted SS for tests

Source	DF	Seq SS	Adj SS	Adj MS	F	P
DIET	1	0.06123	0.06123	0.06123	4.32	0.071
Error	8	0.11339	0.11339	0.01417		
Total	9	0.17462				

Coefficients table

Term	Coef	SECoef	T	P
Constant	4.50799	0.03765	119.74	0.000
DIET				
1	0.07825	0.03765	2.08	0.071
2	*−0.07825*			

While there is some suggestion that Diet 1 is better, the results are not significant. The total degrees of freedom are only 9, so this has become a very small data set, and it is not so suprising that no significant differences have been found. The farmer would have done better to use more pigs, and only to weigh them at the final time point. In other words, this problem might have been better tackled at the design stage. Of course space may have been the limiting factor, but this genuine constraint must be accepted, as it cannot be circumvented by repeated measurements.

In designing experiments, it is important to be clear about which data values can be regarded as independent. This links with the discussion in Chapter 5 about recognising true replicates. The experimental units which are randomly allocated to treatments are both independent, and true replicates. In this case, the individual pigs are randomly allocated to two diets, and the correct dataset contains one datapoint per pig.

Is there any way in which all the information could be combined, yet the analysis is still legitimate? The answer is yes—and this possibility is discussed next.

The multivariate approach

The general approach to multivariate statistics will be described here, with the aim of introducing the principle, and describing when it is likely to be needed. It also provides an illustration of how the inclusion of multiple *Y* variables can be executed without contravening independence.

The principle behind multivariate statistics is to use all the *Y* variables to see if we can distinguish between Diets 1 and 2. To illustrate how this works, consider a simpler case, in which there are just two *Y* variables, `WGHT20` and `WGHT60`. Each pig then provides one data point on the graph, with those on Diet 1 being •, and those on Diet 2 being ▾.

In Fig. 8.6(a), the two diets cannot be distinguished on the basis of `WGHT20`, with both diets giving the full range of readings. They can however be distinguished on the basis of `WGHT60`, with Diet 1 giving higher values.

The alternative illustrated in Fig. 8.6(b) is that the two diets can only be distinguished on the basis of `WGHT20`. In Fig. 8.6(c), the two clusters of points fall

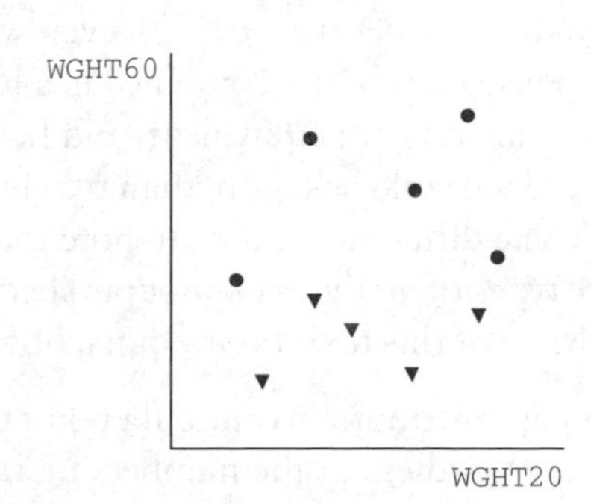

Fig. 8.6(a) `WGHT60` distinguishes the two diets.

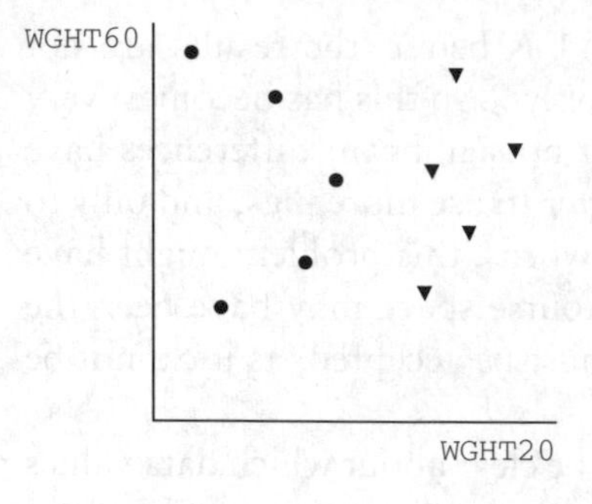

Fig. 8.6(b) WGHT20 distinguishes the two diets.

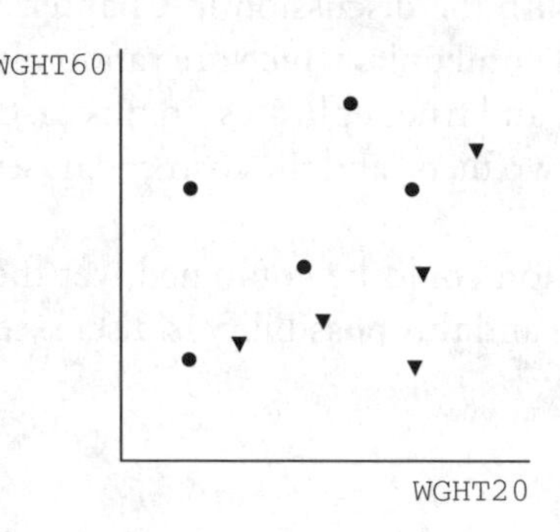

Fig. 8.6(c) A function of WGHT20 and WGHT60 distinguishes the two diets.

on either side of a diagonal line, so a function of both WGHT20 and WGHT60 would best distinguish between them. The fourth possibility (not illustrated), is that a complete mixing of dots and triangles would mean that the two diets could not be distinguished by their WGHT20 or WGHT60 values.

One particular multivariate technique (Discriminant Function Analysis) extends this principle for as many axes as there are *Y* variables in the data set (four in this case). The aim is to see if there is any way of grouping the two sets of data points from the two diets in multidimensional space. The null hypothesis would be that the two clouds of points for Diets 1 and 2 could not be distinguished by any combination of axes.

A 'MANOVA', an analysis of variance with multiple *Y* variables, will test the null hypothesis that there is no difference between the two diets. Box 8.4 summarises the results for the *pigs* dataset.

From this analysis we can conclude that there are no significant differences between the two diets ($p = 0.529$), even when the information contained in all four response variables is combined in a legitimate analysis. In this very simple case, all four tests are equivalent, and hence have the same *p*-value. Once the categorical variable has more than two levels, or there are more variables, the tests become different, leading to potential problems of interpretation.

These types of analyses are not pursued further here as this is not part of the GLM theme of this text. Four points, however, should be noted:

1. One pig contributes to one data point on the graph and in the analysis. This is true regardless of the numbers of axes used, with each data point being

BOX 8.4 Analysis of variance with multiple *Y* variables

MANOVA

Word equation: LOGWT3, LOGWT8, LOGWT20, LOGWT60 = DIET
DIET is categorical

MANOVA for DIET $s = 1$ $m = 1.0$ $n = 1.5$

Criterion	Test statistic	F	DF	P
Wilk's	0.58211	0.897	(4, 5)	0.529
Lawley–Hotelling	0.71790	0.897	(4, 5)	0.529
Pillai's	0.41789	0.897	(4, 5)	0.529
Roy's	0.71790	0.897	(4, 5)	0.529

defined by as many coordinates as there are *Y* variables. In this case, each pig would be defined by one datapoint in four dimensions. The assumption of independence has therefore not been contravened.

2. All the information can be combined in a legitimate analysis.
3. Just as with GLMs, multivariate analyses make assumptions. In GLMs, the error is assumed to be distributed Normally. In multivariate analyses, the *Y* variables are assumed to be distributed multi-Normally. Considering the simplest case of two *Y* variables, a bivariate Normal distribution would be a single peaked hill over the *Y*1 and *Y*2 axes (beyond this it is not possible to visualise!).
4. In general, multivariate analyses have more complicated assumptions, that are more likely to be contravened.

While there are costs to doing these more complex analyses (in terms of assumptions), this complexity may be worthwhile if it is essential to combine data from different *Y* variables. The multivariate approach is needed if evidence from different response variables must be combined in a non-post-hoc way, for example, to show a significant effect of an *X* variable.

8.3 Nested data

A third kind of data set which may contravene independence if not properly handled is that which deals with nested data. This can be illustrated by a study of the calcium content of leaves. Three plants were chosen, from which four leaves were sampled. Then from each of these twelve leaves, four discs were cut and processed. This is illustrated in Fig. 8.7.

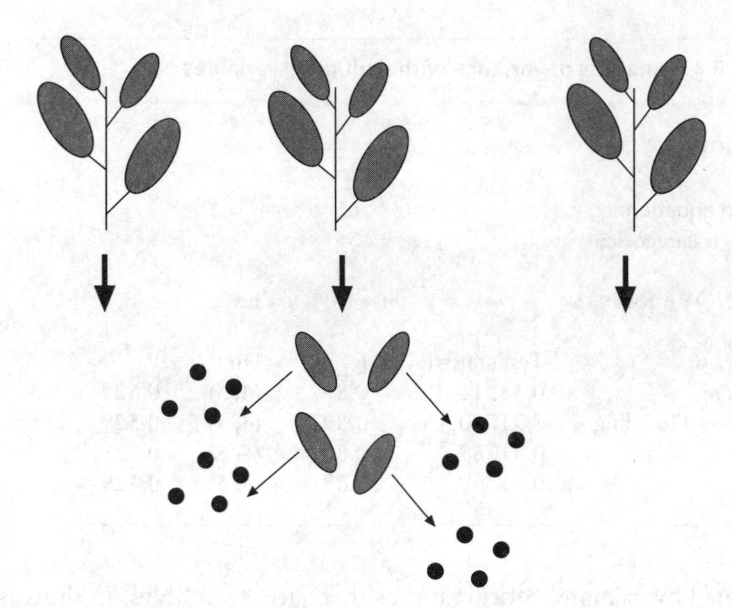

Fig. 8.7 Plants, leaves and discs in a nested design.

This data set will have 48 values of calcium concentration from the leaf discs. To treat these values as 48 independent points however would not be correct—the discs from the same leaves are likely to be closer in value than those from different leaves, and those on the same plant will be more similar to each other than those from different plants. Again, there are natural clusters within the dataset, and this needs to be reflected in the way it is analysed. The clusters occur at two different hierarchical levels; discs within leaves and leaves within plants, so this is rather more complex than simply having repeated measures on the same plant.

The analysis of this kind of hierarchical data is covered in Chapter 12. This example has been mentioned here to draw attention to how its misanalysis would contravene independence.

8.4 Detecting non-independence

The ability to detect non-independence is extremely useful. Non-independence is a common mistake, even in the published scientific literature, and it can have drastic consequences in terms of the conclusions you reach. Moreover, there is no fail safe method. Non-independence cannot always be detected, even if present.

When designing your own experiments, it is important to consider the issues of independence at the earliest possible stage—before collecting data. Recognising true replicates, and randomly allocating your treatments to replicates, is a vital part of this process. When inspecting someone else's analysis, it should be possible to calculate the degrees of freedom expected in the ANOVA table from their description of the experimental design. There are three features which should cause you to be suspicious, even if the experimental design is not available:

1. *Too many data points*. This can be assessed by looking at the error and total degrees of freedom. 1200 counts of leaf area may have been taken from 1200 trees, or 100 leaves from each of 12 trees.
2. *A highly implausible result*. Repeating measurements from a few individuals has the effect of artificially inflating the sample size. Small, insignificant differences can become significant just by over-counting, the implication being that every count is a separate piece of evidence in support of that difference.
3. *Any kind of repeated measure*. Repeated measures can turn up in all sorts of guises, for example, time series and growth curves are effectively repeated measures over time. While there are valid approaches for some of these cases, it is common for invalid methods of analysis to be employed.

These points can be illustrated by considering the following example.

Germination of tomato seeds

An experiment was performed to determine the effects of five different levels of watering on the germination of tomato seeds. The main aim was to determine the level of watering that maximised percentage germination. There were 50 seed trays, and the percentage germination was measured for each tray on three dates and stored in the *tomatoes* dataset. In the first analysis all 150 data points were used (Box 8.5).

BOX 8.5 The WRONG way to analyse tomato germination

General Linear Model

Word equation: PERCGERM = DATE + WATER

DATE and WATER are categorical

Analysis of variance table for PERCGERM, using Adjusted SS for tests

Source	DF	Seq SS	Adj SS	Adj MS	F	P
DATE	2	5713.03	5713.03	2856.51	205.00	0.000
WATER	4	283.16	283.16	70.79	5.08	0.001
Error	143	1992.57	1992.57	13.93		
Total	149	7988.75				

BOX 8.6 The RIGHT way to analyse tomato germination

General Linear Model

Word equation: PG3 = WATER3

WATER3 is categorical

Analysis of variance table for PG3, using Adjusted SS for tests

Source	DF	Seq SS	Adj SS	Adj MS	F	P
WATER3	4	140.97	140.97	35.24	2.21	0.083
Error	45	717.98	717.98	15.96		
Total	49	858.95				

In Box 8.5, the variable are defined as follows. PERCGERM = the percentage germination of a tray on a given date. WATER = the watering treatment for that tray, coded as –2, –1, 0, 1, 2, from lowest to highest watering levels. DATE = the dates of the measurement, coded as 1, 2, 3. The first point to note about the analysis is that the total degrees of freedom are 149 (from 150 – 1). However, the description of the experimental design stated that there were 50 trays, each of which were measured three times. Hence this analysis suffers from repeated measures. On the assumption that each tray of seedlings was randomly allocated to a watering treatment, it is the tray which is the independent unit. If one tray had particularly high germination at time point 1 (just by chance), then it would be likely to have high germination at time points 2 and 3. These three readings would not be independent, but would give the impression that the watering treatment in question was particularly favourable for germination. So while both DATE and WATER are highly significant in this analysis (with $p < 0.0005$ and $p = 0.001$ respectively), we cannot trust these conclusions.

These data have been reanalysed in Box 8.6, using only the 50 measurements from the final date. PG3 is now percentage germination on the final date, and WATER3 is the watering treatment for that tray.

In the new analysis, the total degrees of freedom (49) now reflect the true level of replication in this experiment. However, we have lost any significance of watering treatment with $p = 0.083$. The significance of the earlier result was a direct consequence of contravening independence. (Another way of expressing this is to say that the first analysis was pseudoreplicated. This term was first discussed in Chapter 5).

So in this example the first analysis suggested too many data points, due to repeated measures. This led to a result being significant, but with the correct analysis this significance disappeared. We will return to this data set later, with an even better way of analysing it.

8.5 Summary

- Independence is an assumption which applies to all statistical tests.
- Ignoring the natural subsets when analysing heterogeneous data contravenes the assumption of independence. The possible consequences range from missing significant relationships, to finding spurious relationships.
- Analysing repeated measures is another common way in which independence is contravened. Solutions to this problem include condensing the data to a single summary, or using multivariate statistics.
- Misanalysis of nested data will also contravene independence.
- To detect non-independence, match the experimental design to the analysis.
- Signs of non-independence include too many data points, and highly implausible results. However, non-independence cannot always be detected.
- Possible causes of non-independence include:
  - larvae reared from different batches of eggs
  - measurements made on different days or by different people
  - twins or siblings
  - samples taken from a limited number of plants or habitats
  - animals reared from different litters
  - repeated measures from the same person, place or organism.

8.6 Exercises

How non-independence can inflate sample size enormously

Dr Sharp supervised an undergraduate project on feeding in sheep. She is convinced that male sheep look up more frequently while eating than female sheep (and even has a theory, not relevant here, about why this must be so). The undergraduate spent many uncomfortable weeks in a hide near a field (so as not to disturb the sheep). He recorded the following data for each of twenty one-hour observation periods on each of 3 male and 3 female sheep: (i) total duration of feeding time and (ii) number of look-ups (a look-up is defined as the head rising so that the chin is above the knee, with the eyes open; provided that fewer than four consecutive steps are taken with the head above the chin). The data is recorded in the *sheep* data set as five columns: (i) DURATION of feeding time in minutes; (ii) NLOOKUPS, the number of lookups; (iii) SEX, coded as 1 for female and 2 for male; (iv) SHEEP, coded as 1 to 6 and (v) OBSPER, the number of the observation period from 1 to 20. On each day on which the observations took place, each sheep was observed in a randomly determined order.

BOX 8.7 Look-up rate in feeding sheep

General Linear Model

Word equation: LUPRATE = OBSPER + SEX
OBSPER and SEX are categorical

Analysis of variance table for LUPRATE, using Adjusted SS for tests

Source	DF	Seq SS	Adj SS	Adj MS	F	P
OBSPER	19	0.191918	0.191918	0.010101	2.41	0.003
SEX	1	0.132816	0.132816	0.132816	31.67	0.000
Error	99	0.415244	0.415244	0.004194		
Total	119	0.739977				

Dr Sharp then calculated the lookup rate as NLOOKUPS / DURATION, and did the analysis summarised in Box 8.7.

(1) Criticise the analysis. (Are the datapoints independent?)

(2) Suggest and execute a more appropriate analysis.

(3) Criticise the design.

Combining data from different experiments

Every year Dr Glaikit supervises an undergraduate project, and every year it is the same. Dr Glaikit is convinced that male blackbirds that sing earlier in the year have more young fledged by the end of the year. His students work very hard, listening from early spring for singing by each of the blackbirds; observing the nests; and counting how many young are fledged. This project takes a long time as blackbirds renest until late in the summer. Dr Glaikit chooses a completely different site each year in an effort to find one that will work. The data for five years are in the dataset *birdsong*. For each year there are recorded (i) the date on which each male blackbird first sang for more than twenty minutes in one morning, as days after the first of March (SONGDAY1 etc.), (ii) the number of young fledged from all of the nests incubated in each males territory (YOUNG1 etc.). Dr Glaikit is very disappointed that each year his student fails to find an effect. Nowadays, however, biology students are statistically more wised up. When he offers to let you be the next student to work very hard for no reward, you reply with a counter-offer: if he gives you the data from the last five years, you will analyse it properly.

(1) What does Dr Glaikit not realise about the evidence from the past five years?

(2) How would you analyse it properly?

(3) What happens when you do analyse it properly?

9 Checking the models II: the other three asumptions

This chapter discusses the other three assumptions of GLMs: heterogeneity of variance, Normality of error and linearity/additivity. A dataset which meets independence, but breaks one or more of the other assumptions, can often be corrected by transforming one or more of the explanatory variables, or the response variable. The meanings of these three assumptions, and the consequences of contravening them, are explained with reference to the null data set we created at the beginning of Chapter 8. Model checking techniques are introduced, so that any problems with these assumptions can be diagnosed. Possible solutions are then described; principally those that involve transforming the data.

It is important to get into the habit of checking the assumptions you have made in fitting a particular model. It should become routine, as part of any analysis you do. At the same time however it is important to realise that it is not necessary, and in fact not possible, to get it exactly right. The model checking procedures that we suggest involve looking for patterns in certain plots. If you find yourself scrutinising a plot and wondering if there is a pattern, then it is probably OK. In fact, even weak patterns that are just discernible will probably not cause fundamental problems with your model. This point is picked up in the last exercise of this chapter, where we create datasets which we know to follow the assumptions of a GLM, analyse them, and then subject those analyses to the model checking procedures described here. Even when we know that the data conform to GLM assumptions, the model checking plots do not appear to be perfect.

9.1 Homogeneity of variance

The second assumption of general linear models is that of homogeneity of variance. An alternative way of phrasing this is that we assume the scatter around the model is of equal magnitude throughout the fitted model. The difference between the observed and the predicted value of a datapoint is a **residual.** Returning to the data set created at the beginning of this chapter, the variance of the residuals is set by our chosen value for the error variance. This is the same for all datapoints, regardless of whether the fitted values, as determined by our X variables, are large or small. In real life, however, this is often not the case. One very common example is in weighing or measuring, when it is easier to be more accurate with small objects than large. Balances often have a fixed percentage error rather than a fixed absolute error. The consequences are that absolute errors will rise with the fitted values, leading to inhomogeneous variance in the residuals.

How will this affect the interpretation of our results? The fundamental principle of GLMs is that variance is partitioned between that explained by the model, and that which remains unexplained. The unexplained variance is summarised in one figure—the error mean square. If this is inaccurate, then the F-ratios and significance levels derived from it will be likewise inaccurate. The error mean square then contributes to calculating the standard errors of all parameters estimated in the model. These standard errors are assumed to be equally applicable throughout the model—for both large and small fitted values. It can be seen that if the error mean square is only a measure of 'average variability' in the residuals, then confidence intervals will be too narrow for some parts of the model, and too wide for others.

In summary, when variances are heterogeneous, then there is no single common variance to estimate, but a different variance for each part of the data set.

This can be illustrated by considering a simple example. Fifteen men were put on a diet with the intention of losing weight. The dietician was interested in comparing three different diets to determine the most effective, so he placed five men on each of the three diets. After a month their change in weight was recorded as WGHTCH (see the dataset *diets*). In Box 9.1 and Fig. 9.1, these weight changes are analysed, and the residuals then plotted against the fitted values.

By examining the p-values, it appears that diet does affect weight loss ($p = 0.027$), with Diet 3 being the most effective (the lowest mean of the three in the

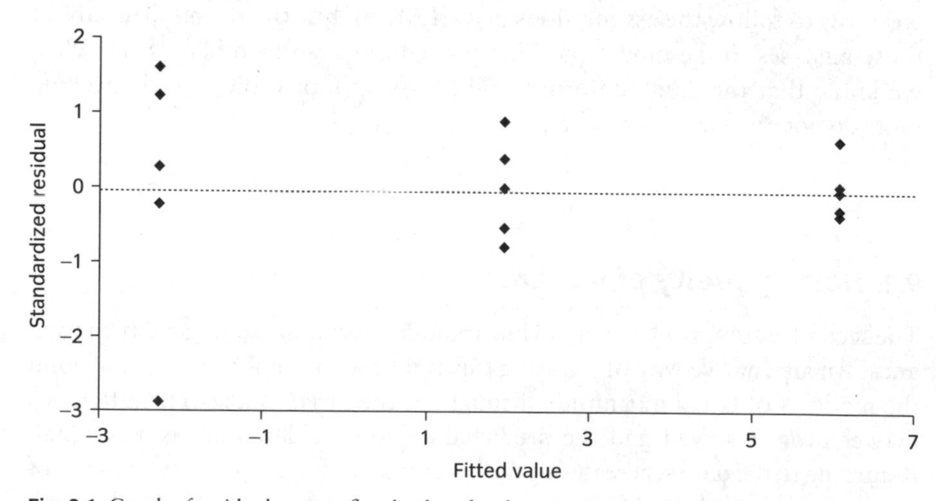

Fig. 9.1 Graph of residuals versus fitted values for the response variable WGHTCH.

BOX 9.1 Weight change with diet

General Linear Model

Word equation: WGHTCH = DIET

DIET is categorical

Analysis of variance table for WGHTCH, using Adjusted SS for tests

Source	DF	Seq SS	Adj SS	Adj MS	F	P
DIET	2	173.56	173.56	86.78	4.97	0.027
Error	12	209.42	209.42	17.45		
Total	14	382.99				

Coefficients table

Term	Coef	SECoef	T	P
Constant	1.924	1.079	1.78	0.100
DIET				
1	0.032	1.525	0.02	0.984
2	4.150	1.525	2.72	0.019
3	*−4.182*			

coefficient table). However, the residual plot indicates that there is something wrong. If these results were to be presented, one conventional way would be to present three means with their standard errors ($\frac{s}{\sqrt{n}}$, with $s = \sqrt{\text{EMS}} = 4.18$). However, the residual plot suggests that the error is far greater for Diet 3 than for Diet 1, so these standard errors would give a misleading impression of the scatter in the raw data. Transformation of the *Y* variable may be able to sort out this problem, and this should be attempted before drawing any conclusions (Section 9.4).

9.2 Normality of error

This assumption refers to the shape of the distribution of the residuals around the model. In our created dataset, the correct distribution of residuals was ensured by drawing them from a Normal distribution. If we wished to deliberately create a dataset which contravened this assumption, then this could be achieved by drawing the random variable from a different distribution. For example a uniform distribution of errors ranging from −1 to 1.

There is sometimes confusion between heterogeneity of variance and Normality of error. The first assumption is concerned with the degree of scatter around

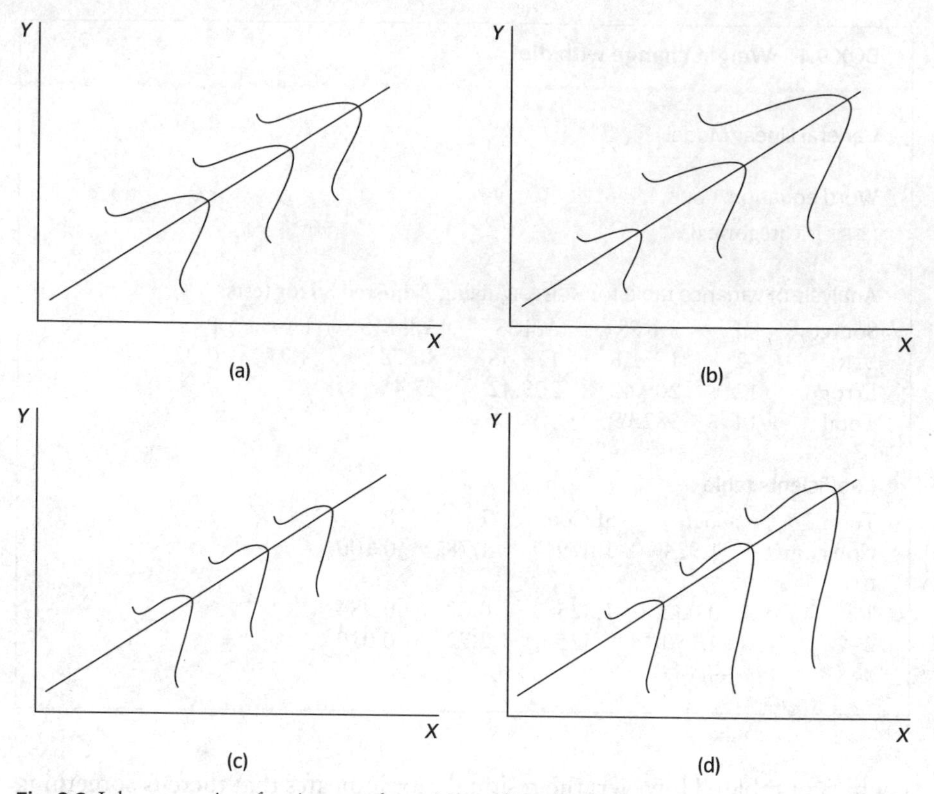

Fig. 9.2 Inhomogeneity of variance and non-Normality of error.

the model (is the width of the distribution of the residuals constant?), and the second is concerned with the shape of the distribution of residuals. This can be illustrated by considering a simple regression model. For example, in Fig. 9.2(a), the distribution of the residuals is sketched in 3D around the regression line —and it is both Normally distributed, and the magnitude of the variance is even throughout the model. In Fig. 9.2(b), the residual variance increases with the fitted values, but is still Normally distributed. In Fig. 9.2(c), the variance around the line is homogeneous, but the distribution is skewed, and finally, in Fig. 9.2(d), the residuals are both skewed and increasing in variance with the fitted values.

In what way does non-Normality of error influence the GLM analysis? In calculating the *F*-ratio, two variances are compared (that variance which is accounted for by the explanatory variable and that which remains unexplained). The *F* distribution is calculated by considering all the possible ratios that are likely to arise under the null hypothesis (in other words, when the null hypothesis is true). In calculating the *F* distribution, it is assumed that the two variances concerned are describing Normal distributions. So the actual shape of the *F* distribution, and therefore the critical *F* value which will determine significance, relies on this assumption of Normality. Once again, if this assumption is not upheld, then our significance levels will be inaccurate.

There are many examples of biological and medical data when the assumption of Normality is unlikely to be upheld. Although this assumption refers to the distribution of residuals, quite often the distribution of the original data and the residuals (once the model has been fitted) are similar. Survival data are a very typical example of non-Normally distributed data. In measuring the longevity of many species, there is high mortality amongst juveniles (pulling the mean down towards zero), and at the other end of the distribution a few individuals live a considerable time. The distribution of survival times will be constrained on one side by zero, but at the other extreme is only constrained by old age. Hence these data are often very right skewed.

9.3 Linearity/additivity

A GLM estimates a linear relationship between the response variables and the explanatory variables. In the case of our constructed data set, the relationship created was a linear one between *Y*, and *X1* and *X2*. So the appropriate word equation is: *Y* = *X1* + *X2*. However, if the data set had been constructed as follows:

$$Y = X1^{2.54} * X2^{0.54} * e^{(1.71 * Noise)}$$

the word equation would no longer be appropriate (* = the multiplication sign). Suppose *LY*, *LX1* and *LX2* are the natural logs of *Y*, *X1* and *X2* respectively. Then taking logs of the previous equation produces

$$LY = 2.54 * LX1 + 0.54 * LX2 + 1.71 * Noise.$$

Notice that this equation shows *LY* to be linear in *LX1* and *LX2*. Thus our assumptions apply to the dataset in which all the variables have been logged. So nonlinear relationships can be analysed using GLMs, as long as an appropriate linearisation exists and is used. Taking logarithms is a very useful way of linearising many multiplicative relationships. This technique is called **transformation**, and is discussed in greater detail in the next section. Other ways in which nonlinearities can be dealt with include fitting interactions (see Chapter 7), or fitting polynomials (see Chapter 10). So it can be seen that ensuring our model is the correct one is one of the major themes for the rest of this book.

9.4 Model criticism and solutions

In the real world we do not know how the data set was constructed, but we can look at the way in which the model we have chosen fits the data. The techniques of model criticism are ways of looking at the data for signs of trouble. There are two ways in which this can be done: formal tests on the residuals, and informal graphic methods. The emphasis here will be on the latter.

The two main variables of interest are the **fitted values**, obtained from the coefficients table, and the **residuals**, which are the differences between the data points and the fitted values. Hence, by definition, the mean of the residuals will be zero. For ease of interpretation, it is best to distinguish between **raw residuals** and **standardised residuals.** Raw residuals are simply the differences between the data and the model, and will therefore be in the same units as the data. Standardised residuals have been transformed to have a standard deviation of 1 (they already have a mean of zero, and are divided by their standard deviation to give a standard deviation of 1). Thus their units can be thought of as 'numbers of standard deviations', and this gives a reference point as to how far from the model any one particular data point is placed.

Histogram of residuals

By plotting a histogram of residuals after the model has been fitted, it is possible to assess visually if the distribution is roughly Normal. Figure 9.3 gives three possibilities.

The symmetric histogram of Fig 9.3(b) suggests that no action need be taken, but the right skew of 9.3(a) and the left skew of 9.3(c) suggests that something needs to be done. For many data sets, right skew will be quite common, so a transformation is needed which effectively pulls together large values, and

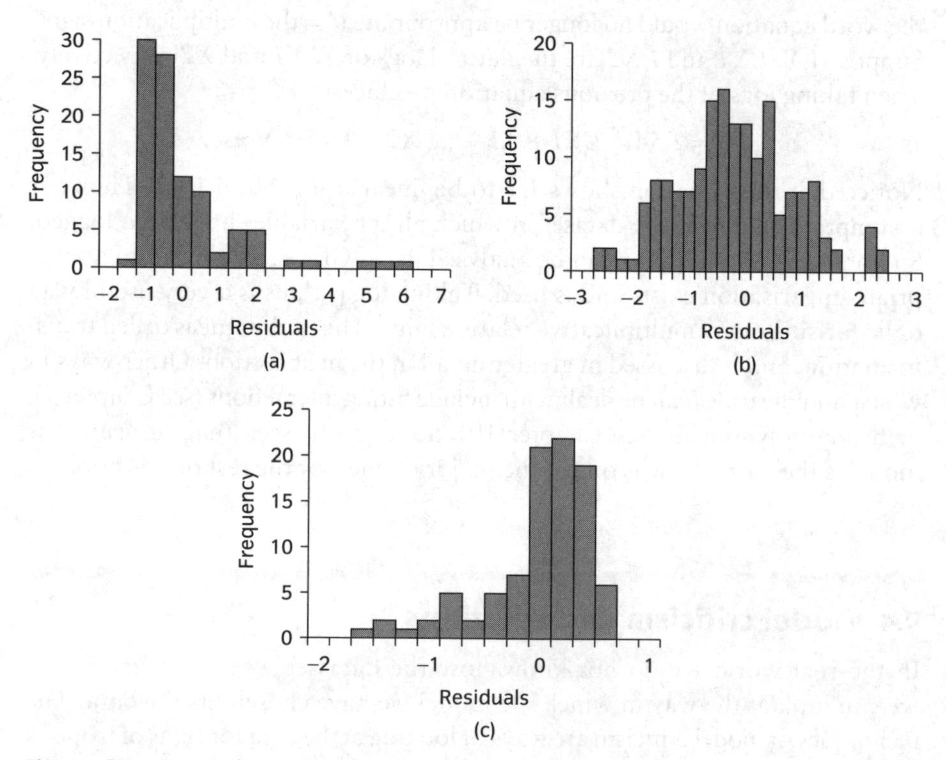

Fig. 9.3 Histograms of residuals.

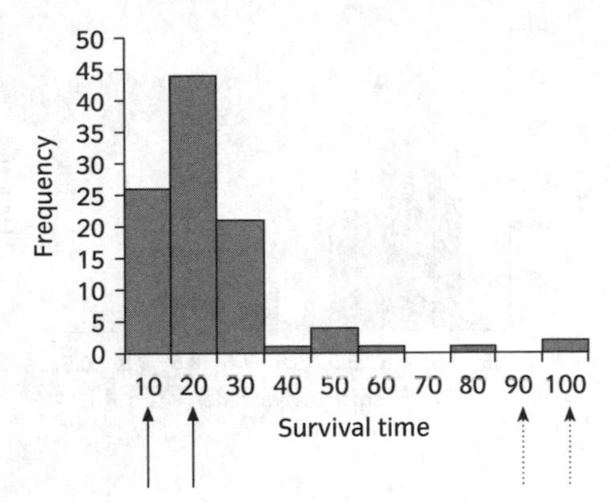

Fig. 9.4 Survival time in days for a cohort of 100 seedlings. The solid arrows are indicating survival times of 10 and 20, and the dashed arrows the survival times of 90 and 100. These arrows follow the same X values as the data is transformed in Fig. 9.5.

stretches out small values. There are three commonly used options: square root, logarithms and inverse.

To illustrate the impact of these transformations on a distribution, consider a dataset of survival times. A cohort of seedlings were followed from germination until death. The mean survival time was 24.8 days, but as can be seen from the distribution in Fig. 9.4, this was because many individuals died during the first few weeks, when they were vulnerable to grazing by snails. After this early period, the mortality rate was not so high, and some individuals survived until the end of the season.

The aim of transforming data is to alter the shape of the distribution, until a symmetrical, Normal shape is produced in the residuals. The three standard transformations mentioned above are useful on right skewed data, because they act differentially on high and low values. For example, taking the square root will have a proportionately greater compressive effect on larger numbers. The end result is that applying a square root transformation to a whole distribution compresses the right hand tail, and stretches out the left hand tail. The two solid arrows indicate the relative positions of two data points at the lower end of the original distribution (10 and 20), and the two dashed arrows indicate relative positions for two high data points (90 and 100) in Fig. 9.4. The relative positions of the same four points are followed in Fig. 9.5, as the data is transformed (square root in 9.5(a), natural log in 9.5(b), and inverse in 9.5(c)). The solid arrows get further apart while the dashed arrows move closer together. This illustrates the relative power of these three transformations; square root being weakest and the inverse the strongest. In the final, inverse, transformation, the two sides of the distribution are reversed, with the left hand side now containing the values that made up the original right hand tail.

In this particular case, the original distribution was so skewed that the strong inverse transformation was required to produce a roughly symmetrical shape. A slightly more formal way of testing this same assumption is given in the next section.

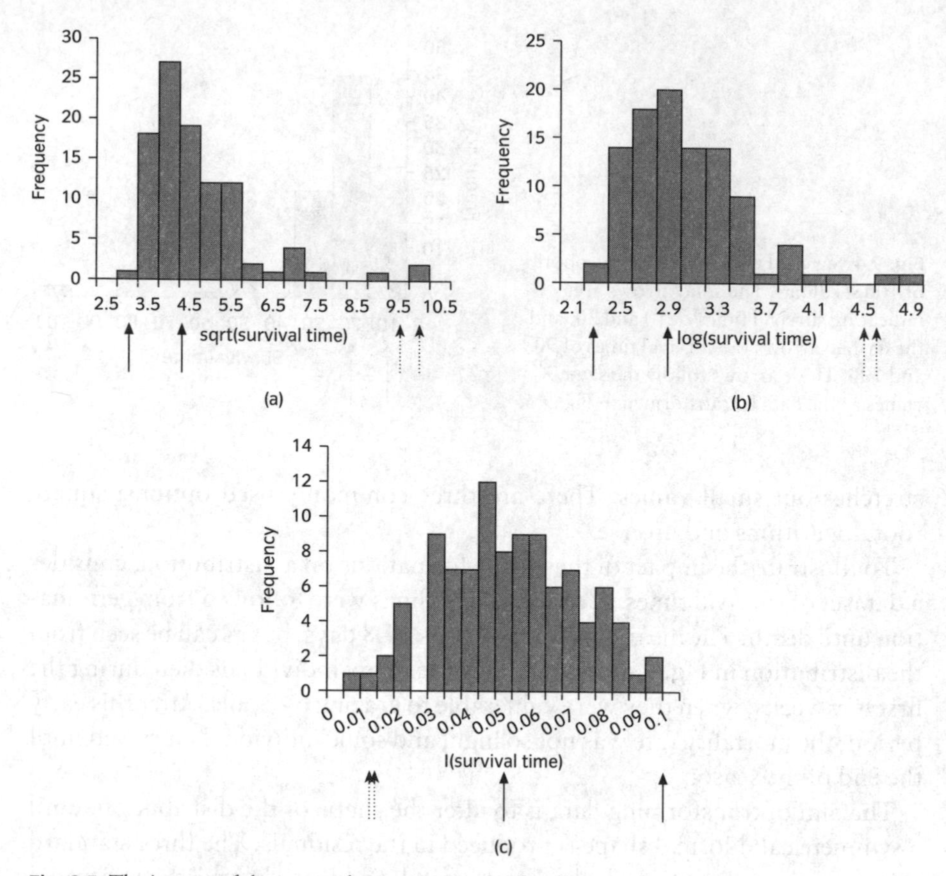

Fig. 9.5 The impact of three transformations on the distribution of survival times.

Normal probability plots

This method allows a more quantitative assessment of the 'Normality' of the distribution of residuals. The dataset *plantlets* contains data on three variables for two species of plant; height, leaf size and longevity. In an analysis which compares the two species, we will see that these three variables produce different residual distributions.

In Box 9.2 and Fig. 9.6, the height of the two species are compared, with a histogram of residuals and a normal probability plot being produced from the analysis.

The residuals in Fig. 9.6 appear to be Normally distributed and originate from a dataset size 50. Imagine these standardised residuals being ranked. If these 50 values have been drawn at random from a Standard Normal Distribution (see revision section), then we can calculate the expected value of the lowest residual—which in this case we would expect to be –2.24 standard deviations from the mean of zero. The lowest standardised residual from our dataset is –2.91. Corresponding expected residuals can be calculated for all 50 points.

BOX 9.2 Analysis height in the plantlets dataset

General Linear Model

Word equation: HGHT = SPECIES

SPECIES is categorical

Analysis of variance table for HGHT, using Adjusted SS for tests

Source	DF	Seq SS	Adj SS	Adj MS	F	P
SPECIES	1	260.83	260.83	260.83	46.28	0.000
Error	48	270.55	270.55	5.64		
Total	49	531.38				

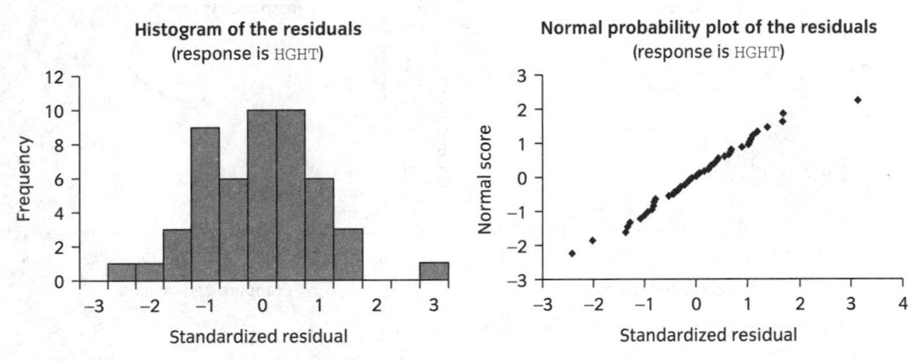

Fig. 9.6 Histogram and Normal probability plot for Normally distributed residuals.

If our residuals really are Normally distributed, then these two sets of values should be closely correlated. When the standardised residuals are plotted against the Normal scores, they produce a straight line. The 'straightness' of such a probability plot can be measured by its correlation coefficient, which is one of the quantitative tests for Normality. Sokal RR and Rohlf FJ (1994) *Biometry: the principles and practice of statistics in biological research*, 3rd edn, Freeman and Co. New York, discuss this subject in more detail, and talk about ways in which you can test for departures from Normality.

What happens to the Normal probability plot when the data deviate from Normality? Considering another variable in this same data set, LONGEV, a rather different pattern of residuals is observed (Box 9.3 and Fig. 9.7). The histogram of residuals has right hand skew, and the Normal probability plot becomes convex. This is because many of the values on the left hand side of the graph are not many standardised residuals away from the mean, therefore the line is pulled above the diagonal. On the right hand side of the graph, some residuals are many more standard deviations away from the mean than expected, causing the line to move below the diagonal.

BOX 9.3 Analysing longevity in the plantlets dataset

General Linear Model

Word equation: LONGEV = SPECIES

SPECIES is categorical

Analysis of variance table for LONGEV, using Adjusted SS for tests

Source	DF	Seq SS	Adj SS	Adj MS	F	P
SPECIES	1	3 300	3 300	3 300	0.30	0.585
Error	48	525 103	525 103	10 940		
Total	49	528 403				

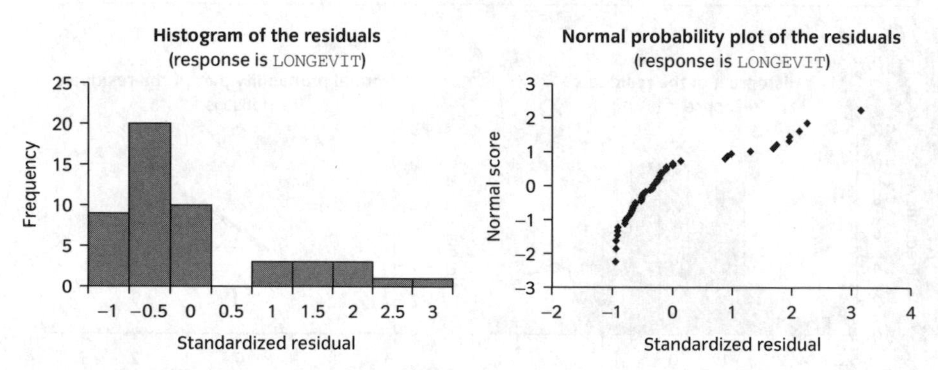

Fig. 9.7 Residuals plots for right hand skewed residuals.

In contrast, in Box 9.4 and Fig. 9.8, the variable LEAFSIZE is compared between species. In this case, the residuals have left hand skew, and the Normal probability plot is concave. The lowest values are much further from the mean (in terms of standardised residuals) than would be expected from a normal distribution, pulling the line below the diagonal, and drawing it out to the left.

These plots may also alert you to the possibility of outliers, as they will appear at the extreme ends of the graph, either above or below the line (depending

BOX 9.4 Analysing leaf size in the plantlets dataset

General Linear Model

Word equation: LEAFSIZE = SPECIES

SPECIES is categorical

Analysis of variance table for LEAFSIZE, using Adjusted SS for tests

Source	DF	Seq SS	Adj SS	Adj MS	F	P
SPECIES	1	0.1152	0.1152	0.1152	0.25	0.618
Error	48	21.9496	21.9496	0.4573		
Total	49	22.0648				

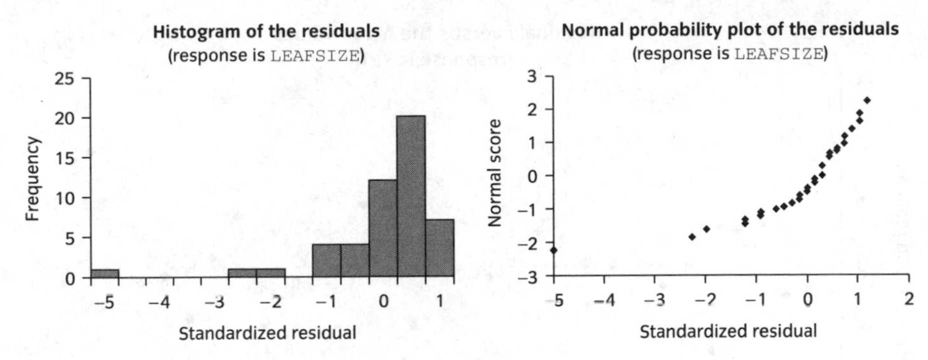

Fig. 9.8 Residual plots for left hand skewed residuals.

upon whether they contribute to right or left skew). For example, in Fig. 9.8, there is one value on the extreme left of the residual histogram—this appears as the lowest dot on the left of the Normal probability plot.

So to summarise, to detect non-Normality, plot histograms of residuals and normal probability plots. To correct for non-Normality, try transforming the *Y* variable by using square root, log or inverse transformations.

Plotting the residuals against the fitted values

This second graphical method indicates if there are any problems with heterogeneity of variance and linearity. If the model fitted the data *exactly*, and there was no error variance at all, then the residuals would all be zero, and they would fall on the horizontal axis. However, there is always a degree of variation around the model, and the assumption of a GLM is that this variance is even. In the following example, three different *Y* variables (Y1, Y2 and Y3) are regressed against X (to be found in the data set *3Ys*).

In Box 9.5 and Fig. 9.9, the standardised residuals are fairly evenly scattered above and below their mean of zero. Most residuals lie between –2 and +2, as would be expected from Normally distributed variables.

BOX 9.5 **Dataset *3Ys*, *Y1***

General Linear Model

Word equation: Y1 = X

X is continuous

Analysis of variance table for Y1, using Adjusted SS for tests

Source	DF	Seq SS	Adj SS	Adj MS	F	P
X	1	345 772	345 772	345 772	1.1E+05	0.000
Error	28	89	89	3		
Total	29	345 861				

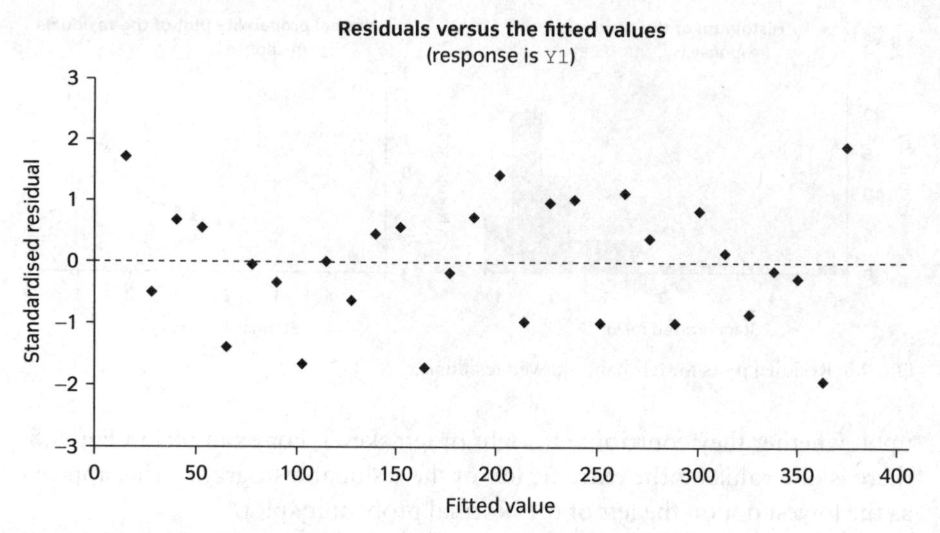

Fig. 9.9 Residuals with homogenous variance against fitted values.

In Box 9.6 and Fig. 9.10 however, the variance in the residuals appears to be increasing with the fitted values. This is by far the more common problem when analysing biological data, as it often becomes more difficult to measure your *Y* variable accurately as values get larger. Measurement error is not the only, and probably not even the major, reason for this pattern. Statistical error often consists of omitted relevant variables (either because they were not measured, or their relevance not realised). These omissions, if they interact multiplicatively as fitted values get larger, will cause increasing error variance.

Less common is a decrease in residual variance with increasing fitted values, as illustrated in Box 9.7 and Fig. 9.11.

The means by which these problems can be solved have, in fact, already been discussed in the last section when correcting for non-Normality. The three commonly used transformations of square root, logarithms and taking the

BOX 9.6 **Dataset *3Ys*, *Y2***

General Linear Model

Word equation: Y2 = X

X is continuous

Analysis of variance table for Y2, using Adjusted SS for tests

Source	DF	Seq SS	Adj SS	Adj MS	F	P
X	1	896 449	896 449	896 449	2.99	0.095
Error	28	8 401 229	8 401 229	300 044		
Total	29	9 297 677				

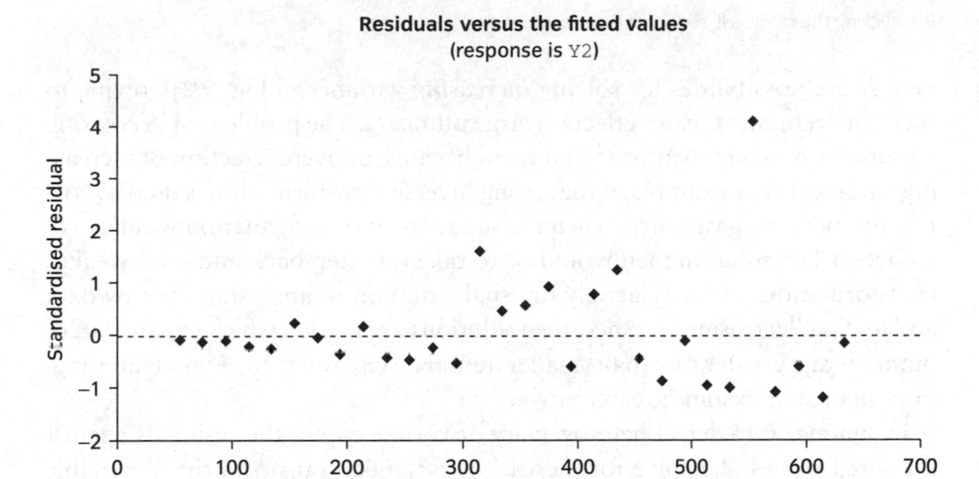

Fig. 9.10 Residuals showing increasing variance with fitted values.

BOX 9.7 Dataset *3Ys, Y3*

General Linear Model

Word equation: Y3 = X

X is continuous

Analysis of variance table for Y3, using Adjusted SS for tests

Source	DF	Seq SS	Adj SS	Adj MS	F	P
X	1	348 148	348 148	348 148	3.0E+05	0.000
Error	28	33	33	1		
Total	29	348 181				

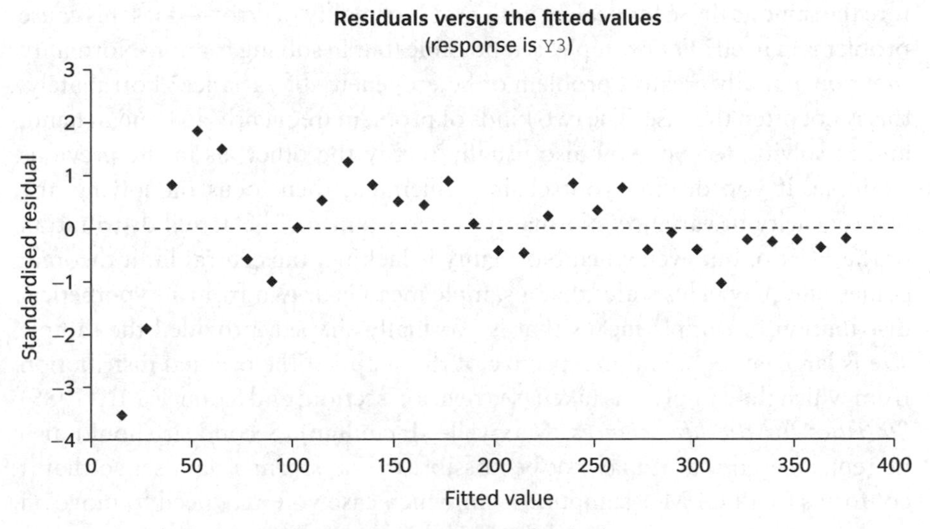

Fig. 9.11 Residuals showing decreasing variance with fitted values.

inverse are possibilities for solving *increasing* variance in Fig. 9.10, owing to their differentially greater effect on large numbers. The problem of *decreasing* variance is most frequently encountered because of overcorrection of increasing variance! (For example, if the strong inverse transformation is used to correct for increasing variance, when the square root transformation would have sufficed.) The solution then would be to take one step back and try a weaker transformation. In the relatively unusual situation of analysing the raw data and finding decreasing variance, then solutions are needed which expand larger numbers at a greater rate than smaller numbers: e.g. squaring, exponentiating, or even raising to some greater power.

In summary, to detect heterogeneity of variance, plot the residuals against the fitted values. To solve for increasing variance, transform the *Y* variable using the square root, log or inverse transforms, and for decreasing variance, square the *Y* variable, or raise it to some higher power. If these relatively simple solutions do not work, then the Box–Cox formula may offer an alternative (see Section 9.6).

Transformations affect homogeneity and Normality simultaneously

To illustrate how spread and shape of distributions are altered simultaneously by transformations, consider the three distributions plotted on the same axes in Fig. 9.12(a). They have means of roughly 2, 10 and 25, and have ranges of (1, 5), (5, 15), (15, 35). Their variances are clearly different, and the two left hand distributions are also skew to the right. In Fig. 9.12(b), the data has been transformed on the $\log_2$ scale, resulting in three distributions that are both symmetrical, and with equal variances.

The solutions suggested for dealing with heterogeneity of variance are therefore the same as those for dealing with non-Normality of error—does this cause problems in itself? For example, is it possible that in solving for non-Normality that you actually create a problem of heterogeneity of variance? Fortunately, this is not often the case. The two kinds of problem frequently go hand in hand, and in solving for one you also usually rectify the other, as in the previous example. If you do find yourself in a dilemma, then focus on solving any heterogeneity in variance. Normality is important for the actual distribution of the *F*-ratio, but even when Normality is lacking, the central limit theorem comes into play. This states that a sample mean is drawn from a hypothetical distribution of sample means that is Normally shaped, provided the sample size is large enough, and irrespective of the shape of the original distribution from which the sample was taken (see revision section, and Samuels ML (1989) *Statistics for the life sciences*, Maxwell–Macmillan). Second, it should also be remembered that it may not be possible to transform a data set so that it conforms to all GLM assumptions—in which case you may need to move on to alternative statistical methods beyond the remit of this text.

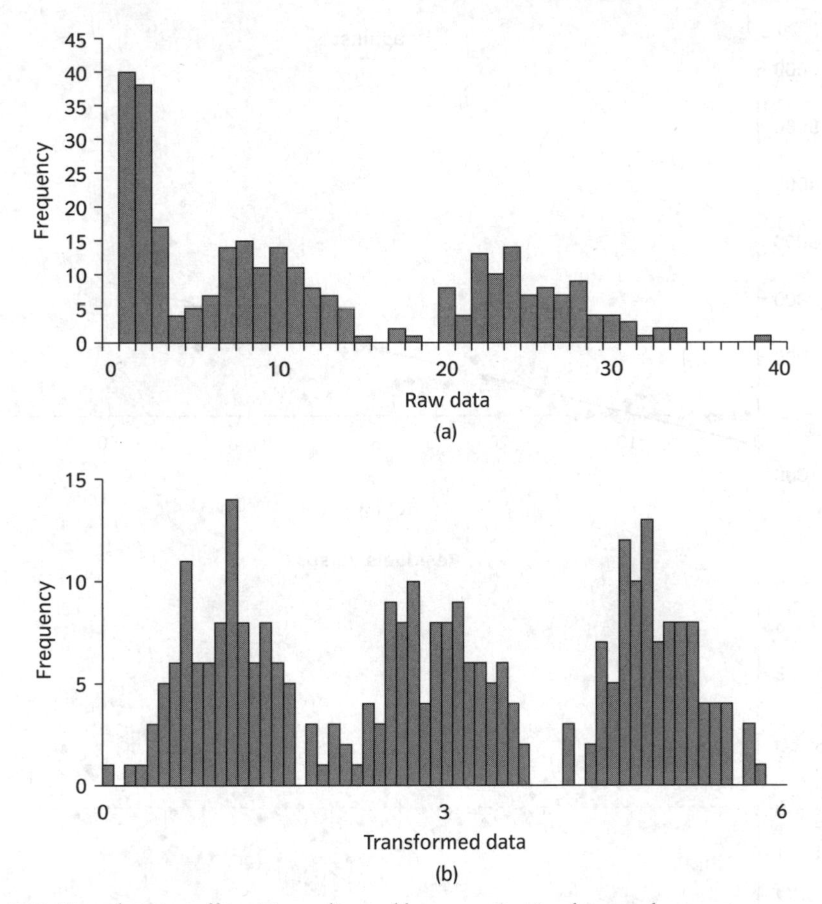

Fig. 9.12 Transforming affects Normality and homogeneity simultaneously.

The plot of residuals against fitted values can also provide indications of nonlinearity. Plotting against the fitted values is essentially looking at the overall fit of a model. If there is more than one continuous explanatory variable, then it would be useful to know which of these is responsible for any nonlinearities —so this topic is discussed in the following section.

Plotting the residuals against each continuous explanatory variable

This plot looks directly at whether Y is linear in any particular X variable. Figure 9.13(a) illustrates a curvilinear relationship between Y and X, with the straight line that would be fitted by the linear model $Y = X$. We consider more formally later in this chapter how to view these plots, but here introduce the subject by looking at some examples.

Figure 9.13(b) shows the residual plot generated from fitting the straight line. The residuals form a distinct **U** shape, with the first 10 residuals all being positive, followed by a clump of negative residuals, and another clump of

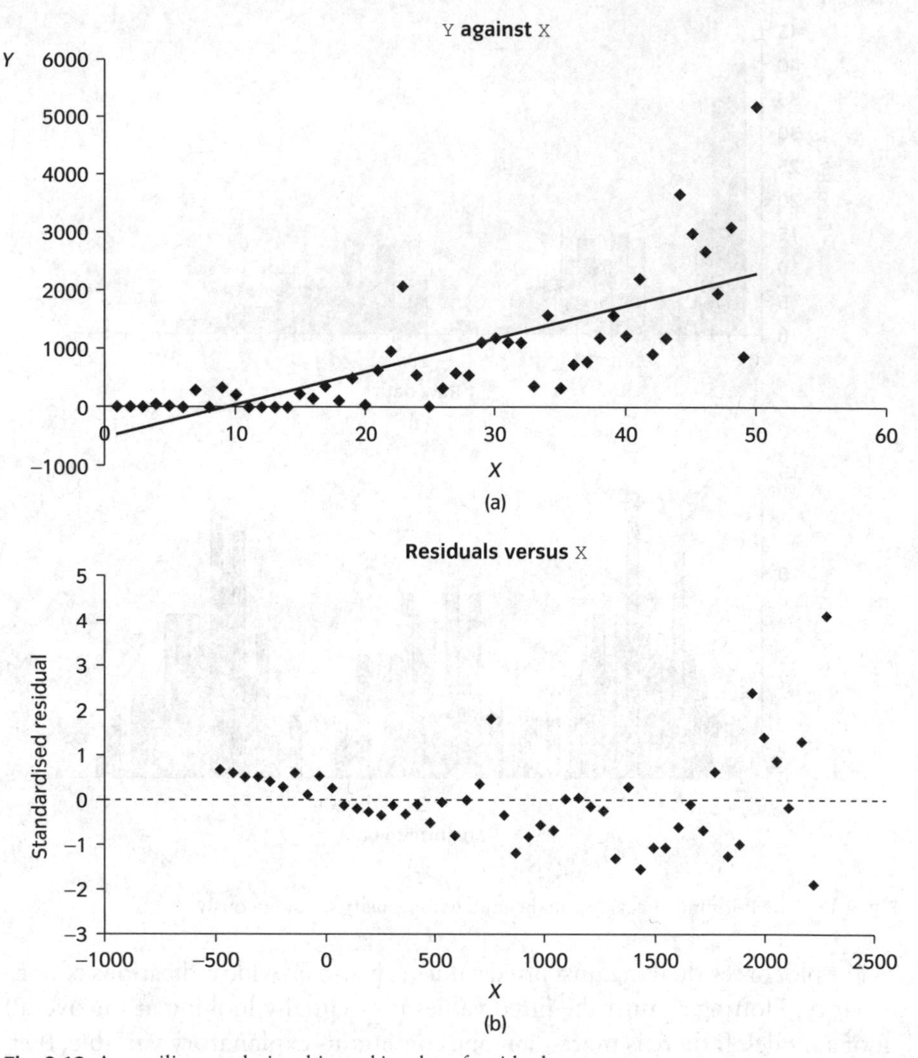

Fig. 9.13 A curvilinear relationship and its plot of residuals.

positive residuals on the right hand side. This corresponds exactly with all the data points being above the model in the first graph, then mostly below etc. This kind of curve is relatively easy to detect. If the original relationship between Y and X is more complicated, then other patterns can arise, one example being given in Fig. 9.14.

Solutions for nonlinearity

When faced with problems of nonlinearity, there are a number of options open to you: interactions, transformations and polynomials. While heterogeneity of variance requires transformation of the Y variable, nonlinearity may be solved by transforming X and/or Y variables.

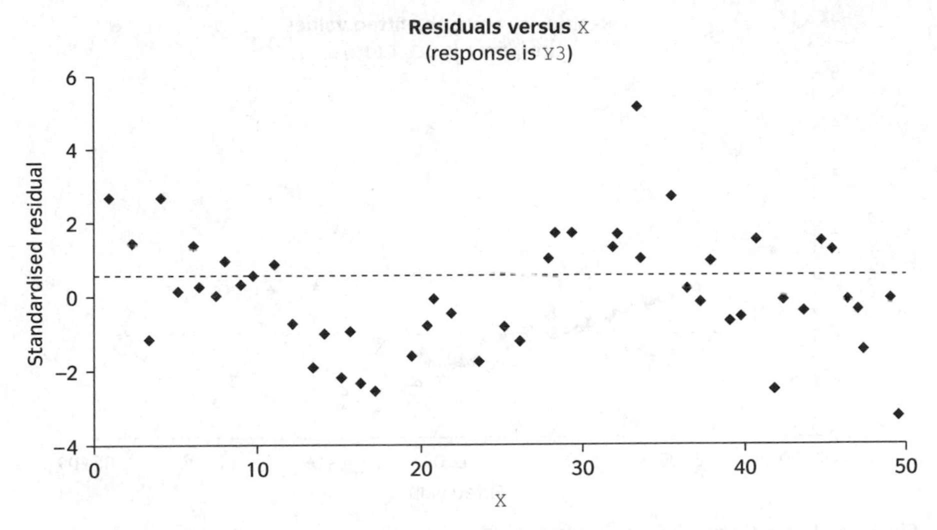

Fig. 9.14 A more complex residual plot.

Interactions and transformations as a means of solving nonlinearity

Nonlinearity can be solved by fitting an interaction between two categorical variables. This is best demonstrated with an example. An experiment was performed to determine the effects of nutrient concentrations on bacterial growth. The two nutrients were sucrose and leucine. The sucrose was added to 48 agar plates at four different concentrations, and the leucine at three different concentrations in a factorial design. The plates were set up over four days and, on each day, every treatment combination was used. DAY was therefore treated as a block. The data are stored in the *bacteria* dataset.

In the first analysis (Box 9.8), DENSITY has been used in its untransformed state, and only the main effects have been fitted. Both SUCROSE and LEUCINE are highly significant ($p = 0.004$ and $p < 0.0005$ respectively).

BOX 9.8 Analysing bacterial growth without interactions

General Linear Model

Word equation: DENSITY = DAY + SUCROSE + LEUCINE

DAY, SUCROSE and LEUCINE are categorical

Analysis of variance table for DENSITY, using Adjusted SS for tests

Source	DF	Seq SS	Adj SS	Adj MS	F	P
DAY	3	1.1570E+19	1.1570E+19	3.8566E+18	0.52	0.674
SUCROSE	3	1.1895E+20	1.1895E+20	3.9651E+19	5.31	0.004
LEUCINE	2	1.4762E+20	1.4762E+20	7.3811E+19	9.88	0.000
Error	39	2.9136E+20	2.9136E+20	7.4709E+18		
Total	47	5.6951E+20				

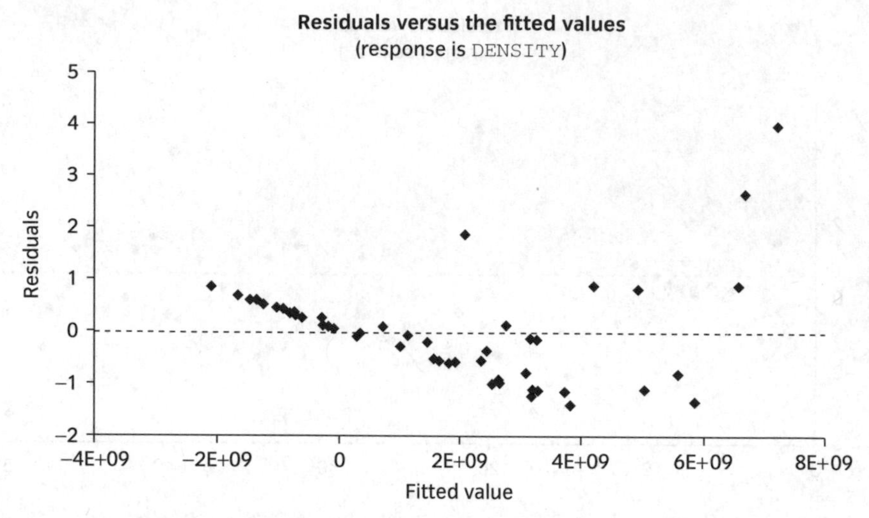

Fig. 9.15 Bacterial growth ignoring interactions.

BOX 9.9 Reanalysis of bacterial growth, including the interaction

General Linear Model

Word equation: DENSITY = DAY + SUCROSE + LEUCINE + SUCROSE * LEUCINE

DAY, SUCROSE and LEUCINE are categorical

Analysis of variance table for DENSITY, using Adjusted SS for tests

Source	DF	Seq SS	Adj SS	Adj MS	F	P
DAY	3	1.1570E+19	1.1570E+19	3.8566E+18	0.81	0.496
SUCROSE	3	1.1895E+20	1.1895E+20	3.9651E+19	8.36	0.000
LEUCINE	2	1.4762E+20	1.4762E+20	7.3811E+19	15.56	0.000
SUCROSE * LEUCINE	6	1.3479E+20	1.3479E+20	2.2464E+19	4.73	0.001
Error	33	1.5658E+20	1.5658E+20	4.7447E+18		
Total	47	5.6951E+20				

A residual plot (Fig. 9.15) however indicates that there are difficulties with this model. The residuals of the lower fitted values are all positive, followed by mainly negative, with a few positive residuals for the highest fitted values. This clearly contravenes nonlinearity.

So the analysis is repeated, including the interaction term between sucrose and leucine, shown in Box 9.9.

In the second analysis, the interaction is very significant ($p = 0.001$). It is therefore necessary that it is included in the model. The residual plot (Fig. 9.16) has improved, but still has problems.

The problems with linearity have disappeared, but there are still problems with heterogeneity of variance. To a certain extent, this is to be expected, as the

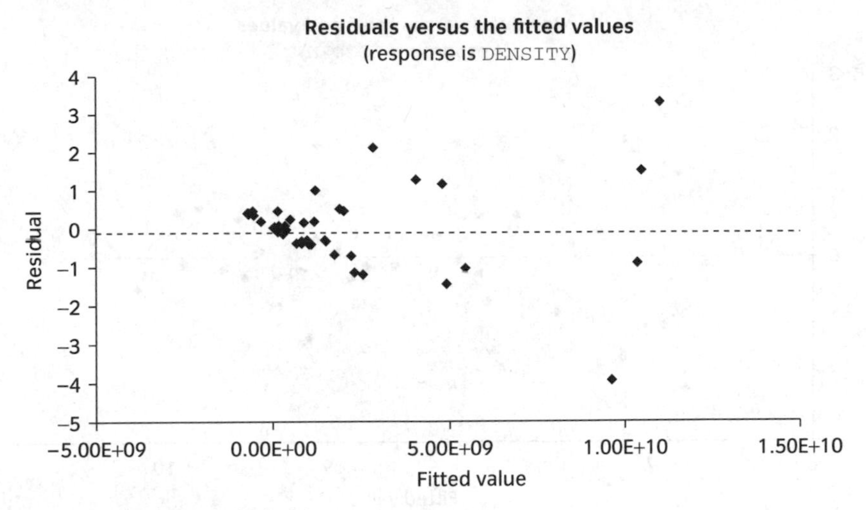

Fig. 9.16 Residual plot including interactions.

BOX 9.10 Reanalysis of bacterial growth with transformation

General Linear Model

Word equation: LOGDEN = DAY + SUCROSE + LEUCINE + SUCROSE * LEUCINE

DAY, SUCROSE and LEUCINE are categorical

Analysis of variance table for LOGDEN, using Adjusted SS for tests

Source	DF	Seq SS	Adj SS	Adj MS	F	P
DAY	3	1.0461	1.0461	0.3487	1.38	0.265
SUCROSE	3	20.8387	20.8387	6.9462	27.55	0.000
LEUCINE	2	15.1785	15.1785	7.5892	30.10	0.000
SUCROSE * LEUCINE	6	1.1489	1.1489	0.1915	0.76	0.607
Error	33	8.3204	8.3204	0.2521		
Total	47	46.5326				

errors in estimating density are likely to rise with density. So a third analysis transforms density to log(density) (Box 9.10 and Fig. 9.17).

Now sucrose and leucine are more significant (higher *F*-ratios in both cases), but the interaction is not significant. The residual plot has also improved, having lost its problems of nonlinearity and heterogeneity of variance.

This example illustrates two very important points. The first is that the fitting of an interaction between two categorical variables can solve problems of nonlinearity. In this particular example, this did not solve all of our problems with the residual plot, and so other measures had to be taken. The second point is that log transforming the *Y* variable is an alternative solution to fitting an interaction to solve for non-additivity in the model, as logarithms convert

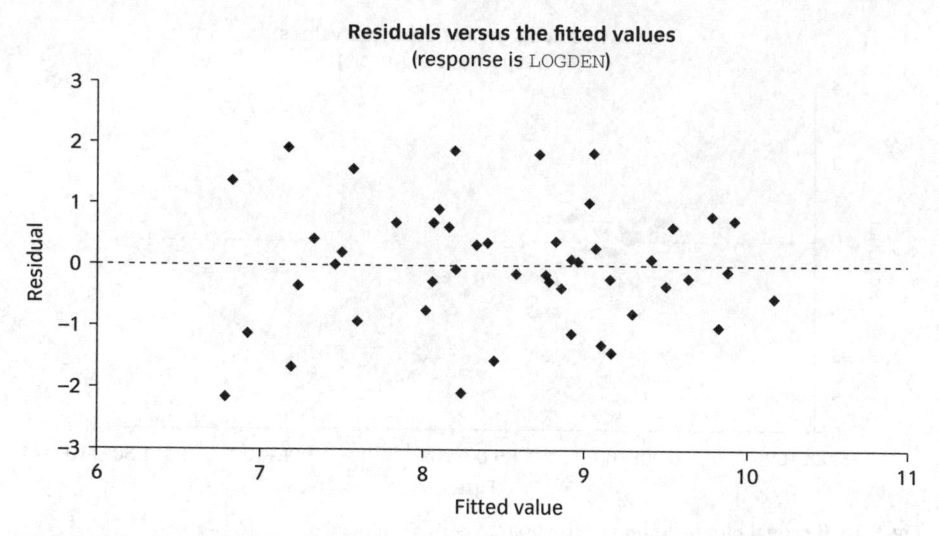

Fig. 9.17 Residual plot including interaction and transformation.

multiplicative relationships into additive ones. In this case, transforming the *Y* variable was appropriate, as the explanatory variables are categorical. Once the *Y* variable had been transformed, the interaction term was no longer needed.

In other cases, where nonlinearity involves continuous variables, it may be more appropriate to log transform both *Y* and *X* variables. One such example is given in the case study of Section 9.5. A third alternative is to fit a polynomial (for example, $X + X^2$), to investigate if there is a curvilinear component to the relationship between *X* and *Y*. This is discussed in more detail in Chapter 11.

Hints for looking at residual plots

One problem with relying on the simpler graphical methods for checking models is that it can appear to be rather subjective. To make it a little more precise, imagine dividing the plot into vertical strips as shown in Fig. 9.18.

Within each strip, the distribution of the points should be roughly the same. If there are differences in the mean between strips (clumps above or below the line), then linearity is in doubt. If there are differences in variance (the range of the residuals) then homogeneity of variance is in doubt. It does not matter whether there are differences in the numbers of points per strip. However, care must also be taken with this method, as the sample size within any one strip is typically quite small (4 to 7 points in the example above). The fewer the number of points, the more likely it is that all will be one side of the zero line. It is also true that one common illusion is to imagine that there is lower variance in strips with fewer points. The residual plot in Fig. 9.18 was drawn from a Normal distribution and so should pass the visual inspection. One helpful technique for judging how worried to be by a given plot is to create a series of Normal plots to compare with it.

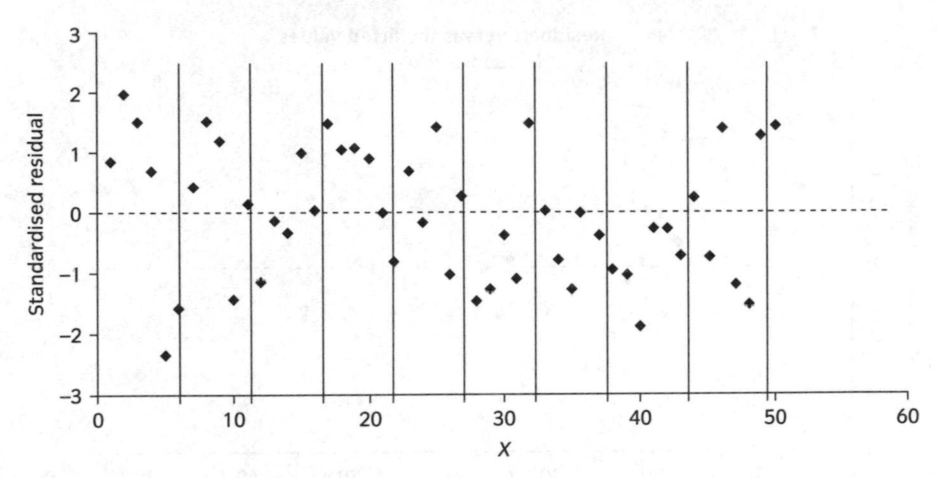

Fig. 9.18 Interpreting residual plots.

9.5 Predicting the volume of merchantable wood: an example of model criticism

The volume of merchantable wood in a tree can only be exactly determined after felling. However, if it can be predicted from measurements that can be taken from a living tree, this would be of great assistance to foresters. In Box 9.11 the forester has analysed the relationship between the allometric measurements of a tree, and the volume of merchantable timber from a dataset of 31 felled trees. She used tree diameter to predict volume, and followed her

BOX 9.11 Untransformed analysis of tree volume

General Linear Model

Word equation: `VOLUME = DIAM`

`DIAM` is continuous

Analysis of variance table for `VOLUME`, using Adjusted SS for tests

Source	DF	Seq SS	Adj SS	Adj MS	F	P
`DIAM`	1	7581.9	7581.9	7581.9	419.47	0.000
Error	29	524.2	524.2	18.1		
Total	30	8106.1				

Coefficients table

Term	Coef	SE Coef	T	P
Constant	−36.945	3.365	−10.98	0.000
`DIAM`	60.791	2.968	20.48	0.000

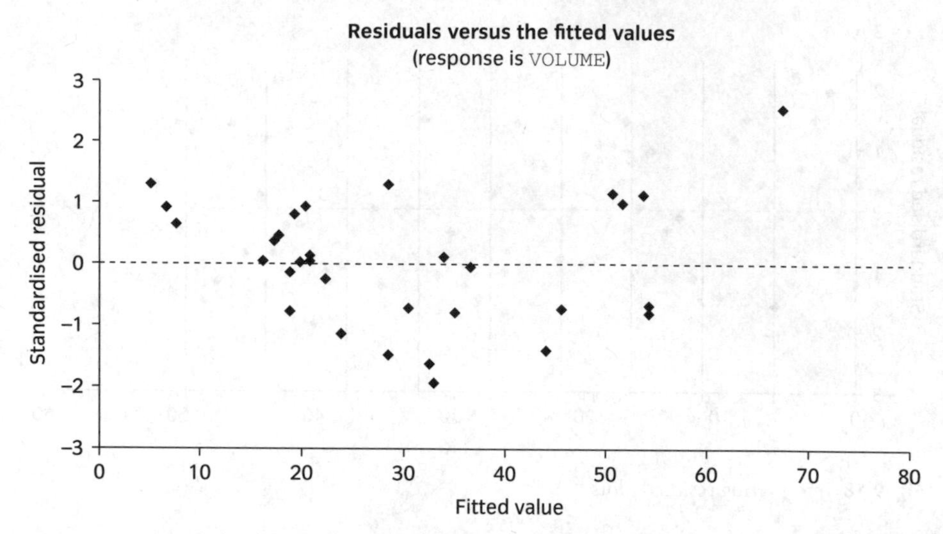

Fig. 9.19 Residual plot for *merchantable timber*.

analysis with a plot of the residuals against the fitted values. (These data are to be found in the *merchantable timber* dataset).

There is clearly a highly significant relationship between volume and diameter, but the residual plot suggests that there are problems with this model. The residuals (Fig. 9.19) form a distinct U shape, suggesting that the assumption of linearity does not hold. The scatter on the right hand side of the graph also appears to be greater than on the left, suggesting heterogeneity of variance. It is not uncommon to find these two problems together.

The residual plot suggests that the relationship between volume and diameter is best described by a curve. Two options for solving this problem include fitting a quadratic term (Chapter 11), or transforming. The latter option is pursued here, because the equation for the volume of a cylinder is:

$$volume = \pi(radius)^2\, height$$

or rephrased in terms of the diameter rather than the radius:

$$volume = \frac{\pi(diameter)^2 height}{4}.$$

So if the volume of a tree can be adequately represented by a cylinder, then the forester would expect a quadratic relationship between *volume* and *diameter*. This multiplicative relationship can be converted to an additive one by taking logarithms.

$$\log(volume) = -0.242 + \log(height) + 2 * \log(diameter)$$

The intercept of -0.242 comes from $\left(\frac{\pi}{4}\right)$. By using the natural log transformation, the relationship can be linearised, and so conform to GLM assumptions.

BOX 9.12 Comparing trees to cylinders

General Linear Model

Word equation: LVOL = LHGHT + LDIAM

LHGHT and LDIAM are continuous

Plot residuals against fitted values

Analysis of variance table for LVOL, using Adjusted SS for tests

Source	DF	Seq SS	Adj SS	Adj MS	F	P
LHGHT	1	3.4957	0.1978	0.1978	29.86	0.000
LDIAM	1	4.6275	4.6275	4.6275	698.74	0.000
Error	28	0.1854	0.1854	0.0066		
Total	30	8.3087				

Coefficients table

Term	Coef	SE Coef	T	P
Constant	−1.7047	0.8818	−1.93	0.063
LHGHT	1.1171	0.2044	5.46	0.000
LDIAM	1.98271	0.07501	26.43	0.000

The forester then tested the hypothesis that the trees could be described by the equation for a cylinder by transforming, and fitting the model of Box 9.12, and noted the residuals plotted in Fig. 9.20.

The first point to notice is that the residual plot is much improved, and is now perfectly acceptable. Both linearity and homogeneity of variance have been achieved by the transformations. (Remember that linearity is tackled by transforming X and/or Y, but for homogeneity of variance, it must be the Y

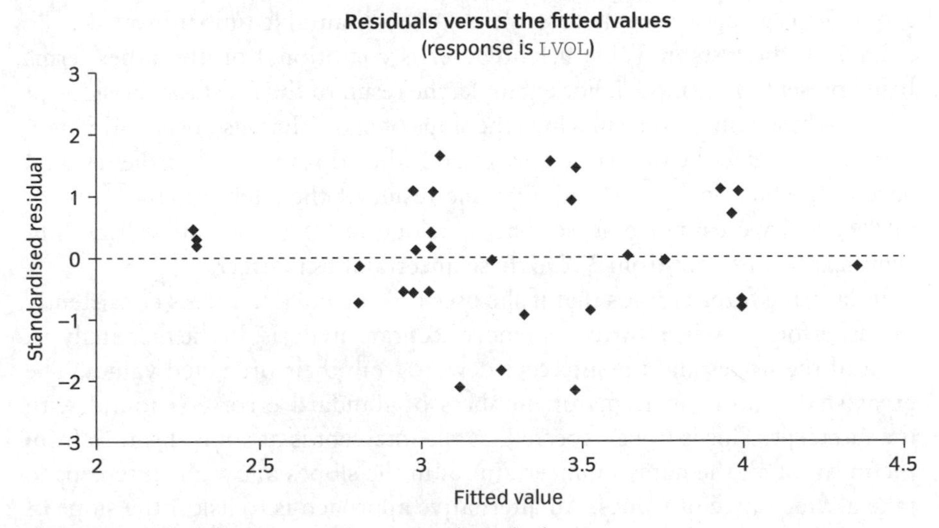

Fig. 9.20 Residual plot for tree volume treated as cylinder.

Table 9.1 Testing the 'trees are cylinders' hypothesis

Parameter	Hypothesis	t-test with 28 degrees of freedom	Significance?
Intercept	$=-0.242$	$\frac{-1.7047-(-0.242)}{0.8818}=-1.659$	$p>0.05$
Slope of log(*diameter*)	$=2$	$\frac{1.98271-2}{0.07501}=-0.2305$	$p>0.05$
Slope of log(*height*)	$=1$	$\frac{1.1171-1}{0.2044}=0.5729$	$p>0.05$

variable that is transformed). Secondly, both log(*diameter*) and log(*height*) are significant explanatory variables in this analysis, with $p<0.0005$. Finally, how do we test the hypothesis that the equation of a cylinder is a good descriptor of a tree? (This revises material first covered in Chapter 3).

The fitted value equation derived from the coefficients table is:

$$\log(volume) = -1.7047 + 1.98271 \log(diameter) + 1.1171 \log(height).$$

So testing the hypothesis can be divided into three parts: (i) is the slope of log(*diameter*) significantly different from 2? (ii) is the slope of log(*height*) significantly different from 1? (iii) is the intercept significantly different from −0.242? The coefficients table not only provides the parameter estimates, but also their standard errors. So these questions can be answered by three t-tests, as shown in Table 9.1.

So the slopes of log(*diameter*) and log(*height*) are not significantly different from that predicted by the equation for a cylinder, and the intercept, while lower than the predicted value, is not significantly lower. It could be concluded that a cylinder is an approximate description of merchantable timber in a tree.

Each of the tests in Table 9.1 however is conditional on the other terms being present in the model. For example, the result of the t-test for the slope of `LDIAM` is based on a model in which the slope of `LHGHT` has also been estimated. If we constrained the slope of `LHGHT` to be 1, then this would alter the residual variance, which in turn would alter the result of the t-test for the slope of `LDIAM`, and we cannot assume that the slope of `LDIAM` would still be 'not significantly different from 2'. Do these uncertainties matter?

In fact, the forester finds that if she uses this formula, her trees consistently underperform, so she returns for a more accurate analysis. In the first analysis, both of the slopes and the intercept deviate from their predicted values. The greatest deviation (in terms of numbers of standard errors) is found with the intercept. This is to be expected, as the intercept is at some distance from the mean of all the data, so uncertainties in the slopes allow the intercept to take a wide range of values. An alternative approach is to ask if the slope of log(*height*) is constrained to 1, and the slope of log(*diameter*) is constrained

BOX 9.13 Estimating the mean deviation of the intercept from the expected cylinder intercept

Calculate a new variable, LDEVINT defined as:

LDEVINT = LVOL – LHGHT – 2*LDIAM

Request descriptive statistics for LDEVINT

Variable	N	Mean	SE Mean
LDEVINT	31	–1.1994	0.0142

to 2, what will then be the value of the intercept, and to what extent does this deviate from the expected value provided by the formula of a cylinder? Given the formula for a cylinder expressed in logs, we have the following:

$$\log(volume) - \log(height) - 2 * \log(diameter) = -0.242.$$

So by using the data set values for height, volume and diameter, we can arrive at a set of values (for the deviation of the intercept), the mean of which should be –0.242, and request descriptive statistics as in Box 9.13.

The estimate we have obtained is considerably lower than the expected value of –0.242. The usual t-test gives:

$$\frac{-1.1994 - (-0.242)}{0.0142} = -67.5$$

which, with 30 degrees of freedom gives a p-value of very much less than 0.001. As the forester expected, the trees are yielding consistently less timber than would be expected from a cylinder of the same height and diameter. This may be because the trunk tapers towards the top of the tree. In fact, the diameter is traditionally measured at breast height, where we would expect it to be close to its maximum in a tree of any age. So it would be helpful to the forester if we could go further than this, and estimate what fraction of a cylinder is represented by a tree. This will be the actual volume of timber divided by the predicted volume, or in logs:

$$\log(fraction\ of\ cylinder\ used)$$
$$= \log(volume) - [-0.242 + \log(height) + 2 * \log(diameter)].$$

The analysis is summarised in Box 9.14.

Finally, by using descriptive statistics, we can provide the forester with a 95% confidence interval for the average amount of useable timber in a tree, expressed as a fraction of the cylinder predicted from the tree's height and diameter. Using the formula for a 95% confidence interval with 30 degrees of freedom we have:

$$-0.9574 \pm 2.042 * 0.0142.$$

BOX 9.14 Estimating the proportion of the cylinder that represents useable timber

Calculate the fraction of timber used by creating a new variable LUSEFRAC by:

$$\text{LUSEFRAC} = \text{LVOL} - (-0.242 + \text{LHGHT} + 2 * \text{LDIAM})$$

Request descriptive statistics for LUSEFRAC

Variable	N	Mean	SE Mean
LUSEFRAC	31	−0.9574	0.0142

The whole analysis has been conducted in natural logs, so it is helpful to backtransform when the results are presented. It is important that the back-transformation is conducted after the confidence interval has been calculated (as it is the logs of the data that have conformed to our model assumptions). This results in the 95% confidence interval

$$(0.373, 0.395).$$

Thus, the forester can conclude that the mean volume of merchantable timber supplied by a tree is between 37.3% and 39.5% of the hypothetical cylinder, with 95% probability. As we discussed in Chapter 2, the prediction interval for an individual tree will be wider than this confidence interval for the mean.

9.6 Selecting a transformation

Transformations have proved to be one of the most useful tools for manipulating our data to conform to the assumptions of a GLM. They can solve problems with both nonlinearity and heterogeneity of variance. If there is no a priori reason for using any particular transformation, then a useful rule of thumb is to start with the weakest (square root—assuming increasing variance with the fitted values, or right skew of the residuals). After transformation and reanalysis, it is important *to go through the model criticism loop again*, to determine if the problem has been solved, or a stronger transformation is required. Only when the residual histograms and plots are satisfactory should you confidently proceed with the analysis.

Certain types of data are likely to require particular transformations. The survival data described in Fig. 9.4 is typical of data that will need the strong inverse transformation to rectify the right hand skew, because it is constrained on the left by zero, and on the right, by old age. It is often recommended to use the negative inverse, because then the signs of coefficients are the same as in the original model. This makes interpretation easier. Allometric data is likely to have a variance that increases with the mean, to the degree that requires the log transformation. It is standard in the field to log allometric data before analysis. If data follow a Poisson distribution (as some count data do), then again the variance will increase with the mean, but will be stabilised by the square root transform. Analysing count data is covered in more detail in Chapter 13. So when analysing these three types of data, we have some prior information as to what to expect.

Which transformation is required all depends upon the rate at which the variance increases with the mean, or in other words, the strength of the variance–mean relationship. The strength can sometimes be quantified as the power k in the relationship

$$\text{variance} \propto \text{mean}^k,$$

with the condition that $k \geq 0$. So we are dealing with right skewed positive variables, which have variances that increase with the mean (fitted values). If the correct transformation is proving elusive, then one possibility to estimate the value of k. This can be done by fitting the model, calculating the residuals, and then plotting the log of the squared residuals against the log of the fitted values. This is effectively fitting the model:

$$\log(\text{residual variance}) = k * \log(\text{mean fitted values}).$$

The slope of the line is therefore an estimate of k.

Once we have an estimate of k, then we could use the **Box–Cox formula** to predict exactly which transformation would solve our problem of heterogeneity of variance. If y is the original response variable, then Y would be the transformed variable, using the formula:

$$Y = \frac{y^{(1-k/2)} - 1}{1 - k/2}.$$

It can be seen that when $k = 0$, then the variance is constant, and the Box–Cox formula suggests no transformation. When the variance is directly proportional to the mean (so that $k = 1$), then the Box–Cox formula suggests a square root transform. As k approaches 2 the limit of the formula is $\ln(y)$, the transformation used for allometric data. When $k = 4$, then it produces an inverse transform, but the subtraction of 1 and division by $1 - k/2$ means that the distribution doesn't become inverted, with the right hand side of the raw data becoming the left hand side of the transformed distribution (as we observed in Fig. 9.5(c)). So this formula suggests the transformations already discussed, depending upon the strength of the variance–mean relationship, but also every possible transformation in-between. So the Box–Cox formula produces a whole series of transformations for correcting right skewed positive variables. In practice, considering the three possibilities of square root, log or inverse is usually sufficient.

The analysis of proportions and percentages present a rather different problem, as they are constrained at one end by zero, and the other by 1 (or 100%). Errors therefore become very asymmetric at the extremes, but are approximately symmetric in the centre. There is one special transform that Normalises this kind of data, called the arcsine transform.

$$Y = \sin^{-1}\sqrt{proportion}$$

or

$$\sin Y = \sqrt{proportion}.$$

Finally, when presenting the results of transformed data, it is often friendly to the reader to transform back into the original units, as in the example of the *merchantable timber* data analysis in Section 9.5. However, if you construct confidence intervals, then these should be calculated with the transformed data first, and backtransformed afterwards.

9.7 Summary

- The four assumptions of general linear models are independence, homogeneity of variance, linearity and Normality of error.
- If these assumptions are contravened, then the significance levels are no longer valid.
- Model criticism techniques were described, focusing on graphical methods.
- To test for non-Normality of error, histograms of residuals and Normal probability plots can be used.
- To test for inhomogeneity of variance, plot the residuals against the fitted values.
- To test for linearity, plot the residuals against the fitted values, and also against any continuous variables in turn.
- Solutions for non-Normality include transforming the data by square root, log and inverse transforms (if the residuals are right skewed), and by squaring or exponentiation (if the residuals are left skewed).
- To correct for variance which increases with the fitted values, again square root, log and inverse transformations are useful, with the square root being the weakest and the inverse the strongest.
- Transformations affect homogeneity and normality simultaneously.
- To correct for nonlinearity solutions include (i) fitting an interaction; (ii) transforming the response and/or the explanatory variables; (iii) fitting polynomials (quadratics, cubics etc.).

- Log transformation and interactions are two alternative means of solving for non-additivity. Model criticism techniques are required to determine which is most suitable.

- If the variance–mean relationship can be directly estimated, the Box–Cox formula is another possible means of determining which transformation might be most suitable.

9.8 Exercises

Stabilising the variance

This example illustrates the effect of three standard transformations on the variance in a dataset. Twenty-four plots of soya beans were divided into three equal groups in a completely randomised design, and each group received a different formulation of a selective weedkiller (stored in the data set *soya beans*). The aim is to choose the formulation of weedkiller which inflicts least damage on the crop itself. The amount of DAMAGE and a code for weedkiller (WDKLR) are entered into a worksheet. Three transformations are used on the data, square root, logarithm and negative inverse. Four graphs are plotted in Fig. 9.21 of the variable DAMAGE against WDKLR in its raw and transformed states.

Basic statistics have also been tabulated for each of the transformations in Box 9.15.

(1) From inspecting the graphs, and the tables, which transformation would you feel is most acceptable?

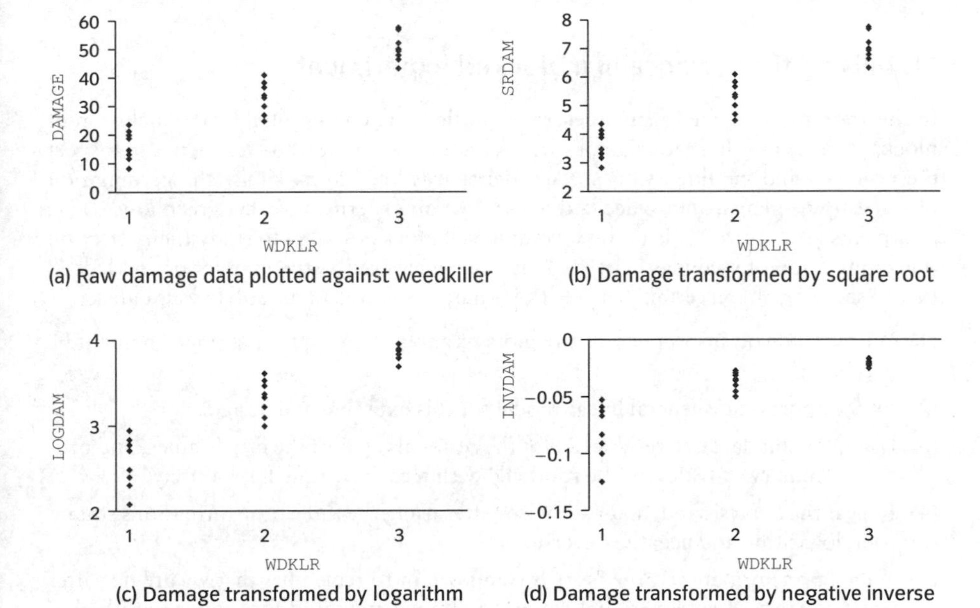

Fig. 9.21 Plots for *Soya Bean* dataset.

BOX 9.15 Basic statistics for *Soya Bean* dataset

Descriptive statistics: DAMAGE described by levels of WDKLR

	N	Mean	Median	StDev
DAMAGE 1	8	13.50	13.50	3.82
2	8	28.38	28.50	5.88
3	8	49.75	48.50	6.98

Descriptive statistics: SRDAM described by levels of WDKLR

	N	Mean	Median	StDev
SRDAM 1	8	3.641	3.669	0.529
2	8	5.301	5.338	0.558
3	8	7.038	6.964	0.492

Descriptive statistics: LOGDAM described by levels of WDKLR

	N	Mean	Median	StDev
LOGDAM 1	8	2.565	2.596	0.297
2	8	3.3260	3.3498	0.2133
3	8	3.8985	3.8815	0.1393

Descriptive statistics: INVDAM described by levels of WDKLR

	N	Mean	Median	StDev
INVDAM 1	8	−0.07998	−0.07500	0.02460
2	8	−0.03667	−0.03510	0.00799
3	8	−0.02044	−0.02062	0.00282

Stabilising the variance in a blocked experiment

In this example the experimental design is a little more complicated by the inclusion of blocks. This means that the original variables are not so useful, and it is better to inspect the residuals and the fitted values. A gardener was keen to maximise the germination of some alpine plants, and so decided to conduct an experiment in her greenhouse. Five treatments (TRT) were applied in six randomised blocks (BLOCK) to study their effect on the germination of seedlings. The response was measured as the number of cotyledons (NCOT) showing after a certain period. These data are stored in the *cotyledons* dataset.

(1) Why would you suspect before any plots or analyses that a transformation might be necessary?

(2) Fit the appropriate general linear model for this experimental design.

(3) Using the standardised residuals, plot the residuals against the fitted values, and produce summary statistics for the residuals within each treatment separately.

(4) Repeat the analysis and model criticism after using the three transformations square root, logarithm and negative inverse.

(5) Which transformation is the best? It is important to remember that we are not after exact equality of variances, but rather to remove any trend for groups with higher means to have higher variances. Exact equality of variances is not attainable by transformation, and variation among the variances of the groups is to be expected.

Lizard skulls

A student measured 81 skulls of a species of lizard. For each skull, the measurements taken allowed the calculation of the following quantities:

JAWL the length of the lower jaw

BVOL an estimate of the volume of the brain

SITE a code from 1 to 9 for the site at which the skull had been discovered.

These variables are stored in the *lizards* dataset. She then conducted an analysis, and used some graphical model checking procedures, an edited version of which is given in Box 9.16 and Fig. 9.22.

BOX 9.16 Analysis of lizard skulls

General Linear Model

Word equation: BVOL = JAWL

JAWL is continuous

Analysis of variance table for BVOL, using Adjusted SS for tests

Source	DF	Seq SS	Adj SS	Adj MS	F	P
JAWL	1	0.52278	0.52278	0.52278	284.65	0.000
Error	78	0.14325	0.14325	0.00184		
Total	79	0.66603				

Coefficients table

Term	Coef	SECoef	T	P
Constant	−0.041583	0.00806	−5.16	0.000
JAWL	0.52456	0.03109	16.87	0.000

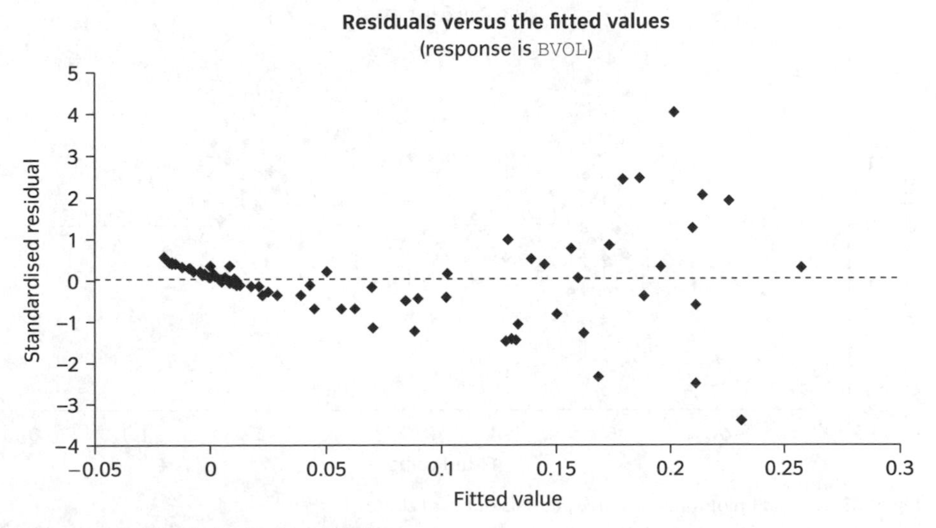

Fig. 9.22 Residual plot for lizard skulls.

BOX 9.17 Alternative analysis for lizard skulls

General Linear Model

Word equation: LBVOL = SITE + LJAWL

SITE is categorical and LJAWL is continuous

Analysis of variance table for LBVOL, using Adjusted SS for tests

Source	DF	Seq SS	Adj SS	Adj MS	F	P
Site	8	193.445	0.301	0.038	0.19	0.992
LJAWL	1	10.405	10.405	10.405	52.52	0.000
Error	70	13.867	13.867	0.198		
Total	79	217.717				

Coefficients table

Term	Coef	SECoef	T	P
Constant	−0.1302	0.5556	−0.23	0.815
SITE				
1	0.0267	0.3156	0.08	0.933
2	0.0644	0.1816	0.35	0.724
3	0.0836	0.1505	0.56	0.580
4	−0.0871	0.1487	−0.59	0.560
5	−0.0164	0.2107	−0.08	0.938
6	0.0099	0.2232	0.04	0.965
7	−0.1204	0.2486	−0.48	0.630
8	0.1036	0.3393	0.31	0.761
9	*−0.0643*			
LJAWL	1.9904	0.2747	7.25	0.000

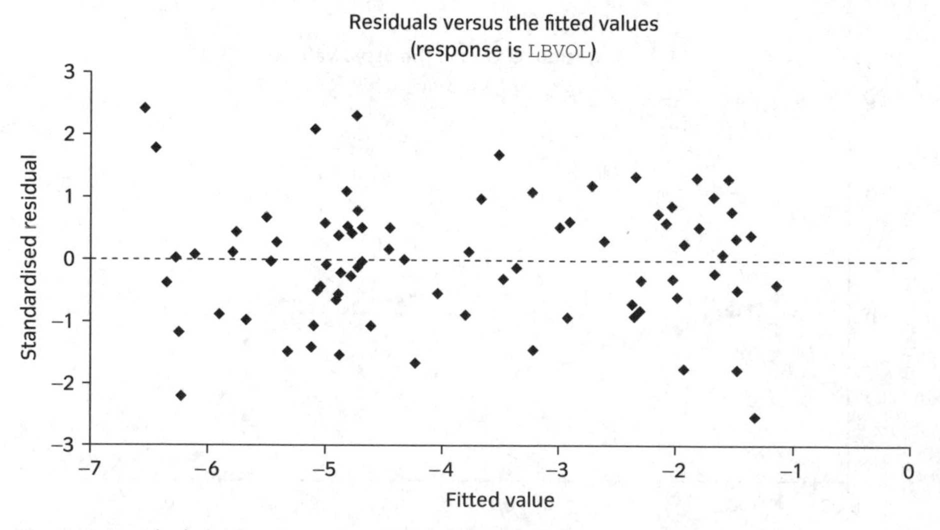

Fig. 9.23 Residual plot for alternative analysis of lizard skulls.

(1) Discuss how well the model fits the data, in light of the plot of residuals.

(2) How could a transformation improve the model? Which transformation might you expect to be most useful?

She then conducted the alternative analysis in Box 9.17 and Fig. 9.23.

(3) Which model do you prefer and why?

(4) Discuss the pattern of residuals in the second graph.

(5) What do you conclude about differences between the 9 sites?

(6) Test the hypothesis that the slope of LBVOL on LJAWL is 3.

Checking the 'perfect' model

In this exercise, datasets which are known to be Normal will be created, and then histograms of these 'Normal' distributions examined.

(1) The dataset *Squirrels* contains the weight of 50 male and 50 female squirrels (in the variables MALE and FEMALE respectively). Plot histograms of weight and of log weight for each of the sexes separately. Do any of these histograms appear to follow a Normal distribution?

(2) Draw a Normal probability plot for each of the four sets of data (female weight raw and logged and male weight, raw and logged). What do you conclude from these probability plots?

(3) Calculate the mean and standard deviation of female squirrel weight and male squirrel weight.

(4) Create a new dataset of female squirrel weights by drawing 50 random numbers from a Normal distribution with the same mean and standard deviation as the female squirrels in the dataset. Repeat this 10 times. For each dataset, draw a histogram of those weights. How do these really Normal histograms compare with the histograms of the original weights and the logged weights?

(5) In a similar manner, create 10 datasets by drawing 50 numbers at random from a Normal distribution with the same mean and standard deviation as the male squirrels in the dataset. Plot a Normal probability plot for each of these created datasets. What conclusions would you draw?

10 Model selection I: principles of model choice and designed experiments

10.1 The problem of model choice

In the last two chapters the focus has been on choosing a model which conforms to the four assumptions of a GLM. However, this is not the only, and not even the first, criterion on which to choose the best model. There may be several models which produce satisfactory residual plots but which contain different combinations of explanatory variables. What criteria should be used to pick the best model from these alternatives?

This problem can be illustrated by considering one of the simplest models—using one continuous variable X to predict the response variable Y (as stored in the *simple polynomial* dataset).

In the first analysis (Box 10.1), a linear model has been fitted, but the residual plot (Fig. 10.1) suggests that the assumption of linearity has been contravened.

BOX 10.1 Fitting and checking a linear model

General Linear Model

Word equation: `Y1 = X1`

`X1` is continuous

Analysis of variance table for `Y`, using Adjusted SS for tests

Source	DF	Seq SS	Adj SS	Adj MS	F	P
`X1`	1	6 663 021	6 663 021	6 663 021	722.89	0.000
Error	78	718 946	718 946	9 217		
Total	79	7 381 967				

Coefficients table

Term	Coef	SECoef	T	P
Constant	−128.08	19.40	−6.60	0.000
`X1`	29.473	1.096	26.89	0.000

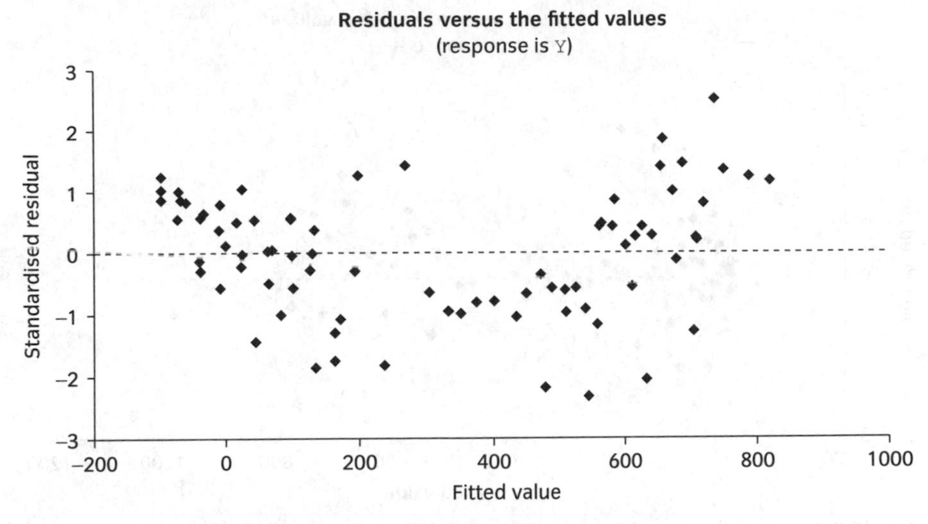

Fig. 10.1 Residual plot for the linear model of Box 10.1.

There is a distinct hole in the residuals above the zero line in the centre of the plot. One way of solving this problem is to fit a polynomial, as mentioned in Chapter 9. Two possible polynomial models are considered—the quadratic (Box 10.2 and Fig. 10.2) and the cubic (Box 10.3 and Fig. 10.3).

Both the quadratic and the cubic models produce satisfactory residual plots. So perhaps the quadratic model is sufficient, and it is not necessary to include

BOX 10.2 Fitting and checking the quadratic model

General Linear Model

Word equation: Y1 = X1 + X1 * X1

X1 is continuous

Analysis of variance table for Y, using Adjusted SS for tests

Source	DF	Seq SS	Adj SS	Adj MS	F	P
X1	1	6 663 021	3 597	3 597	0.60	0.442
X1 * X1	1	256 148	256 148	256 148	42.62	0.000
Error	77	462 798	462 798	6 010		
Total	79	7 381 967				

Coefficients table

Term	Coef	SECoef	T	P
Constant	−7.62	24.21	−0.31	0.754
X1	3.189	4.122	0.77	0.442
X1*X1	0.8525	0.1306	6.53	0.000

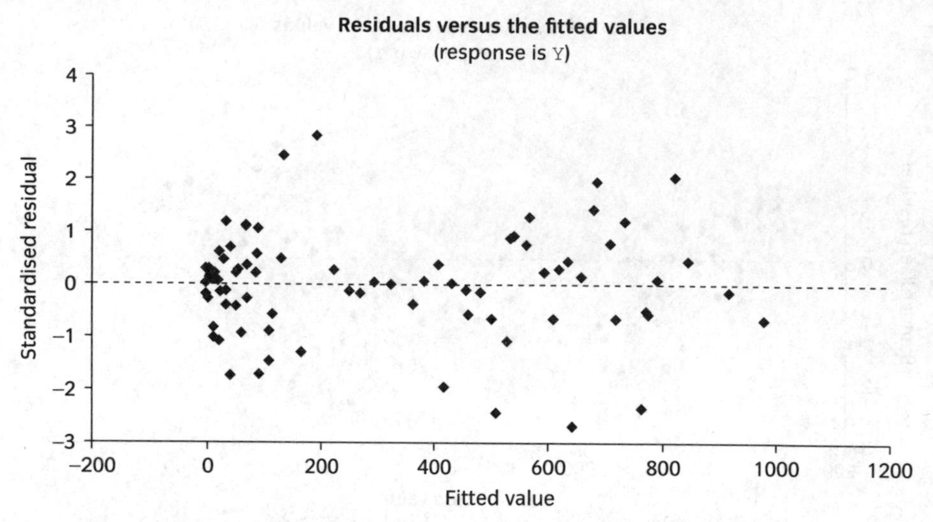

Fig. 10.2 Residual plot for quadratic model.

BOX 10.3 Fitting and checking the cubic model

General Linear Model

Word equation: `Y1 = X1 + X1 * X1 + X1 * X1 * X1`

`X1` is continuous

Analysis of variance table for `Y`, using Adjusted SS for tests

Source	DF	Seq SS	Adj SS	Adj MS	F	P
X1	1	6 663 021	2 505	2 505	0.41	0.523
X1 * X1	1	256 148	4 763	4 763	0.78	0.379
X1 * X1 * X1	1	720	720	720	0.12	0.732
Error	76	462 078	462 078	6 080		
Total	79	7 381 967				

Coefficients table

Term	Coef	SECoef	T	P
Constant	−15.75	33.92	−0.46	0.644
X1	6.179	9.625	0.64	0.523
X1 * X1	0.6169	0.6971	0.89	0.379
X1 * X1 * X1	0.00500	0.01452	0.34	0.732

the extra complication of a cubic term. In this example, the residual plots do suggest strongly that the quadratic model is the best, but in other cases this may not be so easy. We need to be more precise about the reasons why one model should be rejected and another accepted. For example, we should be able to say how strong the evidence is that the quadratic term is needed, and

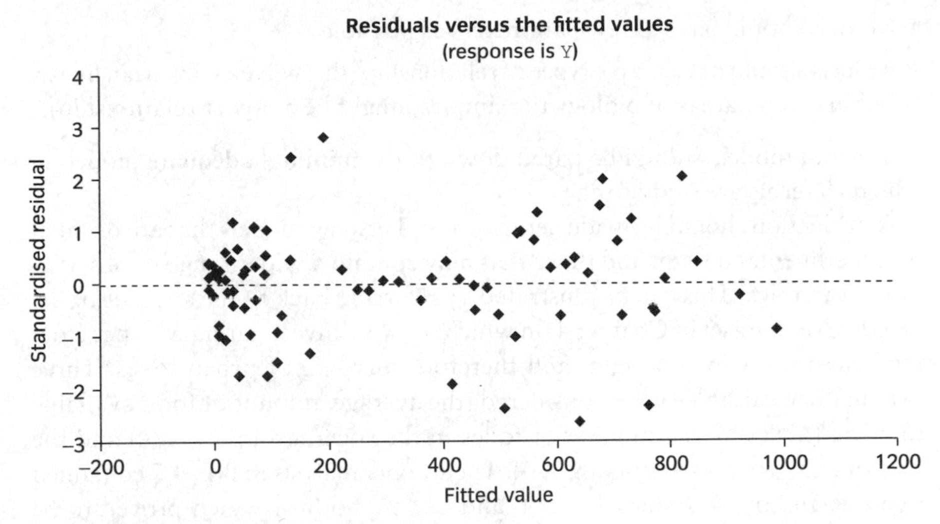

Fig. 10.3 Residual plot for cubic model.

how weak the evidence is that the cubic term is needed. In other words we want p-values. These are normally obtained from the model output. However, when comparing these three outputs, there appears to be a problem. In the first analysis, the linear component of the model is found to be significant ($p < 0.0005$ from an F-ratio of 723). However, in the second output, when the quadratic term has been included, the linear component becomes insignificant, with $p = 0.442$. Similarly, the p-value for the quadratic term is significant in the second analysis ($p < 0.0005$), but not in the third ($p = 0.379$). We need a set of rules about which p-values to accept and why.

This particular example will prove to be straightforward, and in fact inspection of the residuals has already given sufficient clues as to which model is best. Frequently choosing between models is a rather more complex process. We will consider three principles of model choice: economy of variables; multiplicity of p-values; and considerations of marginality. The next section introduces these principles, and applies them to this particular example. Applying these principles to different types of analyses will continue until the end of the following chapter.

10.2 Three principles of model choice

Economy of variables

The first principle of model choice is the simpler the better. The general policy of preferring simple explanations to complex ones translates in statistical terms into principles such as:

- Models should have as few parameters as possible.
- Models should have simple types of relationship (this will usually mean linear, but in certain areas of biology the simplest might be a power relationship).

In general, models should be pared down to the **minimal adequate model**, in which all terms are significant.

A distinction should be made here however between models that are dictated by experimental design, and those that may contain a wider range of observational variables. This can be illustrated by referring back to the analysis of the *urban foxes* dataset in Chapter 4, in which we were investigating which factors influenced the winter weight, and therefore survival, of urban foxes. Three explanatory variables were considered (the average amount of food available to a fox (AVFOOD), the number of foxes in the social group (GSIZE) and the area of that group's territory (AREA)). The second analysis in Box 4.7 contained two explanatory variables: AVFOOD and GSIZE, both of which proved to be significant (with p-values less than 0.0005). A third analysis, Box 4.8, included AREA as an additional explanatory variable, which proved insignificant. In the third model, the significance of AVFOOD was reduced, as AREA and AVFOOD shared informativeness. Consequently, the better model contains only the two explanatory variables AVFOOD and GSIZE. The process of removing unnecessary variables is called **model simplification.**

In contrast, Box 10.4 displays the analysis of an experiment designed to compare four varieties of pea in three blocks (stored in the *pea* dataset). The analysis shows block to be insignificant—does this mean that block should be removed, and the simpler model fitted?

While it is tempting to remove block and present a simpler model just containing variety, this may not be the correct model. This problem was first discussed in Section 7.4, when we were considering how to present the results of

BOX 10.4 Analysing a field experiment with insignificant blocking

General Linear Model

Word equation: YIELD = BLOCK + VARIETY

BLOCK and VARIETY are categorical

Analysis of variance table for YIELD, using Adjusted SS for tests

Source	DF	Seq SS	Adj SS	Adj MS	F	P
BLOCK	2	10.51	10.51	5.25	0.32	0.727
VARIETY	3	348.76	348.76	116.25	7.17	0.002
Error	18	291.83	291.83	16.21		
Total	23	651.10				

a factorial experiment when the interaction was not significant. The same principle applies here. If the 'true relationship' in the population only contains the explanatory variable VARIETY, then the correct model would not include the additional variable BLOCK. However, these occasions cannot be distinguished from those in which we have simply failed to detect the significance of BLOCK. By excluding BLOCK, we are picking a subset of analyses in which BLOCK is not significant (this will include those when it is truly insignificant, and those when we have failed to detect it). This will bias the error mean square, and will therefore affect the validity of the remaining *F*-ratios, and the standard errors of any parameter estimates.

In practice, the error mean square may be changed only slightly. However, the best rule of thumb for designed experiments is to perform the natural analysis that follows from the design. With many observational studies there is no single 'natural analysis', so this simple principle is not available and more difficult decisions may have to be taken.

Multiplicity of *p*-values

This is a good point to revise exactly what a *p*-value means (and this is discussed in some detail in Appendix 1). A *p*-value of 0.05 means that, when the null hypothesis is true, there is a 5% chance of obtaining a significant result and so drawing a false conclusion. So any analysis has a probability of producing a significant result even when there is no relationship between the response and explanatory variable. A result being significant when the null hypothesis is true is referred to as a type I error, and by setting the threshold value of p at 0.05 we attempt to ensure that it happens only 5% of the time. However, if we look at five different relationships for the same purpose, each has a probability of 0.05 of being significant due to chance alone, so the probability of finding at least one of the five significant becomes much higher.

What exactly is meant by looking at several relationships for the same purpose? An astrologer may wish to make much of the fact that those of the star sign 'Scorpio' are significantly shorter than those of another star sign. So from a large database, the height of Scorpio's are compared with those of 'Taurus', and the ten alternatives. Eleven relationships have been investigated, to determine which star sign the taller people belong to. Each of these eleven tests has a 0.05 probability of being significant. If it could be assumed that the tests are independent, the binomial distribution could be used to calculate the combined probability of concluding at least one of the eleven relationships are significant by chance alone (and the answer is approximately 0.43). So if there are no differences in height we are running a 43% chance of making a mistake —unacceptably high to a scientist!

This is quite an obvious example of asking the same question many times, but the problem is not uncommon, particularly in observational datasets.

It will be discussed in more detail in this context in Chapter 11. The same principles can be applied to designed experiments with many terms. For example, a factorial experiment with three variables will have seven p-values in the ANOVA table (three main effects, three two-way interactions and one three-way interaction), and with more complex experiments the number of p-values can increase dramatically. If a higher order interaction is only just significant, and it was not a principle aim of the experiment to test this (or maybe if it is unexpected given previous studies), then it may be sensible to view it with scepticism. Prior to analysis, you may decide to apply a requirement of 1% significance to terms that aren't thought a priori to be of interest. This would allow you to focus more directly on the terms of interest.

Considerations of marginality

Marginality is all about the **hierarchies** involved as soon as we introduce interactions.

Hierarchies must be respected in model formulae

What is meant by a hierarchy in GLM? In a designed experiment with three explanatory variables, A, B and C, there are three levels to the hierarchy: (1) the main effects, A, B and C; (2) the three two-way interactions, A * B, B * C, A * C; (3) the three-way interaction, A * B * C. In an analysis which involves continuous variables, there are also hierarchies when polynomials are fitted. For example, a cubic polynomial has three levels to its hierarchy: (1) the linear component X; (2) the quadratic component X^2; (3) the cubic component X^3. The simplest lower order terms come first in the hierarchy, and the more complex terms come later. The first rule of marginality is that these hierarchies must be respected when composing a model formula, not including an interaction before its component main effects. A legitimate model formula is

$$\mathit{Yield} = \mathrm{A} + \mathrm{B} + \mathrm{C} + \mathrm{A} * \mathrm{B} + \mathrm{A} * \mathrm{C} + \mathrm{B} * \mathrm{C} + \mathrm{A} * \mathrm{B} * \mathrm{C}$$

which can be summarised as A | B | C (exact notation depending upon the statistical package you use: see package specific supplements). An illegitimate formula is

$$\mathit{Yield} = \mathrm{A} + \mathrm{B} + \mathrm{A} * \mathrm{B} + \mathrm{A} * \mathrm{C} + \mathrm{B} * \mathrm{C} + \mathrm{C} + \mathrm{A} * \mathrm{B} * \mathrm{C}.$$

The problem with the second formula is that the main effect C appears in the formula after A * C. In contrast, the following formula would be legitimate:

$$\mathit{Yield} = \mathrm{A} + \mathrm{B} + \mathrm{A} * \mathrm{B} + \mathrm{C} + \mathrm{A} * \mathrm{C} + \mathrm{B} * \mathrm{C} + \mathrm{A} * \mathrm{B} * \mathrm{C}$$

because no interaction appears before its component main effects.

We have insisted that lower order terms must precede higher order terms, but it would be even more egregious to omit the lower order terms altogether.

Significance of interactions includes importance of main effects

Why is it important to respect these hierarchies? The answer is that an interaction contains the main effects it is composed of. Whether or not an interaction is significant tells us if the story the data illustrate is simple or more complicated. For example, referring to the *tulips* dataset of Chapter 7, an interaction such as WATER*SHADE indicates that the effect of SHADE on the number of blooms depends upon the WATER treatment. If this is true, it would then be nonsense to say that neither of the main effects of WATER and SHADE are important in determining the number of blooms. So the interaction actually includes the main effects (as well as other effects of course). Similarly, if the quadratic term X^2 is significant, then this includes a linear component as well. If any power of X is included, then so must all lower powers. In some statistical packages, if you specify a model with only the interaction, then it will automatically be assumed that you are fitting all main effects and interactions.

So the second principle of marginality is that if an interaction is significant, then the main effects must also be important, regardless of their significance levels. This is equivalent to asking sensible questions, and coming up with the simplest possible answer. The questions are: 'Does the impact of shade on the number of blooms depend upon the level of water? If not, does shade have an impact on the number of blooms and does water have an impact on the number of blooms?' By breaking the questions down into main effects and interactions, we can see if it is necessary to have the complication of the interaction, or whether we can get away with just the main effects.

Do not test main effects with a sum of squares adjusted for the interaction

Remember the definition of an adjusted sum of squares: it is the explanatory power of a variable when the influence of all other variables in the model have been taken into account. So because an interaction contains a main effect, the sum of squares for the main effect when the interaction has been taken into account does not represent a sensible question. For example, it would be as if we were asking if water had an influence on the number of blooms, over and above the fact that water and shade in combination influence bloom number.

So considerations of marginality have given us a set of rules by which we can pick valid and relevant p-values to test for significance of different components of the model when hierarchies are involved. This completes the third of our set of three principles of model choice. The next section returns to the polynomial example presented earlier in this chapter to apply these three principles and elucidate them by example.

Model choice in the polynomial problem

The analyses in Boxes 10.1–10.3 present three models. By applying the three principles of model choice it is now possible to make a formal decision between them.

Economy of variables

The aim is to chose the simplest model that adequately describes the data. So we don't wish to include unnecessary powers of `X1`, but neither should there be any pattern in the residuals. This suggests that the model `Y1 = X1 + X1 * X1` is the most appropriate. The quadratic term is needed because it is significant in the quadratic model, while the cubic term is not because it is insignificant in the cubic model.

Multiplicity of p-values

It may be tempting to fit many powers of `X1` on an exploratory basis. If there are a priori reasons for expecting certain powers to be important, then they should be included at the outset. In some cases, interpretation of the residual plots may be difficult, and it may be that although one power is insignificant, a higher power is also tested. If inspecting higher powers on a purely exploratory basis such as this, it is best to limit it to two powers above the last significant power, and be sceptical if the higher of the two is only borderline in its significance. In the context of our example, the cubic term is insignificant, and it would be acceptable to fit a term to power 4. However, the residual plot is satisfactory, so in many cases the researcher would look no further.

Considerations of marginality

In each of the ANOVA tables of Boxes 10.1–10.3, only one p-value is valid—the value for the highest power in the model formula. So in the first analysis, `X1` gives a p-value of <0.0005, in Box 10.2, `X1 * X1` gives a p-value of <0.0005, and in Box 10.3, `X1 * X1 * X1` gives a p-value of 0.732. The fact that the p-values for `X1` and `X1 * X1` in the final cubic model are not significant (0.523 and 0.379) is of no concern, because they are inadmissible—they are based on SS that have been adjusted for higher order terms. These significance tests are based on sums of squares that have been adjusted for the higher order interaction, the cubic term, and are therefore not asking sensible questions.

So in conclusion, the best model will be `Y1 = X1 + X1 * X1`. Having decided on this model, then the parameter estimates should be taken from the relevant coefficient table. In this case:

$$\text{Y1} = -7.62 + 3.189 * \text{X1} + 0.8525 * \text{X1} * \text{X1}.$$

There is an alternative to fitting all three models separately. If all F-ratios were based on sequential sums of squares (as in Box 10.5), and the model was fitted with all terms in the correct order, then the rules of marginality would be satisfied. Each term would then only be adjusted for earlier terms in the model, and if they were all of lower power, then this would be acceptable.

None of the rules of marginality has been contravened. The F-ratio for `X1 * X1 * X1` has been adjusted for `X1` and `X1 * X1`, but the ratio for `X1 * X1` is only adjusted for `X1`. With this one analysis we can determine the best model. However, it is still true that to obtain the parameters for the best model, we would need to refit, dropping the `X1 * X1 * X1` term. One disadvantage of using

BOX 10.5 ANOVA tables using sequential rather than adjusted sums of squares

General Linear Model

Word equation: Y1 = X1 | X1 | X1

X1 is continuous

Analysis of variance table for Y, using Sequential SS for tests

Source	DF	Seq SS	Adj SS	Seq MS	F	P
X1	1	6 663 021	2 505	6 663 021	1095.90	0.000
X1 * X1	1	256 148	4 763	256 148	42.13	0.000
X1 * X1 * X1	1	720	720	720	0.12	0.732
Error	76	462 078	462 078	6 080		
Total	79	7 381 967				

Coefficients table

Term	Coef	SECoef	T	P
Constant	−15.75	33.92	−0.46	0.644
X1	6.179	9.625	0.64	0.523
X1 * X1	0.6169	0.6971	0.89	0.379
X1 * X1 * X1	0.00500	0.01452	0.34	0.731

this method is that the residuals have not been examined for the simpler models. This can assist in model choice, or alert you to other problems, such as heterogeneity of variance. It is also important to remember why polynomials are so useful—they tell us something about the shape of the relationship. In Section 10.5, a case study is analysed which makes use of polynomials for this purpose.

10.3 Four different types of model choice problem

Choosing between models is a problem that arises with all types of analyses. Problems are divided here into four broad groups, three of which are considered in this chapter, and the fourth rather more tricky group is discussed in Chapter 11.

Group 1

The first of these groups are orthogonal designed experiments. In these cases, the model is prescribed by the experimental design itself. If the experiment includes interactions, then considerations of marginality will apply. Owing to the orthogonal design, these experiments are particularly easy to interpret, because the sequential and adjusted sums of squares will be exactly equal. This means that the third consideration of marginality is not relevant. An example of model choice with an orthogonal experiment is given in Section 10.4.

Group 2

The second group are designed experiments which have lost orthogonality, perhaps through losing a few datapoints. This will be reflected in slight differences between the sequential and adjusted sums of squares. If this loss is slight, then it usually does not cause problems with the analysis—but it may require a few alternative model fits to ensure this. An example is given in Section 10.4.

Group 3

The third group includes designed experiments with continuous variables, including polynomials. Once again, marginality is the main consideration in model choice. An example is given in Section 10.5.

Group 4

The fourth group is of models in which many (usually continuous) variables are fitted. These are often referred to as multiple regression problems. In these cases, economy of variables and multiplicity of p-values are the main concerns of model choice. Deducing which variables are best included in the model will usually involve more than purely statistical issues, but also practical issues and previous studies. The extent to which the variables share information will be an important consideration during model simplification. If there are great differences between the sequential and adjusted sums of squares for the different variables, this will alert you to possible complications during model simplification. These types of analyses are the subject of Chapter 11.

One of the first steps in model choice is to decide what kind of problem you are dealing with. This can be answered by considering (a) the explanatory variables (Are they categorical or continuous? Are there interactions?); and (b) the differences between sequential and adjusted sums of squares (Is it an orthogonal experiment? Has orthogonality been lost? Are there continuous variables that affect each other's informativeness?). In the next section, orthogonal, or near orthogonal designed experiments are considered, and in the final section, designed experiments with continuous variables.

10.4 Orthogonal and near orthogonal designed experiments

Model choice with orthogonal experiments

Farmers of potato crops often experience problems with potatoes rotting while in storage. An experiment was conducted to find the conditions under which

BOX 10.6 Analysis of a factorial experiment

General Linear Model

Word equation: ROT = BAC | TEMP | OXYGEN

BAC, TEMP and OXYGEN are categorical

Analysis of variance table for ROT, using Adjusted SS for tests

Source	DF	Seq SS	Adj SS	Adj MS	F	P
BAC	2	651.81	651.81	325.91	13.91	0.000
TEMP	1	848.07	848.07	848.07	36.20	0.000
OXYGEN	2	97.81	97.81	48.91	2.09	0.139
BAC * TEMP	2	152.93	152.93	76.46	3.26	0.050
BAC * OXYGEN	4	30.07	30.07	7.52	0.32	0.862
TEMP * OXYGEN	2	1.59	1.59	0.80	0.03	0.967
BAC * TEMP * OXYGEN	4	81.41	81.41	20.35	0.87	0.492
Error	36	843.33	843.33	23.43		
Total	53	2707.04				

to keep potatoes to minimise the rate at which rotting occurs. The variables were oxygen (3 levels: OXYGEN), temperature (2 levels: TEMP) and bacterial inoculation (3 levels: BAC). This last variable concerns the amount of bacterial inoculum injected into the potatoes at the beginning of the experiment (to ensure that some rotting occurred during the course of the experiment). It was coded as 1, 2 and 3, but these categories are also ordinal (as is also the case for oxygen and temperature). There were three replicates of each treatment combination, completing the factorial design. The analysis of this experiment is given in Box 10.6 (with the data stored in the *potatoes* dataset). The first point to note is that the sequential and adjusted sums of squares are equal, confirming that the experiment is exactly orthogonal. This greatly assists our interpretation, because inferences about each of the terms in the model formula can be drawn independently of the other terms in the model. So all p-values in the ANOVA table are valid, and we can draw conclusions about the main effects even though the model formula contains interactions.

The first term that should be considered in the ANOVA table is the highest order interaction (BAC * TEMP * OXYGEN). This is not significant ($p = 0.492$), so we can turn our attention to the three second order interactions. Of these, only BAC * TEMP is significant ($p = 0.05$), and so the main effects of BAC and TEMP must also be important (owing to the second consideration of marginality). The evidence that the main effects of BAC and TEMP combine non-additively is borderline however, so we may decide not to accept the significance of the interaction. In this case, there is overwhelming evidence that both the main effects matter.

This interpretation is straightforward. This emphasises the usefulness of orthogonal designs. We shall return to this example in Section 10.5, and see how this particular analysis can be taken further.

Model choice with loss of orthogonality

In many experiments, although the original design was intended to be orthogonal, mishaps during the execution of the experiment lead to a few missing values. This can result in us needing to fit more than one model to be sure of the interpretation of the results.

In this example, an experiment was performed to look at the influence of three nutrients on the growth of cacti. The three nutrients, nitrate, phosphate and water, were added at four levels to cactus plants in a factorial design. After three weeks, the dry weight of the cacti were measured. Unfortunately, during the drying process, some labels were lost, making the original orthogonal and balanced design non-orthogonal and unbalanced. This becomes apparent though a comparison of the sequential and adjusted sums of squares, which now have slight differences (Box 10.7). (These data are stored in the *cactus plants* dataset).

These differences introduce two complications. Taking H2O as an example, its sequential sum of squares is adjusted for nitrate (NI), while its adjusted sum of squares is adjusted for nitrate and phosphate (PH) (in fact everything else in the model). So now its sum of squares depends upon which other

BOX 10.7 Loss of orthogonality affects the equality of adjusted and sequential sums of squares

General Linear Model

Word equation: DRYW = NI | H2O | PH

NI, H2O and PH are categorical

Analysis of variance table for DRYW, using Sequential SS for tests

Source	DF	Seq SS	Adj SS	Seq MS	F	P
NI	3	38.6797	37.9039	12.8932	109.39	0.000
H2O	3	0.2332	0.5108	0.0777	0.66	0.579
PH	3	26.7578	26.0708	8.9193	75.67	0.000
NI * H2O	9	2.0007	1.7625	0.2223	1.89	0.062
NI * PH	9	0.7641	0.8178	0.0849	0.72	0.689
H2O * PH	9	1.2027	1.2618	0.1336	1.13	0.346
NI * H2O * PH	27	1.7371	1.7371	0.0643	0.55	0.964
Error	104	12.2579	12.2579	0.1179		
Total	167	83.6334				

terms have been controlled for—assessing its importance can no longer be done without reference to its place in the model formula. Secondly, its adjusted sum of squares is now also adjusted for terms that include H2O in higher order interactions. As discussed in Section 10.2, this contravenes the rules concerning hierarchies, and leads to the ANOVA table which bases its p-values on adjusted sums of squares as asking a question that is no longer sensible. (For example, does water influence the growth of cacti when the combined effects of water and nitrate have been taken into account?)

These complications require two courses of action: (1) the ANOVA table must be constructed with the sequential rather than adjusted sums of squares; (2) the model should be fitted with the explanatory variables in different orders, to see if this really does affect our interpretation. In this example, because there are three different factors, there are six different models (NI | H2O | PH, H2O | PH | NI, PH | NI | H2O, PH | H2O | NI, H2O | NI | PH, NI | PH | H2O). Three examples are given in Box 10.8.

The p-values for each term are slightly different between these three analyses, but they do not differ greatly. Given that the sequential and adjusted sums of squares are not very different in Box 10.8, this is what we would have expected. It is always important however to check this by fitting the models based on both the sequential and the adjusted sums of squares.

In this example, the loss of orthogonality has not hindered the interpretation of the experiment, and this is usually the case if there are not many missing values, or they are spread fairly evenly across the different treatments. The situation might have been different if the missing values result in whole treatment

BOX 10.8(a) **Reanalysing the growth of cacti due to loss of orthogonality**

General Linear Model

Word equation: DRYW = H2O | PH | NI

NI, H2O and PH are categorical

Analysis of variance table for DRYW, using Sequential SS for tests

Source	DF	Seq SS	Adj SS	Seq MS	F	P
H2O	3	0.4339	0.5108	0.1446	1.24	0.304
PH	3	25.6528	26.0708	8.5509	72.55	0.000
NI	3	39.5841	37.9039	13.1947	111.95	0.000
H2O * PH	9	1.2508	1.2618	0.1390	1.18	0.316
H2O * NI	9	1.9310	1.7625	0.2146	1.82	0.073
PH * NI	9	0.7859	0.8178	0.0873	0.74	0.671
H2O * PH * NI	27	1.7371	1.7371	0.0643	0.55	0.964
Error	104	12.2579	12.2579	0.1179		
Total	167	83.6334				

BOX 10.8(b) Reanalysing the growth of cacti due to loss of orthogonality

General Linear Model

Word equation: DRYW = PH | NI | H2O

NI, H2O and PH are categorical

Analysis of variance table for DRYW, using Sequential SS for tests

Source	DF	Seq SS	Adj SS	Seq MS	F	P
PH	3	25.2647	26.0708	8.4216	71.45	0.000
NI	3	39.8666	37.9039	13.2889	112.75	0.000
H2O	3	0.54395	0.5108	0.1798	1.53	0.212
PH * NI	9	0.8976	0.8178	0.0997	0.85	0.576
PH * H2O	9	1.2780	1.2618	0.1420	1.20	0.300
NI * H2O	9	1.7920	1.7625	0.1991	1.69	0.101
PH * NI * H2O	27	1.7371	1.7371	0.0643	0.55	0.964
Error	104	12.2579	12.2579	0.1179		
Total	167	83.6334				

BOX 10.8(c) Reanalysing the growth of cacti due to loss of orthogonality

General Linear Model

Word equation: DRYW = PH | H2O | NI

NI, H2O and PH are categorical

Analysis of variance table for DRYW, using Sequential SS for tests

Source	DF	Seq SS	Adj SS	Seq MS	F	P
PH	3	25.2647	26.0708	8.4216	71.45	0.000
H2O	3	0.8220	0.5108	0.2740	2.32	0.079
NI	3	39.5841	37.9039	13.1947	111.95	0.000
PH * H2O	9	1.2508	1.2618	0.1390	1.18	0.316
PH * NI	9	0.9248	0.8178	0.1028	0.87	0.553
H2O * NI	9	1.7920	1.7625	0.1991	1.69	0.101
PH * H2O * NI	27	1.7371	1.7371	0.0643	0.55	0.964
Error	104	12.2579	12.2579	0.1179		
Total	167	83.6334				

combinations being lost. It may then be that the significance of some factors depend upon the order of the variables in the model formula. In these cases, only those terms that are consistently significant can be relied upon, and it may be necessary to repeat the experiment.

10.5 Looking for trends across levels of a categorical variable

In some experiments, there is an option as to whether the variables should be treated as categorical or continuous. If the coding for that variable is ordinal (as in the potato analysis of Box 10.6), then it can make sense to look for trends in the same way as for other continuous variables. The categorical option will tell us if there are significant differences between the different levels of that variable, but will give no indication of trend (apart from by a visual inspection of the means). There may be distinct advantages to treating such variables as continuous, as will be illustrated in the following example.

Returning to the potato rot experiment, Box 10.9 presents a table of means for the significant terms.

These means are then translated into an interaction diagram in Fig. 10.4. Because means have been fitted (rather than lines), the points have been joined together. If we had treated BAC as a continuous variable we could have drawn straight lines.

Figure 10.4 suggests that the relationship between ROT and BAC is linear for Temperature 2, and perhaps a hint that it may be curvilinear at Temperature 1. BAC was treated as categorical in the first analysis—but it is ordinal, so it may be treated as continuous. This will allow us to ask the following four questions: (1) Is there a linear trend in the relationship between ROT and BAC?; (2) Is the slope of this trend the same at the two temperatures?; (3) Is there a curvilinear component to the relationship between ROT and BAC?; (4) Is this curvilinear component the same at the two temperatures? Each of these questions is answered by a specific component of the ANOVA shown in Box 10.10.

Because BAC is now being treated as a continuous variable, we are no longer dealing with a fully orthogonal design. Consequently, the ANOVA table of Box 10.10 has been constructed using sequential rather than adjusted sums

BOX 10.9 Degree of rotting in potatoes, according to TEMP and BAC treatments

Least squares means for ROT

BAC *	TEMP	Mean	SE of Mean
1	1	3.556	1.562
1	2	7.000	1.562
2	1	4.778	1.562
2	2	13.556	1.562
3	1	8.000	1.562
3	2	19.556	1.562

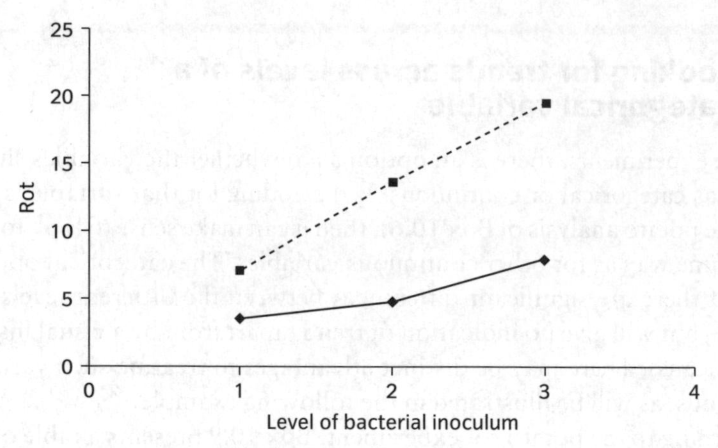

Fig. 10.4 An interaction diagram for the potato rot experiment. (Solid line for TEMP 1, dashed line for TEMP 2).

BOX 10.10 Reanalysing the potato rot experiment, looking for trends

General Linear Model

Word equation: ROT = OXYGEN + TEMP | BAC | BAC

OXYGEN and TEMP are categorical. BAC is continuous

Analysis of variance table for ROT, using Sequential SS for tests

Source	DF	Seq SS	Adj SS	Seq MS	F	P
OXYGEN	2	97.81	97.81	48.91	2.35	0.106
TEMP	1	848.07	4.68	848.07	40.79	0.000
BAC	1	650.25	5.78	650.25	31.27	0.000
TEMP * BAC	1	148.03	15.43	148.03	7.12	0.010
BAC * BAC	1	1.56	1.56	1.56	0.08	0.785
TEMP * BAC * BAC	1	4.90	4.90	4.90	0.24	0.630
Error	46	956.41	956.41	20.79		
Total	53	2707.04				

of squares, to obtain valid p-values for all terms. As always, the first term to be examined is the highest order interaction: TEMP * BAC * BAC. As BAC is now continuous, BAC * BAC is a quadratic term. So the three way interaction is asking whether the quadratic term is different for TEMP level 2 compared to TEMP level 1 (as we suspected that the lower line of Fig. 10.4 might be curved while the upper one is straight). This is insignificant ($p = 0.630$). BAC * BAC is also insignificant ($p = 0.785$), so there is no curvilinear component to the relationship between ROT and BAC at all. Moving on to TEMP * BAC, this is significant ($p = 0.01$), and therefore the main effects of TEMP * BAC must also be important.

The ANOVA table in Box 10.10 was constructed using the sequential sums of squares, owing to the loss of orthogonality once BAC was treated as continuous. However, the sequential and adjusted sums of squares for OXYGEN remain the same—so the loss of orthogonality has not been complete. There are three levels of OXYGEN, and within each level all combinations of TEMP and BAC are represented. Consequently, OXYGEN remains orthogonal to TEMP and BAC.

At first glance, the suggestion could be that little has been gained by reanalysing this data set using BAC as a continuous rather than categorical variable, because the same terms (TEMP * BAC and therefore the main effects) are significant. However, there are two distinct advantages to this analysis over that of the factorial experiment: (1) we have asked questions about the shape of the relationship between ROT and BAC; (2) the second analysis is more sensitive than the first.

Shape

When examining the interaction diagram, drawn as a result of the factorial analysis, it was clear that as BAC levels increased, then so did ROT. It was not clear however whether this increase was linear, or curvilinear (especially for Temperature 1). As a result of the second analysis it is now apparent that there is a trend and there is no evidence for nonlinearity. While the slopes are significantly different for the two temperatures, there is also no evidence for a curvilinear component in either case. It is only by fitting BAC as a continuous variable that such questions about shape can be answered.

Sensitivity

While the same terms are significant, in the second analysis they are *more* significant (the p-values for TEMP * BAC being 0.05 in the first analysis and 0.01 in the second). Why is this, and will this always be the case?

In Fig. 10.5, the two methods of analysis have been compared. The sum of squares for BAC in the first analysis is 651.81. In the second analysis, this sum of squares has been partitioned into the linear component (BAC), and the curvilinear component (BAC * BAC). The degrees of freedom have been partitioned likewise. Because most of the variability in BAC is explained by the linear component (650.25), with very little being partitioned to the quadratic component, yet the degrees of freedom have been halved, the F-ratio for BAC in the second model is much greater (though $p < 0.0005$ in both cases). Similarly, the interaction of the first model (TEMP * BAC) has been partitioned into the linear differences and the quadratic differences between the two temperatures. Again, most of the variability is explained by fitting two lines with different slopes, and practically none by allowing the lines to curve differently for each temperature. Consequently the F-ratio for TEMP * BAC is also larger (and the p-value smaller) in the second analysis.

The variable BAC has been decomposed into its linear and quadratic components. This is called a polynomial decomposition of the sum of squares, and

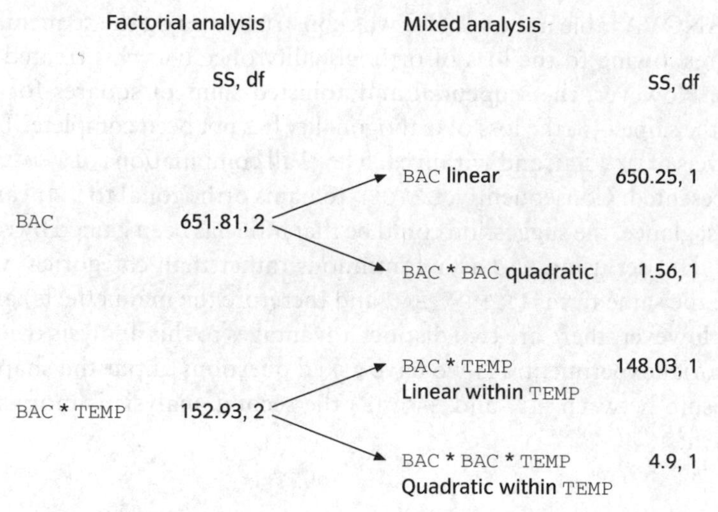

Fig. 10.5 Partitioning of variation between linear and quadratic components.

it allows questions to be asked about the shape of relationships. If there is a trend, as in this case, it also results in a more sensitive test, and so more significant p-values. Fitting lines asks the question 'Is there a consistent trend?' whereas fitting means just asks 'Are there differences?'. The nature of the analysis should also be reflected in the way in which the results are presented. In Fig. 10.4, the means for ROT at different levels of BAC and TEMP were presented, and the means joined by a line for each temperature. Now that BAC has been fitted as a continuous variable, then the model can be represented legitimately as two straight lines (Fig. 10.6).

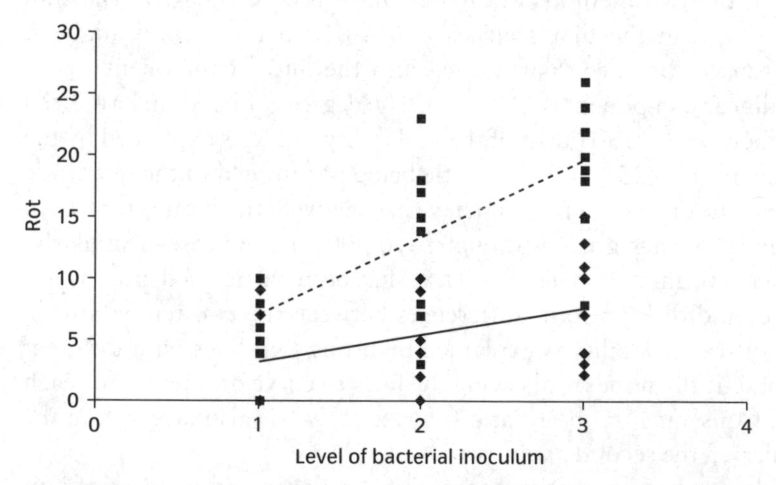

Fig. 10.6 Presenting the results for rotting potatoes. (Solid line is Temp 1, dashed line is Temp 2).

The sum of squares for BAC was decomposed into linear and quadratic components. In fact, it would not have been possible to fit any higher powers in this example, as the variable BAC had only two degrees of freedom in the first instance. This limits the number of powers that could be fitted, and the general rule is that if an ordinal categorical variable has n levels, then it may be decomposed into $n - 1$ powers as a continuous variable. In fact, if an ordinal categorical variable has a greater number of levels, the potential for increased sensitivity by using it as a continuous variable is even further enhanced. A factor with n levels has $n - 1$ degrees of freedom. If most of the variation is explained by a linear trend, the mean square could be enhanced by a factor of $n - 1$ (as the mean square for the linear component of that factor is obtained by dividing by 1 rather than $n - 1$). This could easily move an F-ratio from insignificant to highly significant.

In the last two sections, the process of model selection has been illustrated for designed experiments, where considerations of marginality have been the key concern in model choice. We have also discussed how to cope with the loss of orthogonality, and how to look for trends across categorical variables. In the next chapter, we move on to observational data sets, when the number of possible models increases greatly.

10.6 Summary

- Three principles of model choice were introduced: economy of variables, multiplicity of p-values; considerations of marginality.
- The application of these three principles was illustrated with a problem involving polynomials.
- The relative importance of the three principles will depend upon the type of data set being considered. There are two main types of problem: (1) those in which the sequential and adjusted sums of squares are equal (designed orthogonal experiments), or nearly equal (loss of orthogonality); (2) those in which there are great differences between sequential and adjusted sums of squares.
- For orthogonal (and nearly orthogonal) experiments, the key principle is that of marginality. Model choice for orthogonal experiments need only involve the ANOVA table based on adjusted sums of squares. Loss of orthogonality means that p-values should be derived from sequential sums of squares, and the explanatory variables should be fitted in all possible orders (within the constraints of marginality).
- Ordinal categorical variables have the option of being fitted as continuous variables. If there is a consistent effect (for example, linear or quadratic), this method will be more sensitive. A second advantage is that it allows a polynomial decomposition of the sum of squares for that variable, so enabling us to ask questions about the shape of the relationship.

10.7 Exercises

Testing polynomials requires sequential sums of squares

In this example two sets of data are analysed, Y against X and Y against XS (all stored in the X *and* XS dataset). The second explanatory variable, XS, has been calculated from X by subtracting 0.2. So X and XS are essentially the same variable, but measured on slightly different scales. They are both plotted against Y in Fig. 10.7, and, as expected, the shape of the relationship is the same.

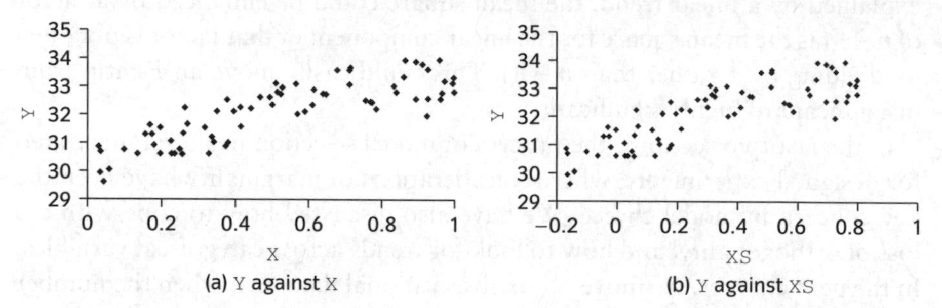

Fig. 10.7 Plots to explore fitting polynomials.

BOX 10.11(a) **Analysis based on** X

General Linear Model

Word equation: Y = X | X | X

X is continuous

Analysis of variance table for Y, using Adjusted SS for tests

Source	DF	Seq SS	Adj SS	Adj MS	F	P
X	1	58.906	0.719	0.719	2.46	0.121
X * X	1	4.305	0.151	0.151	0.52	0.475
X * X * X	1	0.542	0.542	0.542	1.85	0.178
Error	68	19.876	19.876	0.292		
Total	71	83.629				

General Linear Model

Word equation: Y = X | X | X

X is continuous

Analysis of variance table for Y, using Sequential SS for tests

Source	DF	Seq SS	Adj SS	Seq MS	F	P
X	1	58.906	0.719	58.906	201.53	0.000
X * X	1	4.305	0.151	4.305	14.73	0.000
X * X * X	1	0.542	0.542	0.542	1.85	0.178
Error	68	19.876	19.876	0.292		
Total	71	83.629				

BOX 10.11(b) Analysis based on XS

General Linear Model

Word equation: Y = XS | XS | XS
XS is continuous

Analysis of variance table for Y, using Adjusted SS for tests

Source	DF	Seq SS	Adj SS	Adj MS	F	P
XS	1	58.906	10.502	10.502	35.93	0.000
XS * XS	1	4.305	0.029	0.029	0.10	0.753
XS * XS * XS	1	0.542	0.542	0.542	1.85	0.178
Error	68	19.876	19.876	0.292		
Total	71	83.629				

General Linear Model

Word equation: Y = XS | XS | XS
XS is continuous

Analysis of variance table for Y, using Sequential SS for tests

Source	DF	Seq SS	Adj SS	Seq MS	F	P
XS	1	58.906	10.502	58.906	201.53	0.000
XS * XS	1	4.305	0.029	4.305	14.73	0.000
XS * XS * XS	1	0.542	0.542	0.542	1.85	0.178
Error	68	19.876	19.876	0.292		
Total	71	83.629				

Two analyses are conducted for each data set: one in which the adjusted sums of squares are used to calculate the *F*-ratio, and one in which the sequential sums of squares are used. These are shown in Box 10.11.

(1) The shape of a curve does not change on subtracting a constant from X. Compare the adjusted SS for X and XS, and also the sequential SS for X and XS. Which type of SS are not generally useful under these circumstances?

(2) Which of the ANOVA tables in Box 10.11(b) is to be preferred and why?

(3) In what circumstances (not involving interactions) are tests based on Adjusted SS to be preferred?

Partitioning a sum of squares into polynomial components

A factorial experiment was conducted to investigate the yield of barley. Thirty-six plots were divided into four blocks. Three varieties were compared at three different row spacings. These data are stored in the *barley* dataset, in variables BYIELD, BSPACE, BVARIETY and BBLOCK.

(1) Treating the variables as categorical, conduct a GLM of the factorial experiment, request the means and draw an interaction diagram. Your interaction diagram should resemble the one given in Fig. 10.8.

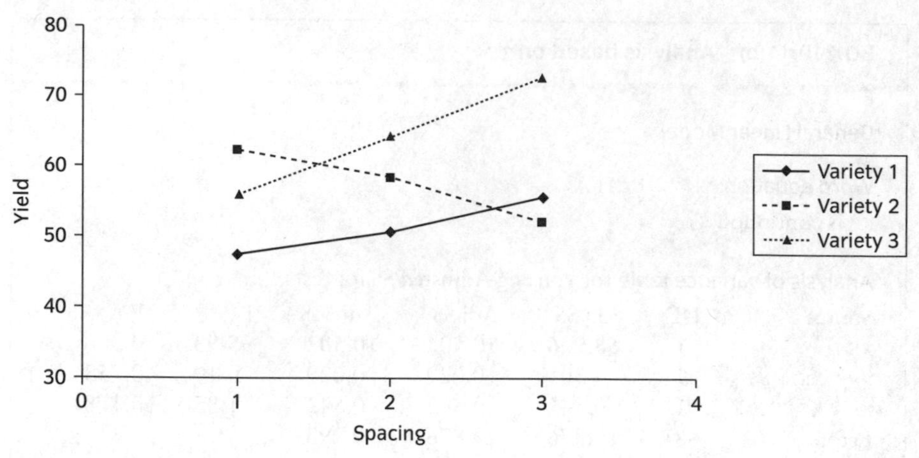

Fig. 10.8 Interaction diagram from the *barley* dataset.

Table 10.1 Table of sums of squares

	Adjusted sum of squares	Sequential sum of squares	Degrees of freedom
Linear term for BSPACE in the second analysis.			
Quadratic term BSPACE * BSPACE in the second analysis.			
Sum of the above.			
Categorical variable BSPACE in the first analysis.			

(2) What conclusions do you draw from your analysis about the influence of spacing on yield? What additional questions might you ask after drawing the interaction diagram?

(3) Now conduct a second GLM analysis treating BSPACE as a continuous variable and including a quadratic term. Complete Table 10.1.

(4) For which of the three columns does the sum of the polynomial components equal the values as a categorical variable?

(5) How is the shape of the graph of BSPACE against BYIELD reflected in the polynomial decomposition?

(6) How do your conclusions differ from those you drew after the first analysis?

11 Model selection II: datasets with several explanatory variables

In a typical multiple regression problem, there is one Y variable, and a large number of possible X variables. (This is in contrast to designed experiments, where the X variables are largely limited to the subset of variables manipulated by the researcher.) The aim of the analysis is to choose a selection of X variables which best describe or predict Y. The principles of model selection are the same (economy of variables, multiplicity of p-values and marginality), but the problem is much greater as there are many more models to choose from.

If we consider the situation when we have only six X variables, there are, in fact, 63 possible models (6 with one X variable, 15 with two X variables, etc.), and this is ignoring the possibility of any interactions between them. If just two-way interactions are allowed, there are over 30 000 possible models! It is not practical to examine all these possibilities, and so it is unusual to consider any interactions in multiple regression problems (unless a specific hypothesis involving an interaction is being tested). Thus considerations of marginality are less frequently encountered. However, we are still left with a considerable problem in choosing the best fit model.

The second consequence of dealing with datasets with several continuous variables is that the variables will not be orthogonal. Consequently, the informativeness of one variable is likely to be influenced by which other variables are included in the model. This concept was first introduced in Chapter 4, where two kinds of influence were discussed. The first kind of relationship was when two variables share information, so that when one is included in the model, the other can add little predictive power. The second kind of relationship is when one variable increases the informativeness of the other. These two relationships were distinguished by comparing the patterns of sequential and adjusted sums of squares in models that contained two variables. If the adjusted sum of squares for the first variable was lower than its sequential sum of squares, then it shared information with the second variable. In contrast, if its adjusted sum of squares was higher, then the second variable increased its informativeness. These patterns were useful in deducing the relationships between two or three explanatory variables, and they will also be useful for models containing many more variables.

So in summary, given that interactions are rarely utilised in multiple regression problems, the two most important principles are therefore the economy of variables, and multiplicity of p-values.

11.1 Economy of variables in the context of multiple regression

In the context of polynomials and designed experiments, the principle of making the model as simple as possible was fairly straightforward. Extra polynomial powers would only be added as necessary, as indicated by residual plots and p-values. With designed experiments, all potential explanatory variables are dictated by the experimental design. In multiple regression problems, there is the potential for a much wider range of X variables to be included. For each variable there is a statistical cost. One continuous variable uses up one degree of freedom and, if not significant, there is a risk of increasing the error mean square. However, there are often real world costs as well. Even though a variable may contribute significantly to prediction of Y, it may be that its measurement is expensive in terms of resources. While it is not possible to discuss real world costs in a general context, in practice it will often temper our choice of the best model.

A second consideration may be the number of variables used to predict Y. In comparing two models, it may be important to take into consideration not only the amount of variance explained, but also the number of X variables used to do the explaining. Many formalised methods have been suggested for measuring how economically a model fits a dataset, and packages report varying numbers of them. As an introduction to this area, we will look more closely at two approaches.

R-squared and adjusted *R*-squared

In Chapter 2, when simple regression was first introduced, the statistic R^2 was used as a measure of the fraction of variance explained by a model. This was calculated as:

$$R^2 = \frac{\text{Total SS} - \text{Residual SS}}{\text{Total SS}}$$

where SS = sum of squares. The value of R^2 will lie between 0 and 1, and the higher R^2 is, the greater the fraction of variance explained by the model. However, a full comparison of two models also needs to take account of the number of explanatory variables used. One disadvantage of R^2 as calculated here is that as more variables are added to the model, the greater R^2 will be. It will rise asymptotically towards 1, regardless of whether the variables added are significant or not.

One solution to this problem is to calculate the adjusted R^2 as follows:

$$R^2_{adj} = \frac{\text{Total MS} - \text{Residual MS}}{\text{Total MS}}$$

where

$$\text{Total MS} = \frac{\text{Total SS}}{\text{Total DF}}.$$

The residual MS is an estimate of the error variance, and represents the scatter around the regression line. The lower this scatter, the higher the R^2_{adj}.

Given that

$$\frac{\text{Residual SS}}{\text{Total SS}} = 1 - R^2$$

from above, then

$$R^2_{adj} = 1 - (1 - R^2)\left(\frac{\text{Total DF}}{\text{Residual DF}}\right).$$

The purpose of this algebra is to understand what happens to R^2_{adj} as the number of terms in the model increases. If we add unimportant terms to the model, while R^2 may rise, the residual degrees of freedom will drop, so on average R^2_{adj} will plateau below 1. (We can only be confident what will happen to R^2_{adj} on average, as individual terms close to significance may or may not be important, and could cause a slight increase or decrease in R^2_{adj}). It therefore combines information on the amount of variance explained, and the number of variables used. To illustrate this (see Fig. 11.1), consider two models which explain the same proportion of variance. Model 1 has a raw R^2 of 0.8 with three explanatory variables. Model 2 has the same raw R^2, but has five explanatory variables.

The raw R^2 and the total degrees of freedom will be the same for both models, but Model 1 has used up fewer degrees of freedom, and so the residual degrees of freedom are greater. Consequently, Model 1 has the larger adjusted R^2, as it explains the variance more economically.

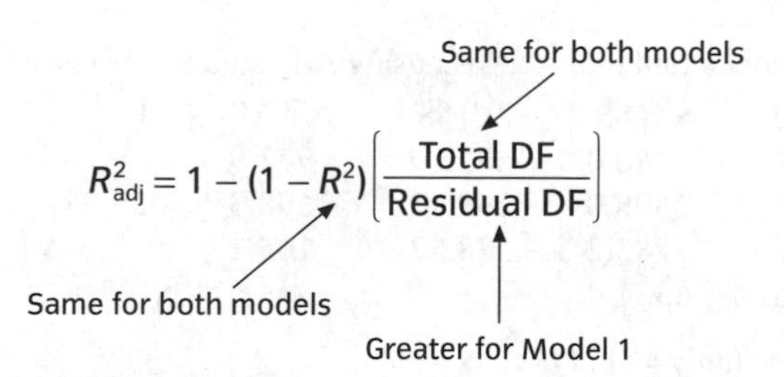

Fig. 11.1 Comparing adjusted R^2 for two models.

Table 11.1 *Peru* dataset to study the influence of changing altitude on blood pressure

AGE	Age in years
YEARS	Number of years since migration from high altitude
WGHT	Weight in kg
HGHT	Height in mm
CHIN	Chin skin fold in mm
FOREARM	Forearm skin fold in mm
CALF	Calf skin fold in mm
PULSE	Pulse rate in beats per minute
SYSTOL	Systolic blood pressure
DIASTOL	Diastolic blood pressure

Migrating to a lower altitude

This example illustrates how R^2 and adjusted R^2 change as more variables are added to a model. The data set originates from a study conducted by anthropologists on the effects of a change in environment on blood pressure. A group of 39 men of varying ages who had moved from a high altitude region at various times in the past to a much lower altitude were monitored for blood pressure and a variety of other body measurements described in Table 11.1 (and stored in the *Peru* dataset).

Previous studies suggested that the two most important variables were likely to be the length of time since the migration had occurred, and the weight of the individual. These two variables proved to be significant, and the regression

BOX 11.1(a) Multiple regression for blood pressure

Model 1

General Linear Model

Word equation: SYSTOL = YEARS + WEIGHT

YEARS and WEIGHT are categorical

Analysis of variance table for SYSTOL, using Adjusted SS for tests

Source	DF	Seq SS	Adj SS	Adj MS	F	P
YEARS	1	50.0	972.9	972.9	9.26	0.004
WEIGHT	1	2698.3	2698.3	2698.3	25.68	0.000
Error	36	3783.2	3783.2	105.1		
Total	38	6531.4				

$R^2 = 42.1\%$ $R^2(\text{adj}) = 38.9\%$

BOX 11.1(b) Multiple regression of factors influencing blood pressure after a change in altitude: Model 7

General Linear Model

Word equation: SYSTOL = YEARS + WEIGHT + AGE + HEIGHT + CHIN + FOREARM + CALF + PULSE

YEARS, WEIGHT, AGE, HEIGHT, CHIN, FOREARM, CALF and PULSE are all continuous.

Analysis of variance for SYSTOL, using Adjusted SS for tests

Source	DF	Seq SS	Adj SS	Adj MS	F	P
YEARS	1	50.0	697.6	697.6	6.41	0.017
WEIGHT	1	2698.3	2201.7	2201.7	20.22	0.000
AGE	1	27.9	97.4	97.4	0.89	0.352
HEIGHT	1	61.4	263.6	263.6	2.42	0.130
CHIN	1	366.9	249.3	249.3	2.29	0.141
FOREARM	1	42.7	59.2	59.2	0.54	0.467
CALF	1	14.7	16.2	16.2	0.15	0.703
PULSE	1	3.0	3.0	3.0	0.03	0.870
Error	30	3266.7	3266.7	108.9		
Total	38	6531.4				

$R^2 = 50.0\%$ $R^2(\text{adj}) = 36.6\%$

model gave an R^2 of 0.421, and an adjusted R^2 of 0.389. However, to check if any of the remaining variables were important, one of the anthropologists fitted a whole series of models, each incorporating an extra variable, calculating the R^2 and the adjusted R^2 in each case. The first and last of these models are shown in Box 11.1.

None of the additional six variables are significant, yet the R^2 has risen from 0.421 to 0.5. The adjusted R^2, on the other hand, has dropped from 0.389 to 0.366, indicating that the earlier model explained the variance more economically. In fact, whether a variable will increase or decrease the adjusted R^2 depends on whether its F-ratio is greater or less than 1. If the F-ratio is exactly 1, then the fraction of error SS accounted for by that variable is exactly the same as the fraction of error DF also removed by that variable, leaving the error MS unaltered (and therefore the adjusted R^2 unaltered).

Prediction intervals

Prediction intervals provide a second means of assessing how economically a model fits the data. The concept of prediction intervals was introduced in Chapter 2 for a simple regression equation. When predicting Y from one X

variable, there are two sources of uncertainty: (1) the scatter around the line; (2) the error in estimating the true line. For a simple regression two parameters are estimated, an intercept and a slope. Each of these terms contributed a source of uncertainty to the prediction interval equation discussed in Section 2.5. Reproduced here is the prediction interval for Y at $X = X'$:

$$PI = \hat{Y} \pm t_{\text{crit}} s \sqrt{\frac{1}{m} + \frac{1}{n} + \frac{(X' - \bar{X})^2}{SS_x}}$$

where

$$SS_x = \Sigma(X - \bar{X})^2.$$

The term $1/m$ represents the scatter around the true line ($m = 1$ when predicting a single data point, 2 when predicting the mean of two data points etc.). The term $1/n$ represents the uncertainty in the estimate of the intercept (n = the number of data points). The third term represents the uncertainty in the slope.

In a multiple regression problem, several X variables are now used to predict Y. For each extra explanatory variable, an extra model parameter is estimated, and this will affect the formula for a prediction interval in two ways. The first is that each added parameter contributes a term to the equation describing the uncertainty. In multiple regression, the extra term is normally a slope, and so the extra term will be of the form $(X' - \bar{X})^2/SS_x$. The second point is that each extra parameter reduces the residual degrees of freedom and so increases t_{crit}. In both cases this will increase the width of the prediction interval as the number of X variables rises. However, if the extra variable is significant, and leads to a reduction in the error mean square, then this will act towards reducing the width of the interval ($s = \sqrt{\text{EMS}}$). If prediction is the primary aim of the exercise, then the relative costs and benefits of adding a variable in terms of improving or worsening prediction should be considered. Adding an irrelevant variable to the model will on average worsen prediction.

Returning to the dataset investigating changes in blood pressure due to altitude, a prediction interval was calculated for an individual of weight 87 kg who had migrated to lower altitude 40 years prior to the study. In Box 11.2, both a prediction interval and a confidence interval are calculated using Model 1, containing only the two significant explanatory variables.

BOX 11.2 A prediction interval using a model with two explanatory variables

Regression equation: SYSTOL = 50.3 – 0.572 YEARS + 1.35 WEIGHT

Predicted values and intervals

Fitted value	StDev fitted value	95.0% CI	95.0% PI
145.25	6.06	(132.96, 157.54)	(121.10, 169.40)

BOX 11.3 A prediction interval using a model with five explanatory variables

Regression equation: SYSTOL = 117 − 0.573 YEARS + 1.87 WEIGHT − 0.253 AGE − 1.27 FOREARM − 0.0530 HEIGHT

Predicted values and intervals

Fitted value	StDev fitted valuc	95.0% CI	95.0% PI
146.71	7.32	(131.81, 161.60)	(120.92, 172.49)

The first point to note is the difference between the confidence interval and the prediction interval. The confidence interval represents the scatter around the line, while the prediction interval includes the additional uncertainty in the line itself. So the prediction interval is wider than the confidence interval. In Box 11.3, a prediction interval is requested for the same individual for a model containing five rather than two explanatory variables.

The prediction interval for the same individual in the second model is wider, even though extra information has been added to the model. Because this information did not add significantly to the explanatory power of the model, this resulted in increasing the uncertainty in the prediction. The simpler model therefore provides the more accurate predictions.

At first, this appears to be something of a paradox—that by adding extra information we end up being less accurate in our predictions. However, any dataset always contains a mixture of signal and noise. Variables that truly represent the underlying relationship add signal as well as noise, while other variables just add noise, so making our predictions less precise. A simple way to see that adding variables can make things worse is to imagine a situation with one X variable, just five datapoints and a true linear relationship. An analysis of this dataset is given in Box 11.4, showing the relationship to be linear, with no quadratic component.

Provided that all the values of X are different from each other, we can add polynomial terms until we can fit every single point exactly. This has been done in Fig. 11.2, with a quartic function that goes through all five points. However, this polynomial wobbles enormously within the range of the dataset. Even if we wished to interpolate, the exact fit goes way above any existing point. Slight extrapolation also gives severely negative values that are quite at odds with a reasonable view of the data. Clearly the exact fit is not the best fit!

On first encounter, prediction intervals seem a useful tool to represent the uncertainty in predicted values of Y. However, they should be used with caution for several reasons. They will only contain the true value of Y on the expected percentage of occasions (e.g. 95%) if the model is correct in the first place. A missing polynomial or interaction term or a missing variable would prevent this from being the case. Such inaccuracies would become greater

BOX 11.4 The 'five points' example

General Linear Models

Word equation: Y = X

X is continuous

Analysis of variance table for Y, using Adjusted SS for tests

Source	DF	Seq SS	Adj SS	Adj MS	F	P
X	1	3.3674	3.3674	3.3674	24.97	0.015
Error	3	0.4046	0.4046	0.1349		
Total	4	3.7720				

Coefficients table

Term	Coef	SE Coef	T	P
Constant	−12.925	3.279	−3.94	0.029
X	6.294	1.260	5.00	0.015

Word equation: $Y = X + X^2$

X is continuous

Analysis of variance table for Y, using Sequential SS for tests

Source	DF	Seq SS	Adj SS	Seq MS	F	P
X	1	3.3674	0.0343	3.3674	17.42	0.053
X * X	1	0.0181	0.0181	0.0181	0.09	0.789
Error	2	0.3866	0.3866	0.1933		
Total	4	3.7720				

Coefficients table

Term	Coef	SE Coef	T	P
Constant	−34.48	70.62	−0.49	0.674
X	22.92	54.40	0.42	0.715
X * X	−3.20	10.46	−0.31	0.789

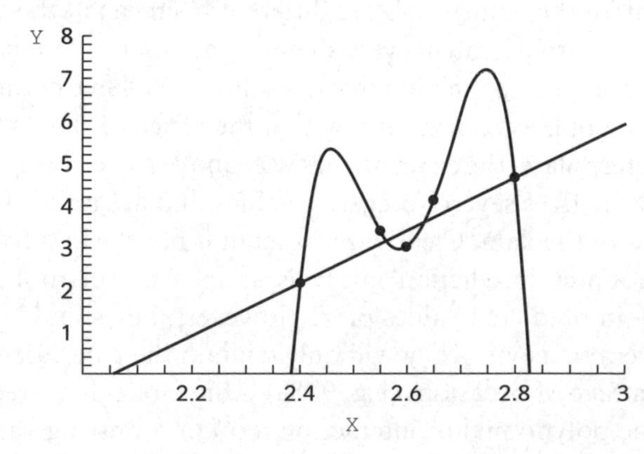

Fig. 11.2 The exact fit is not the best fit.

towards the extreme values of the explanatory variables, and would be greatly magnified if extrapolation was attempted beyond the range of the data set. Finally, the prediction intervals are not independent, so it would be misleading to use a series of prediction intervals to sketch out the confidence envelope for a whole line. Nevertheless, prediction intervals may be compared to determine how economically a model fits the data. Our purpose here is also to provide a clear demonstration that the addition of a variable can worsen the predictive power of a model.

11.2 Multiplicity of *p*-values in the context of multiple regression

The enormity of the problem

The more X variables we choose to investigate, the more likely it is that at least one will be significant by chance alone. Significance when the null hypothesis is true is a Type I error. By setting our threshold p-value at 0.05, we accept that there is a 5% chance that a test statistic will be significant through chance processes alone. The risk of making a Type I error will rise if we investigate several p-values for the same purpose. If, for simplicity, it is assumed that a set of k p-values are independent, then the risk of a Type I error can be quantified. For each p-value, let this risk be Λ. The probability that one or more will be significant by chance alone can be described as:

Probability of a Type I error
= 1 – Probability that none are significant.

By the multiplicative law of probability for independent events, the chance that all k are not significant is just $1 - \Lambda$ times itself k times, i.e. $(1 - \Lambda)^k$. So the probability that one is significant by chance alone = $1 - (1 - \Lambda)^k$. So, for example, with six p-values and Λ set to 0.05, the probability of making a Type I error becomes about 26%. (*Post hoc* comparison methods and Bonferroni corrections etc. are based on this kind of calculation). In reality, p-values are not independent, and this will alter this risk (and also make it hard to quantify). This is obviously a serious problem in multiple regression, and whenever analysing these kinds of problems, it is important to carefully consider how to minimise this risk.

Possible solutions

As we mentioned in the introduction to this chapter, it is rare to consider any interactions in multiple regression problems unless testing specific hypotheses, or solving problems in nonlinearity, as this greatly increases the number of

p-values considered. Beyond this, there are three further means of reducing the risks of Type I errors.

Focus don't fish

1. Include as few X variables as possible in the first place (i.e. reduce k of the formula above). If two X variables are likely to be highly correlated (for example, the length of the right leg and the length of the left leg), then only include one.
2. Use any information at your disposal to statistically eliminate variables known to be important from previous studies, and undertake to include them at the outset, regardless of their significance. For example, previous studies in the literature have shown that age is a significant factor in the incidence of cancer. In any one study, the range of ages may not be sufficient for age to be significant, but it may be wise to decide to include age at the outset (regardless of its p-value). This would statistically eliminate any age effects without adding to the problem of multiplicity of p-values, as the aim of the study is not to test the influence of age itself.
3. It is also important to consider the aim of the study. If the aim is to predict Y on future occasions, then it is sensible to use only those X variables that are likely to be available and easy to measure on those occasions.

In general, the best approach is to cut down on the number of unnecessary questions asked, because in that way we obtain sharper answers to the questions of primary interest.

Stringency

A second approach is to reduce our threshold p-value (often referred to as α) for rejecting the null hypothesis (normally set at 0.05), so making the conditions for significance more stringent. For example, reducing α to 0.01 for a study with 6 variables brings the probability of making a Type I error to 0.059, using the formula from the binomial distribution discussed above. Although the six variables are unlikely to be independent, this formula can be useful as a rule of thumb. This is an example of altering the conventional significance levels so that for one analysis, the probability of making a Type I error is acceptable. This kind of approach is also useful if you cannot avoid doing multiple pair-wise comparisons using t-tests. This is called a Bonferroni correction.

Reverse the process—combine sums of squares

The essence of the general linear model is to partition variance into fine components, to determine which factors are responsible for explaining significant amounts of variation. For example, total SS is partitioned into treatment SS and error SS. In turn, the treatment SS is partitioned into main effects and interactions. The SS of ordinal categorical variables can be partitioned using polynomial decomposition. With multiplicity problems, it is sometimes

Table 11.2 Variables in the *specific gravity* data set

WOODSG	The specific gravity of the timber of a tree, measured after felling.
NFIBSPR	The number of fibres in a given cross-sectional area of spring wood.
NFIBSUM	The number of fibres as above, but for summer wood.
SPRING	The fraction of wood in the sample core that was spring wood.
LASPRING	The light absorption of spring wood.
LASUMMER	The light absorption of summer wood.

expedient to do exactly the reverse, and combine sums of squares and degrees of freedom to ask one question, instead of two or three. This is best illustrated with an example.

An important component of the commercial value of wood is its density, which can be effectively measured by its specific gravity. The *specific gravity* dataset was collected with the aim of producing a model that could predict the specific gravity of a tree from a small sample. Twenty trees were felled, and their specific gravity measured directly (WOODSG), and cores also taken to measure five other variables as described in Table 11.2.

One option is to simply fit all five explanatory variables, but with five variables, multiplicity is likely to be a problem. Examining this data set there also appears to be some redundancy, e.g. the light absorption of spring and summer wood are included separately. One alternative is to decide to reduce the number of questions from five to four, by asking whether light absorption in general helps to predict the specific gravity. In Box 11.5 the full model is fitted, but an *F*-ratio

BOX 11.5 Analysing the specific gravity dataset

General Linear Model

Word equation: WOODSG = NFIBSPR + NFIBSUM + SPRING + LASPRING + LASPRING

NFIBSPR, NFIBSUM, SPRING, LASPRING and LASPRING are continuous

Analysis of variance table for WOODSG, using Adjusted SS for tests

Source	DF	Seq SS	Adj SS	Adj MS	F	P
NFIBSPR	1	0.01574	0.04450	0.04450	1.38	0.260
NFIBSUM	1	0.12459	0.00043	0.00043	0.01	0.910
SPRING	1	1.08211	1.04567	1.04567	32.44	0.000
LASPRING	1	0.00110	0.08563	0.08563	2.66	0.125
LASUMMER	1	0.29132	0.29132	0.29132	9.04	0.009
Error	14	0.45124	0.45124	0.03223		
Total	19	1.96610				

for light absorption can be calculated by combining the sequential sums of squares for the last two terms.

The combined sums of squares for light absorption is (0.00110 + 0.29132) = 0.292, and the combined degrees of freedom are 1 + 1 = 2. This gives a mean square of 0.146, and an *F*-ratio of (0.146/0.03223) = 4.53 with 2 and 16 degrees of freedom. This in turn gives a *p*-value of 0.028 (refer to the package specific supplements for how to calculate the *p*-value from the *F*-ratio). The decision as to whether or not you would reject the null hypothesis would depend upon the level of stringency at which you had set your *p*-values.

The two variables whose SS were combined were at the end of the model formula. This is important, because the two sequential sums of squares of the last two variables are adjusted for all other variables in the model, and our aim was to base the test on the adjusted sum of squares.

11.3 Automated model selection procedures

When faced with a large number of explanatory variables, and the task of selecting a subset of those variables, it is not uncommon for automated model selection procedures to be used. Two simple forms are forwards and backwards stepwise regression. These methods are described here, with the aim of highlighting the pitfalls and problems involved in their use.

How stepwise regression works

Suppose there are five *X* variables in the dataset. For forwards stepwise regression, the starting point is the null model, with no explanatory variables fitted. The first step involves fitting five models, each with one of the explanatory variables ($Y = X1$, $Y = X2$ etc.). Of these five models, the chosen one is that in which the *X* variable has the highest *F*-ratio or lowest *p*-value. If this was, for example *X3*, then the second step would be to include this variable, and fit the other four as second explanatory variables (i.e. fit the models $Y = X3 + X1$, $Y = X3 + X2$ etc.). The next strongest candidate would then be chosen, and added to the model. This would continue, until none of the remaining variables had *F*-ratios sufficiently large or *p*-values sufficiently small to merit inclusion in the model (an *F*-ratio of 4 or a *p*-value of 0.05 is often taken as the cut off point). At this point, the process comes to an end, and the best model has been found.

Backwards stepwise regression follows the same principle, but the starting point is now the full model: i.e. $Y = X1 + X2 + X3 + X4 + X5$. In the first step, the *F*-ratio or *p*-value of each variable is examined, and the variable with the lowest *F*-ratio or highest *p*-value (as determined by the adjusted sum of squares) is removed from the model. This process is repeated until even the weakest variable has an *F*-ratio sufficiently large or a *p*-value sufficiently small to retain it in the model.

Using *F*-ratios or *p*-values as the criteria for inclusion or exclusion will amount to the same thing if the degrees of freedom of the variables are the same (for example, if all variables are continuous).

Stepwise regression will now be used to solve a model selection problem. This example will also illustrate some of the pitfalls in using automated selection procedures.

The stepwise regression solution to the whale watching problem

A firm specialising in marine cruises runs whale watching trips, with the promise of a 50% refund if no whales are seen. To improve their profit margins, they want to be able to predict when they are likely to see whales by 8 a.m. each morning, so that if it is unlikely, they can cancel the trip. Over the first three years of the business they ran 180 trips, and a data base was built up containing a number of variables collected during each trip. The aim of the analysis was to determine which of these variables might be important in predicting the number of whales spotted per minute per trip. The variables are listed in Table 11.3. The response variable was expressed as the number of whales spotted during each trip, corrected for the length of trip, and transformed to conform to the assumptions of GLMs.

Table 11.3 Variables in the whale watching data set

Explanatory variables	
TRIPID	A code that numbers the trips from 1 to 180
YEAR	Code for year (1, 2, 3)
MONTH	Codes for month (5, 6, 7, 8, 9)
DAY	Day of the month
NPASS	The number of passengers on the trip
CLOUD8AM	Cloud cover in harbour at 8am, coded 0 for no cover to 8 for total cover
RAIN8AM	Rain in harbour at 8am, coded as 0, 1, 2 and 3 for none, light, moderate, heavy
VIS8AM	Visibility in harbour at 8am, in kilometres
RAIN	Rain experienced on trip coded 0 to 3 as before
VIS	Visibility during the trip, in kilometres
DURNTOT	Total duration of voyage in minutes
Response variable	
LRGWHALE	$\log\left(0.01 + \dfrac{\text{Number of whales seen that trip}}{\text{Time spent in whale zone}}\right)$

BOX 11.6 Forwards stepwise regression of the whale watching dataset

Stepwise regression

The Y variable is LRGWHAL and the programme is given a choice of 11 predictors

Step	1	2	3
Constant	−4.525	−4.555	−4.641
VIS	0.1252	0.1041	0.1056
T-value	13.91	7.63	7.79
P-value	0.000	0.000	0.000
VIS8AM		0.029	0.037
T-value		2.05	2.56
P-value		0.042	0.011
RAIN			0.146
T-value			2.18
P-value			0.031
R^2	45.67	46.65	47.73
R^2 (adj)	45.44	46.18	47.04

The boss of the firm decided to use stepwise regression to solve this problem, using the *F*-ratio of 4 as the cut-off point for inclusion of a variable in the model. The output from the forwards stepwise regression is given in Box 11.6.

The output from a stepwise regression is given in columns. In step 1, VIS was fitted, the second variable to be added was VIS8AM, and the third RAIN. At this point the program stopped, as no further variables gave significant *F*-ratios when added to the model. With stepwise procedures, a number of measures of model fit are included in the output—only R^2 and adjusted R^2 are shown here, but others will be described in the package specific supplements.

Has this automated procedure given us a sensible and useful model? In this particular example, the boss has thrown all eleven variables into the stepwise program. Because he has failed to focus on the most relevant variables, this output will certainly suffer from problems with multiplicity of *p*-values. While the first *p*-value is highly significant ($p < 0.005$ for VIS), the evidence for VIS8AM is less convincing at $p = 0.042$. At least one out of ten variables will be significant about 40% of the time, so we should treat this second result with scepticism. That was a statistical argument about interpreting the *p*-value. Non-statistical arguments are also important.

Would a model containing VIS and VIS8AM suit our purpose? VIS is a measure of visibility during the trip, so while it is likely to be a good predictor of the number of whales seen, it cannot be measured in advance, whereas VIS8AM can be. The boss failed to take into account which variables are available for prediction at the right time, compared to those that may explain the relationship retrospectively.

There are a number of problems with relying on such automated procedures, some of which are well illustrated by this example:

1. The temptation is to rely on purely statistical considerations in choosing the best model. Automated procedures may seem to the naïve user to have statistical sophistication, and often lead him to neglect to use other relevant information. This can be overcome to some extent by including some variables in the model at the outset.
2. The temptation is also to throw in every conceivable variable, in the hope that 'the program will sort them out'. But this is fishing, not focussing, and causes problems of multiplicity as discussed in Section 11.2.
3. Some packages have the particular restriction that all explanatory variables used in the stepwise procedure are treated as continuous. Here, YEAR is categorical, so this is inappropriate and will be discussed further later.
4. It is also tempting to adhere to the usual statistical criterion for inclusion in the model (an *F*-ratio of 4). When using several variables, this should be more stringent. One option is to make the criterion different at different stages, but there is no precise way in which to decide how this is done.
5. Finally, forwards and backwards procedures can easily give different best models.

We now consider point 5 in more detail. This is a result of the way in which continuous variables share information. For example, consider three X variables, $X1$, $X2$ and $X3$. In Fig. 11.3 the model $X1 + X2 + X3$ is represented in three different ways. In all three cases, the magnitude of the sum of squares represented by fitting $X1 + X2 + X3$ is represented by the full line (and is therefore the same in Fig. 11.3(a)–(c)). In Fig. 11.3(a), the length of the line from 0 to $X1$ denotes the sum of squares by fitting $X1$ alone (similarly for $X2$ and $X3$ in Fig. 11.3(b) and (c) respectively). The second point in Fig. 11.3(a), labelled $X2 + X3$, denotes the sum of squares explained by fitting $X2 + X3$ (or alternatively working downwards,

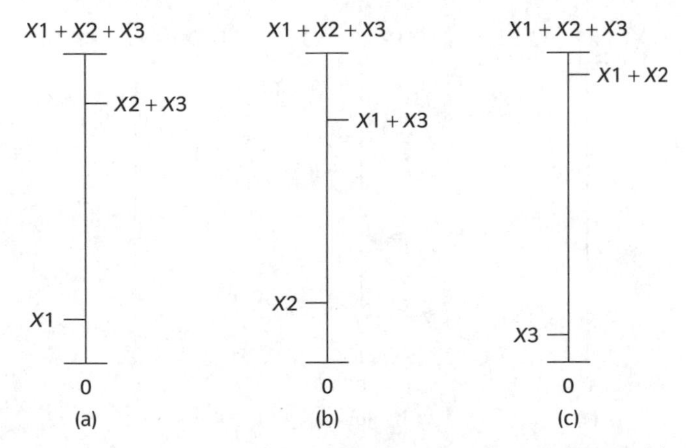

Fig. 11.3 Forwards and backwards regression would produce the same best model.

the difference between $X1 + X2 + X3$ and $X2 + X3$ illustrates the explanatory power lost when $X1$ is removed from the full model). Likewise for $X1 + X3$ and $X1 + X2$ in Fig. 11.3(b) and (c). So which 'best models' would be produced by forwards and backwards stepwise regression?

Forwards stepwise regression would choose $X2$ as the first variable to include, as the distance $0 \rightarrow X2$ is the greatest. The next step would be to jump from $X2$ in Fig. 11.3(b) to either $X2 + X3$ in Fig. 11.3(a) or $X2 + X1$ in Fig. 11.3(c). The latter jump is the greatest, so $X1$ would be entered into the model. Finally, the program would consider if it also needed to include $X3$, but the explanatory power gained is not great (let us define it as insignificant), so the program would stop. Forwards stepwise regression has produced the model $X1 + X2$.

Backwards stepwise regression would start with the full model and drop the least significant term first—in this case $X3$ (as Fig. 11.3(c) has the shortest distance at the top). Dropping either $X1$ or $X2$ would lead to a large loss of explanatory power, so the program would stop here. The same model has been produced by forwards and backwards procedures in this instance.

Figure 11.4(a)–(c) illustrates an alternative scenario. In this case, forward stepwise regression would choose $X2$ first, followed by $X1$ as before, and then stop. However, backwards stepwise regression would drop $X2$ first (Fig. 11.4(b) has the shortest distance at the top), and then stop. In this case the direction of the stepwise regression has influenced the final result. In the second example, $X2$ shares information with $X1 + X3$, so while it is the most significant in the three models containing one variable, it also makes the least difference to the explained sum of squares when removed from the full model.

In summary, many packages contain such automated procedures, so it is important to be aware of them. However, once you have become familiar with the problems and issues surrounding model choice, it is usually preferable to fit a series of General Linear Models, and be more actively involved in the model choice process yourself. The whale watching data set is revisited to illustrate this alternative approach.

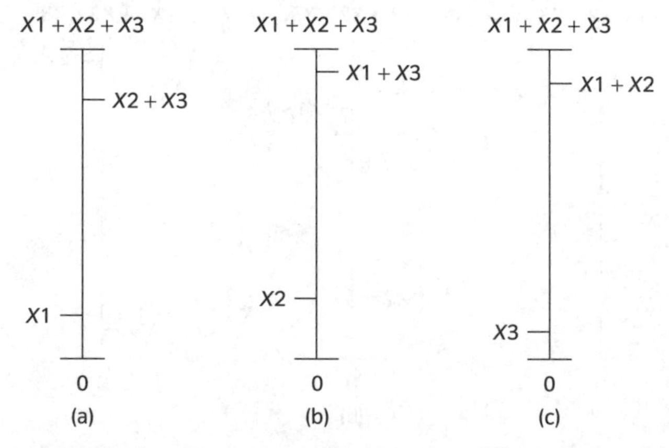

Fig. 11.4 Forwards and backwards regression would produce different best models.

11.4 Whale Watching: using the GLM approach

An employee was given the task of reanalysing the dataset, with the aim of making recommendations about when to cancel trips. The first analysis she conducted is illustrated in Box 11.7.

The employee has been quite restrained in this first analysis. By selecting only five variables for inclusion in the initial model, she has attempted to minimise any problems with multiplicity of p-values. She has focused on those variables that could have potential predictive value. For example, cloud cover, rainfall and visibility at 8 a.m. are included, but not rain and visibility during the course of the trip (as by then it would be too late to cancel). As a result of this initial analysis, visibility at 8 a.m. is highly significant ($p < 0.0005$), and there are clearly good and bad years ($p < 0.022$).

The next step is to simplify the model. Can we do this in one step, by simply picking the variables that are significant in this first model? In other words, can we be sure that knowing VIS8AM, there is no extra information to be gained from RAIN8AM or CLOUD8AM? Unfortunately the answer is no, we have to be more cautious than that.

The differences between the sequential and adjusted sums of squares indicate two points. The first is that there is a considerable drop between the sequential and adjusted sums of squares for both CLOUD8AM and RAIN8AM, indicating that a number of the explanatory variables share the same information. The second point is that the sequential sums of squares are also quite high, suggesting that alone, or in a reduced model, one or other or both of these variables could be significant. Part of the challenge of model simplification in such multiple

BOX 11.7 First GLM analysis of whale watching dataset

General Linear Model

Word equation: LRGWHAL = YEAR + MONTH + CLOUD8AM + RAIN8AM + VIS8AM

YEAR and MONTH are categorical, CLOUD8AM, RAIN8AM and VIS8AM are continuous

Analysis of variance table for LRGWHAL, using Adjusted SS for tests

Source	DF	Seq SS	Adj SS	Adj MS	F	P
YEAR	2	6.7923	2.9032	1.4516	3.88	0.022
MONTH	4	6.1211	3.5533	0.8883	2.38	0.053
CLOUD8AM	1	4.4622	0.1864	0.1864	0.50	0.481
RAIN8AM	1	9.7026	0.5748	0.5748	1.54	0.216
VIS8AM	1	23.8503	23.8503	23.8503	63.80	0.000
Error	222	82.9960	82.9960	0.3739		
Total	231	133.9245				

BOX 11.8 Second GLM analysis of whale watching dataset

General Linear Model

Word equation: LRGWHAL = YEAR + MONTH + CLOUD8AM + VIS8AM

YEAR and MONTH are categorical, CLOUD8AM and VIS8AM are continuous

Analysis of variance table for LRGWHAL, using Adjusted SS for tests

Source	DF	Seq SS	Adj SS	Adj MS	F	P
YEAR	2	6.7923	2.6981	1.3490	3.60	0.029
MONTH	4	6.1211	3.6100	0.9025	2.41	0.050
CLOUD8AM	1	4.4622	0.0293	0.0293	0.08	0.780
VIS8AM	1	32.9781	32.9781	32.9781	88.00	0.000
Error	223	83.5708	83.5708	0.3748		
Total	231	133.9245				

regression problems is to tease apart those variables that are not necessary in the model from those which have valuable predictive power.

So in solving this problem, the order in which the variables appear in the model could be important. The drop in the adjusted sum of squares compared to the sequential sum of squares must be due to terms that follow that variable in the model. As far as RAIN8AM is concerned, this drop must be due to the inclusion of VIS8AM. On the other hand, VIS8AM is significant, even when the other four variables are known. This suggests that VIS8AM should be retained, and that RAIN8AM should be dropped.

The effect of removing RAIN8AM on the significance of the remaining variables cannot however be predicted. The corresponding drop in the adjusted SS compared to the sequential SS for CLOUD8AM could be due to either RAIN8AM, VIS8AM or both. If it is largely due to RAIN8AM, then when it is removed, it is possible that CLOUD8AM becomes significant. Thus model simplification should proceed by removing one term at a time, the next step being illustrated in Box 11.8.

The variable CLOUD8AM remains insignificant on removal of RAIN8AM, so it is now appropriate to remove it as well. This leads to the third model of Box 11.9, in which VIS8AM remains highly significant, YEAR significant at $p = 0.028$, and MONTH has become significant (at $p = 0.045$). The question is now whether we should simplify the model any further.

There are considerations other than statistical ones that should be considered at this point. MONTH is only just significant in the reduced model ($p = 0.045$), but it is also a cheap variable to be included. It is very easy to enter the code for the current month, involving no resources in its measurement. Biologists studying the whales might also expect some seasonal variation. A second consideration is that there are clearly significant differences between the years, but at the beginning of the season, there will be insufficient data to indicate if it is a good year

BOX 11.9 Third analysis of the whale watching dataset

General Linear Model

Word equation: LRGWHAL = YEAR + MONTH + VIS8AM

YEAR and MONTH are categorical, VIS8AM is continuous

Analysis of variance table for LRGWHAL, using Adjusted SS for tests

Source	DF	Seq SS	Adj SS	Adj MS	F	P
YEAR	2	6.7923	2.7000	1.3500	3.62	0.028
MONTH	4	6.1211	3.6917	0.9229	2.47	0.045
VIS8AM	1	37.4110	37.4110	37.4110	100.24	0.000
Error	224	83.6001	83.6001	0.3732		
Total	231	133.9245				

Term	Coef	StDev	T	P
Constant	−4.44327	0.07182	−61.87	0.000
YEAR				
1	−0.08330	0.06238	−1.34	0.183
2	0.15006	0.05583	2.69	0.008
3	*−0.06676*			
MONTH				
5	0.1177	0.1230	0.96	0.339
6	0.01618	0.07803	0.21	0.836
7	0.18729	0.07194	2.60	0.010
8	−0.07378	0.08984	−0.82	0.412
9	*−0.24739*			
VIS8AM	0.10217	0.01020	10.01	0.000

or a bad one. Towards the end of the season, as data for that particular year accumulates, then the variable YEAR will have greater predictive power. So even though it is significant, its usefulness will depend upon the time in the season.

The General Linear Model approach has produced a rather different model from the stepwise regression approach illustrated earlier. One notable difference is the complete absence of the variable YEAR in the explanatory variables chosen by the automated procedure. The explanation for this is very simple, as the stepwise regression procedure treated all variables as continuous. Looking at the chosen model of Box 11.9, the coefficient table illustrates how the number of whales spotted varies from year to year. The deviations from the mean are −0.083, 0.150, −0.067 (the last one being calculated so that deviations from the grand mean sum to zero—see Section 3.2). Year 2 was the best year, with years 1 and 3 being below average, as can be seen in Fig. 11.5.

In the stepwise regression, YEAR was treated as continuous, while in the GLM, it was categorical. Given the way in which the number of whales spotted changes over the three year period, it can be seen that any straight line fitted

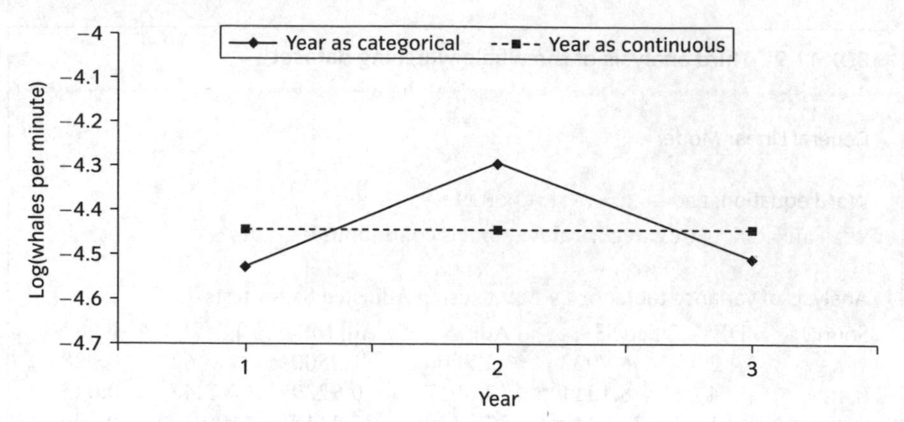

Fig. 11.5 Differences in whale spotting over three years.

through these points would have a slope of zero. This is a good example of when it is not appropriate to treat a categorical variable as continuous, and then only test for linear trends. This emphasises the importance of not only looking at the significance of variables in an analysis, but also the direction and patterns of their effects. This will ensure that the type of analysis conducted is both the most appropriate and the most sensitive.

In summary, this GLM analysis employed the principles of model choice in the context of a multiple regression. By limiting the number of variables included in the initial model on an exploratory basis, the employee attempted to focus on relevant and possibly useful variables, and so minimise problems of multiplicity. Model simplification worked towards economising on the variables to be incorporated in the model, but included other considerations in addition to statistical ones.

11.5 Summary

This chapter has discussed multiple regression models in which the best model is chosen from a subset of possible explanatory variables.

- Interactions are rarely fitted in multiple regression problems. Considerations of marginality are therefore not often encountered in this context.
- We have discussed two of the many measures of how economically a model fits the data, namely (1) adjusted R^2 and (2) prediction intervals.
- Multiplicity of p-values is a great problem in multiple regression. Possible solutions include: (1) focusing on the questions of primary interest rather than indulging in speculative fishing for significance; (2) making the conditions for significance more stringent; (3) combining sums of squares to reduce the number of questions asked.

- The dangers of automatic model selection procedures are: (1) they rely purely on statistical considerations, whereas it is usual for other factors to also be important; (2) they tempt you to rely on default criteria for selection, but it is possible to increase stringency as appropriate; (3) they tempt you to throw in all possible variables, rather than focus on specific question; (4) forwards and backwards procedures can produce different 'best' models. There are circumstances in which stepwise regression could be useful, e.g. when you have lots and lots of X variables in a purely exploratory study, with no prior expectations or constraints.
- An example of model simplification using the GLM approach is given, employing the principles of model choice.

11.6 Exercises

Finding the best treatment for cat fleas

The efficacy of two proprietary treatments for cat fleas was compared by a journalist for a pet magazine. In a survey, households with cats were asked to chose a 'focal cat', and to report information which has been stored in the *cat fleas* dataset as the following variables:

NCATS the number of cats in the household

CARPET whether the focal cat was (CARPET = 1) or was not (CARPET = 2) allowed in rooms with carpets

FLEAS the average density of fleas on the focal cat over the year

TRTMT a code of 1 or 2 for the two proprietary treatments being compared

HAIRL the length in mm of the hair on the focal cat's back

Initially, the journalist performed the simplest analysis using LOGFLEAS as the response variable to ask if the two treatments differed. He then conducted other analyses with both FLEAS and LOGFLEAS as response variables. These are displayed in Box 11.10.

BOX 11.10(a) Cat flea analysis with minimal model: Analysis 1

General Linear Model

Word equation: LOGFLEAS = TRTMT

TRTMT is categorical

Analysis of variance table for LOGFLEAS, using Adjusted SS for tests

Source	DF	Seq SS	Adj SS	Adj MS	F	P
TRTMT	1	1.6118	1.6118	1.6118	1.68	0.199
Error	87	83.5915	83.5915	0.9608		
Total	88	85.2033				

BOX 11.10(b) Cat flea analysis with full model for FLEAS: Analysis 2

General Linear Model

Word equation: FLEAS = TRTMT + HAIRL + NCATS + CARPET

TRTMT and CARPET are categorical, HAIRL and NCATS are continuous

Analysis of variance table for FLEAS, using Adjusted SS for tests

Source	DF	Seq SS	Adj SS	Adj MS	F	P
TRTMT	1	1 412	18 438	18 438	2.72	0.103
HAIRL	1	17 463	4	4	0.00	0.981
NCATS	1	188 947	188 994	188 994	27.90	0.000
CARPET	1	32 498	32 498	32 498	4.80	0.031
Error	84	568 947	568 947	6 773		
Total	88	809 267				

BOX 11.10(c) Cat flea analysis with full model for LOGFLEAS: Analysis 3

General Linear Model

Word equation: LOGFLEAS = TRTMT + HAIRL + NCATS + CARPET

TRTMT and CARPET are categorical, HAIRL and NCATS are continuous

Analysis of variance table for LOGFLEAS, using Adjusted SS for tests

Source	DF	Seq SS	Adj SS	Adj MS	F	P
TRTMT	1	1.6118	5.7573	5.7593	9.03	0.004
HAIRL	1	1.1523	0.1642	0.1642	0.26	0.613
NCATS	1	22.5843	22.5915	22.5915	35.41	0.000
CARPET	1	6.2605	6.2605	6.2605	9.81	0.002
Error	84	53.5944	53.5944	0.6380		
Total	88	85.2033				

(1) What conclusions do you draw from Box 11.10a?

(2) Calculate R^2 for Analyses 2 and 3 (Box 11.10b and c).

(3) From the evidence given above, and drawing from your answer to Question 2, explain whether you would expect model criticism techniques to show the second or third model to be the better model.

(4) Have extra explanatory variables helped or hindered in the treatment comparison? Are all these variables needed? Using the *cat fleas* dataset, produce the final model you would use to predict most efficiently the average density of fleas on the focal cat's back. Compare the raw and adjusted R^2 of this model with that of Analysis 3.

Multiplicity of *p*-values

Construct a null dataset using the following routine:

Pick 30 numbers at random from a normal distribution with mean 10.5 and a standard deviation of 2.4, and place them in a column. (The exact code you need to do this will depend upon the package, and is given in the supplements). Repeat this procedure 11 times.

This dataset could represent measuring eleven variables on each of 30 subjects. Given that the data have been created by drawing values at random, we know that they are quite unrelated. Let us imagine however that this is a multiple regression problem, and analyse the data with the first column as the response variable, and the remaining 10 columns as predictors. For each analysis, an *F*-ratio test may be conducted for the regression as a whole, testing the grand null hypothesis that all the variables are irrelevant. An *F*-ratio test may also be conducted for each explanatory variable in the word equation.

Repeat this analysis with new random variables 10 or 20 times.

(1) Record the *p*-value for the whole regression. How often is the whole regression significant?

(2) Record the significance of the most significant variable for each analysis. How do the *p*-values compare to those recorded for Question 1? Which set of *p*-values is providing more useful information and why?

(3) Record R^2 and adjusted R^2 for each analysis. Given that you know your dataset contains unconnected variables, which of these measures provides you with the most useful information?

(4) Given the four sets of output that you have recorded (two sets of *p*-values and two types of R^2), what pattern would they show as we fitted models with 1, 2, 3, 4, etc. random *X* variables?

12 Random effects

This chapter serves as an introduction to random effects for biologists. It will provide an understanding of what random effects are, and how to recognise them, and we present some simple examples. There is a wealth of theory and many parts of packages that deal with this subject in much greater depth, and in more sophisticated ways, than we do here.

Random effects are easier to interpret if the design is orthogonal. If we lose orthogonality, some or all of the *F*-tests become approximate, because the *F*-ratio we calculate no longer has an *F*-distribution under the null hypothesis. An approximation to the actual distribution under the null hypothesis is found by finding the 'nearest' *F*-distribution, which may have non-integer DF in the numerator. So you may come across an *F*-ratio which has 4.32 and 12 DF.

Dealing with random effects requires familiarity with four new concepts. Three examples will follow, which should encompass the most common situations in which random effects will be encountered by biologists.

12.1 What are random effects?

Distinguishing between fixed and random factors

Random effects are categorical variables whose levels are viewed as a sample from some larger population, as opposed to fixed effects, whose levels are of interest in their own right. Random effects are often variables whose levels are biological individuals, such as humans, pigs or plants. An example would be plants of a particular species which may vary in their level of calcium. To answer this question, we would go out and take a random sample of ten plants. Each plant would have several samples taken from it to measure calcium concentration (CACONC). In comparing the CACONC within and between plants, we could determine if those plants differed in CACONC. The word equation to answer this question would read:

$$\text{CACONC} = \text{PLANT}.$$

In drawing conclusions from this analysis, we would wish to extrapolate from our sample of 10 plants, to plants of the same species in a wider population.

This contrasts with the kinds of analyses covered in earlier chapters, where we have been dealing with **fixed effects**. For example, in investigating the impact of a certain fertiliser on the calcium content of plants, ten plants might be allocated to two treatments. One reading would be taken per plant, and the data would be analysed by:

CACONC = TREATMENT.

The conclusions would relate to whether these specific treatments influenced the calcium concentration of plants. In the first analysis, PLANT is a random factor and multiple readings were taken per plant, while in the second, TREATMENT is a fixed factor, and individual plants have been treated as replicate units (with only one measurement being taken per plant).

There are two properties which distinguish fixed and random effects. The first is **desired inference**. Will the statistical test extrapolate your conclusions to some wider population, or draw conclusions only relevant to the particular experimental conditions? In the first example, plants were defined as a factor in the word equation, and would have been coded as 1 to 10 for the ten different plants. However, we are not particularly interested in those ten plants, but consider them as a random sample of plants of that species. If PLANT proved to be a significant factor, we would conclude that plants of that species differ significantly in their calcium content. The randomness of the sample is very important in principle—there must be an unbiased link between the sample and the population for the conclusions to be valid. In the second experiment, however, the fertiliser treatments (coded as 1 and 2) refer to two specific fertilisers. If the analysis proved significant, we would conclude that those particular fertilisers cause a difference in the calcium content of plants. Our conclusions would refer only to those two particular fertilisers, and not to fertilisers in general. However, we often do extrapolate (or interpolate, which may be more justified) when we have fixed effects for explanatory variables. Take an experiment for which TEMP is an explanatory variable as an example —if we find a difference at 5, 10 and 15 degrees Celsius, then we would conclude that there is likely to be a difference at 12 degrees. In this case however it is the researcher extrapolating from the model. With random effects, it is the model itself which extrapolates to a wider population, in estimating the expected variance between individuals in that population.

The second property to distinguish between fixed and random effects is that of **repetition**. If you performed a similar experiment, what would have to be the same to make the experiment a repeat? In the case of fixed effects, such as fertilisers, the experiment must have the same fertilisers, so the levels of the fixed effect must be the same. To use different fertilisers would be a new experiments, not a repeat of the old one. In the case of comparing the calcium concentration of plants, it is the population that must be the same for the experiment to count as a repeat. The plants can be different, but if the population was different, then that would be asking a different question.

So when considering whether a factor is fixed or random, ask the questions: (1) to what do the conclusions apply?; (2) what needs to be repeated exactly to ask the same question? Random factors are often biological individuals, but can also be farms, schools etc.

Why does it matter?

The word equations that we have fitted to data so far follow the general form:

$$Y = \textit{fixed effects} + \textit{error}.$$

The variance in Y is partitioned between that which can be explained by fixed effects, and that which remains unexplained. The error term is the only term on the right hand side of the equation to have variance (that is, it is the only term that would vary in repetitions of the study). As discussed in Chapter 8, it is also the term which determines the independence of each datapoint. One of the assumptions of GLMs is that the error contributions of each data point are independent. However, if a model contains a random factor, it may look like this:

$$Y = \textit{fixed effects} + \textit{random effects} + \textit{error}.$$

The random factor now also has variance. If this factor is completely unimportant, then the parameters that relate this factor to Y are all zero, and this term falls out of the equation. However, if the random factor is important, the individuals are different. The variation due to this factor will now depend upon which individuals have been randomly sampled from the population, and will also vary in repetitions of the study. Now we need to partition the variation into that which is due to the random factor, and that which is due to error. It is as if we have two error terms in the model, and this will complicate hypothesis testing.

Four new concepts will be introduced to help you understand what happens when a model has more than one error term.

12.2 Four new concepts to deal with random effects

Components of variance

In all the models we have used so far, there has been only one variance to estimate—the error variance. But for each random factor in a model, there is also a variance associated with that factor. For example, in experiments investigating how birds forage for food, ten starlings were trapped in Wytham

Woods. They were each set five different foraging tasks to see how fast they would learn. The model fitted could then be:

$$\text{LEARNTIME} = \text{STARLINGS}.$$

This model will tell us if starlings learn at different rates. However, the variance in learning speed amongst starlings can also be estimated. There are two reasons why an estimate of this second variance would be useful: (1) the magnitude of this variance would be informative in its own right; (2) a comparison of its magnitude with that of the error variance would also be useful. If starling variance is much greater than error variance, then this would indicate that the starlings are very variable material. Any future experiments would benefit from using a larger number of starlings. If starling variance is low compared to error variance, then our techniques for measuring one learning time may not be very accurate, and to improve the experiment these measurement techniques require modification.

So in summary, we now have two components of variance, that due to error and that due to the random factor, and we wish to estimate both.

Expected mean square

The concept of the expected mean square is already a familiar one. In an ANOVA table with fixed effects, the mean square is calculated by dividing the sum of squares by the appropriate degrees of freedom. The resulting mean square may be thought of as the variation attributable to that factor per degree of freedom. The reason why it is mentioned again here is that there is a link between components of variance and expected mean squares. In fact simple additive formulae connect variance components and expected mean squares.

The expected mean square of any factor has two components to it: (1) the variation attributable to that factor (e.g. differences between fertilisers, differences between starlings); (2) the variation attributable to error. For a typical fixed effect (such as the fertiliser experiment of Chapter 1) this may be expressed as:

$$\text{Expected mean square for fertiliser} = 10\frac{\gamma_1^2 + \gamma_2^2 + (-\gamma_1 - \gamma_2)^2}{3-1} + \sigma_E^2$$

where γ_i = the true deviation of the mean yield obtained under Fertiliser i from the grand mean yield, and $3-1$ are the degrees of freedom for fertiliser. This is the mean square for fertiliser, as first discussed in Chapter 1. The parameters γ_i used in this equation are fixed values, and our null hypothesis was that $\gamma_1 = \gamma_2 = 0$.

In contrast, the expected mean square for a random factor allows us to estimate the variation between, for example, our ten starlings. The ten starlings were measured on each of five occasions. The variation in learning time between the starlings is an estimate of the expected variation between starlings in the source population σ_S^2 (from which they were picked). The formula for the expected mean square between starlings is as follows:

$$\text{Expected mean square for starlings} = 10\sigma_S^2 + \sigma_E^2$$

where σ_E^2 = the error variance. So the parameter being estimated in this case is a variance (σ_S^2) of a wider population rather than a fixed effect (γ_i) of the particular levels in the experiment. Here our null hypothesis is $\sigma_S^2 = 0$.

Nesting

Fixed factors are usually **crossed**, while random factors are frequently **nested**. In a factorial design with 3 crop varieties and 3 fertilisers, every combination of fertiliser and variety will be represented, as illustrated in Table 12.1.

Table 12.1 Crossed experimental design

	Variety 1	Variety 2	Variety 3
Fertiliser 1	✡	✡	✡
Fertiliser 2	✡	✡	✡
Fertiliser 3	✡	✡	✡

This is often not the case however with random factors. In an experiment with six blue tits, there are two different foraging schedules, three birds being tested under each schedule. The aim of the experiment is to discover if the learning rate is faster under Schedule 1 or 2. BLUETIT is a random effect, and SCHEDULE is a fixed effect. The birds may be coded as 1 to 6. However, now not all combinations are represented—each blue tit is combined with only one schedule, as the blue tits are nested within schedule. See Table 12.2.

Table 12.2 Nested experimental design

	Blue tit 1	Blue tit 2	Blue tit 3	Blue tit 4	Blue tit 5	Blue tit 6
Schedule 1	✡	✡	✡			
Schedule 2				✡	✡	✡

Alternatively, packages usually allow a coding of nested variables in which numbers are re-used, in this case numbering blue tits 1 to 3 within SCHEDULE 1, and re-using those numbers for the remaining blue tits as 1 to 3 within SCHEDULE 2. This can be confusing if you forget there are actually six blue tits!

Table 12.3 Blue tits nested within two fixed factors

		Reward Rate	
		1	2
Schedule	1	Blue tits 1, 2 & 3	Blue tits 4, 5 & 6
	2	Blue tits 7, 8 & 9	Blue tits 10, 11 & 12

To incorporate a nested factor into a model formula brackets are used. The variable that is nested comes first, and the variable it is nested inside is given inside the brackets. For example:

LEARNRATE = SCHEDULE + BLUETIT (SCHEDULE).

The last term should be read as BLUETIT nested within SCHEDULE. This can become more complicated. For example, if there were also two different reward rates (RWDRATE) as well as two different schedules, and these were in a factorial design, the model formula would read:

LEARNRATE = SCHEDULE | RWDRATE + BLUETIT (SCHEDULE RWDRATE).

This would indicate that BLUETIT was nested within SCHEDULE and REWARD RATE as shown in Table 12.3. There would still be several readings for each blue tit.

Appropriate denominators

In a fixed effects ANOVA, the variation per degree of freedom for the fixed effect is compared to the variation per degree of freedom for error, to obtain the *F*-ratio. The appropriate denominator is always the error mean square. If a random factor is involved however it may be that the error mean square is not the appropriate denominator for some of the terms in the ANOVA table.

For example, if the model is a nested one, there will be different **appropriate denominators** for different levels in the nested design. This is best illustrated with an example. The calcium concentration is measured for a number of leaves on a number of plants. For each plant, five leaves are taken, and from each leaf, three discs are cut. The calcium content is measured separately for each disc. This experiment has a nested design, with discs at the lowest level of nesting, then leaves are nested within plants. In this experiment, two questions are asked: (1) Does the calcium content vary between leaves of a plant?; (2) Does the calcium content vary between plants? To answer the first question, we need to compare the variance within a leaf (between discs from the same leaf) to the variance between leaves (from the same plant). To answer the second, we need to compare the variance between leaves within a plant to the variance between leaves on different plants. These two variance ratios have two different denominators, and there will be a different denominator for each level of nesting. Unfortunately, there may also be some more complex analyses involving random effects, where an appropriate denominator cannot be found, so a synthetic denominator will be constructed. Under these circumstances the *F*-ratio becomes approximate.

These four concepts will now be illustrated by three examples.

12.3 A one-way ANOVA with a random factor

This example illustrates a random factor in its simplest form. The calcium content has been measured in just four leaves, by taking four discs per leaf (Fig. 12.1). The leaves are taken as a random sample from a larger population of leaves, and so are considered a random factor. The data are stored in the *leaves* dataset. This example illustrates how to analyse a random factor, and introduces the link between the components of variance and the expected mean squares. The analysis is illustrated in Box 12.1.

The input

The model formula is very simple: `CACONC` = `LEAF`. However, as well as specifying the model, we need to specify which factors are random (just `LEAF` in this example). The tables of expected mean squares and variance components are given in the output, and you may need to request these.

The ANOVA table

The ANOVA table with random factors is just the same. In this case, the conclusion would be that there are significant differences between the leaves in their calcium content ($p = 0.002$). Further information can be obtained from the 'expected mean squares' table which follows.

The expected mean squares

This second table lists the same sources as in the ANOVA table, prefixed by a reference number. In this example, there are only two sources, `LEAF` and

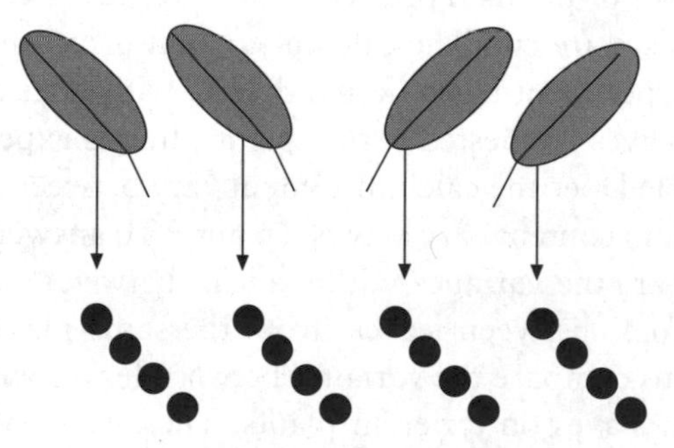

Fig. 12.1 Experimental design for one way ANOVA with leaf as a random factor.

BOX 12.1 One-way ANOVA with a random factor

General Linear Model

Word equation: CACONC = LEAF

LEAF is a random factor

Analysis of variance table for CACONC, using Adjusted SS for tests

Source	DF	Seq SS	Adj SS	Adj MS	F	P
LEAF	3	0.28086	0.28086	0.09362	8.70	0.002
Error	12	0.12910	0.12910	0.01076		
Total	15	0.40996				

Expected Mean Squares using Adjusted SS

Source	Expected Mean Square for each term
1 LEAF	(2) + 4.0000(1)
2 Error	(2)

Error terms for tests using Adjusted SS

Source	Error DF	Error MS	Synthesis of Error MS
1 LEAF	12.00	0.01076	(2)

Variance components using Adjusted SS

Source	Estimated value
LEAF	0.02072
Error	0.01076

error. It also provides a formula which links the variance components with the expected mean squares. In this case, if the variance component for error is denoted σ_E^2 and that for LEAF as σ_L^2, then the link between the expected mean square for LEAF and the variance components is:

$$4\sigma_L^2 + \sigma_E^2,$$

that is: 4 × LEAF *variance component* + *error variance component*. An estimate of each variance component is given below.

The error terms for tests

This table tells us which denominator was used in the *F*-ratio test. In this case, it was the error mean square (just as if LEAF had been a fixed effect). The null hypothesis being tested is that there is no variation between the leaves other than that due to sampling error. If this is the case then $\sigma_L^2 = 0$, and the ratio $\frac{4\sigma_L^2 + \sigma_E^2}{\sigma_E^2}$ would be equal to 1 (as all other *F*-ratios have been under the null hypothesis).

The variance components

The variance component for error (σ_E^2) is the same as the error mean square, and this represents the variation between discs from the same leaf. The variance component for leaf (σ_L^2) is 0.02072. Using the expected mean squares formulae, we can link the variance components to the expected mean squares as follows:

$$\text{Expected mean square for LEAF} = 4 \times 0.02072 + 0.01076 = 0.0936.$$

The LEAF component of variance is multiplied by 4 because there are four discs per leaf. The piecing together of these components is illustrated in Box 12.2, an extract of the output from Box 12.1.

Additional conclusions

In this particular example, we find that the variance in leaf to leaf calcium concentration is 0.0207, while disc to disc variation is estimated at 0.0108.

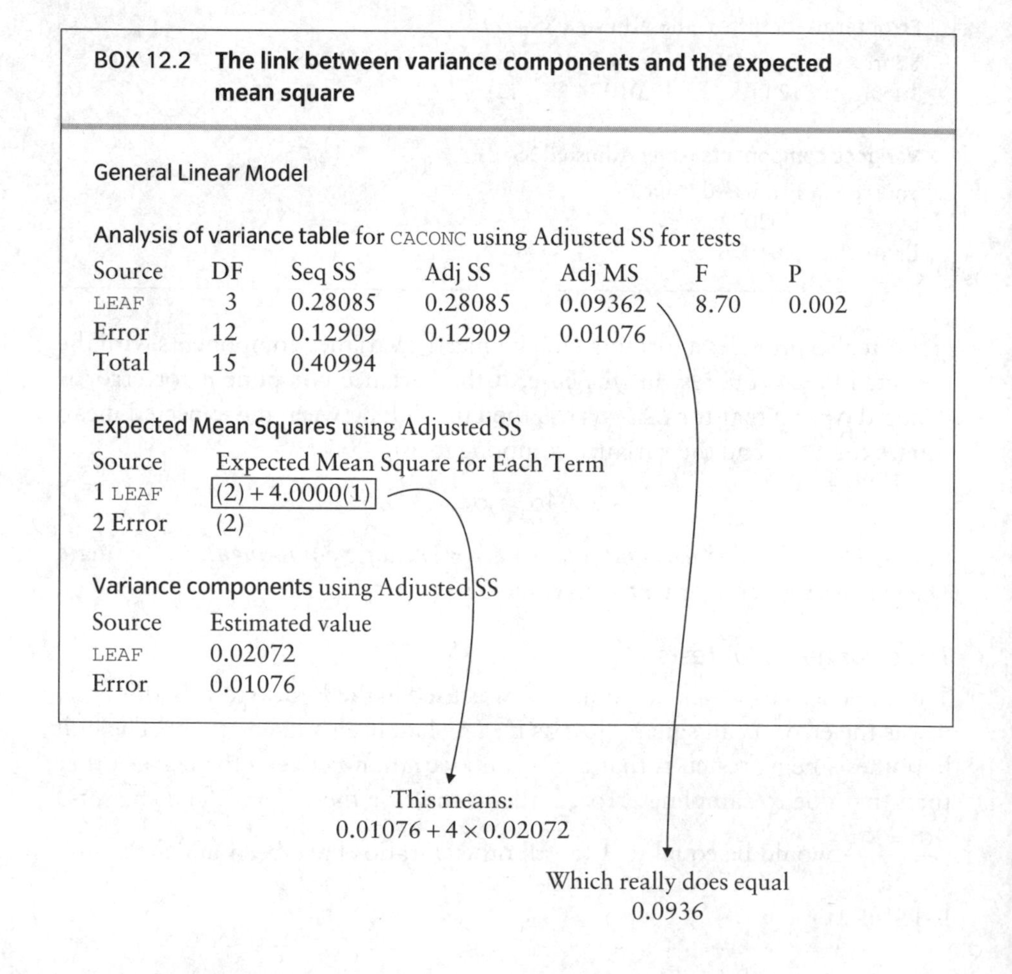

BOX 12.2 The link between variance components and the expected mean square

General Linear Model

Analysis of variance table for CACONC using Adjusted SS for tests

Source	DF	Seq SS	Adj SS	Adj MS	F	P
LEAF	3	0.28085	0.28085	0.09362	8.70	0.002
Error	12	0.12909	0.12909	0.01076		
Total	15	0.40994				

Expected Mean Squares using Adjusted SS

Source	Expected Mean Square for Each Term
1 LEAF	(2) + 4.0000(1)
2 Error	(2)

Variance components using Adjusted SS

Source	Estimated value
LEAF	0.02072
Error	0.01076

The variation between leaves is therefore greater than the variation within leaves. This information would suggest that in future experiments an increase in effort may be best directed in increasing the number of leaves analysed.

So to summarise this example, a null hypothesis has been tested as before, but additional information about the wider population has also been obtained. This information about the variability of the population is of value in itself, but also assists in designing future experiments.

12.4 A two-level nested ANOVA

This example is an extension of the last, with an extra layer in the experimental design. The data are stored in the *leaves nested within plants* dataset and analysed in Box 12.3.

Nesting

Once again, calcium content of leaves is being measured, but on this occasion there are four different plants. From each plant, three leaves are taken, and from each leaf, two discs, as illustrated in Fig. 12.2.

As discs are at the lowest level of nesting, variation between the discs will be taken as error variation (as before). Moving up to the next level of nesting, discs from groups A and B have two reasons to vary—because they come from different leaves, but also because of sampling error. Comparing discs in groups A and C, there are now three reasons to vary: plant to plant variation, leaf to leaf variation and error. These additional components of variance are reflected in the expected mean squares at each level.

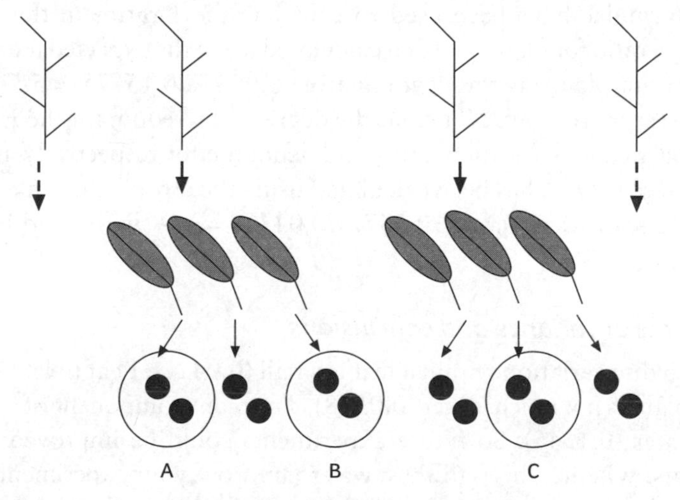

Fig. 12.2 A three-tiered nested experimental design.

Link between expected mean squares and components of variance

At the lowest level of nesting, there is only one component of variance, error, σ_E^2. The expected mean square for error is therefore equal to σ_E^2. At the next level, however, the LEAF2 mean square will contain two components of variance, leaf (σ_L^2) and error. There are two discs per leaf, so the linking formula is:

$$\text{LEAF2 expected mean square} = 2\sigma_L^2 + \sigma_E^2.$$

At the highest level of the plant, there are six discs per plant, so the linking formula becomes:

$$\text{PLANT2 expected mean square} = 6\sigma_P^2 + 2\sigma_L^2 + \sigma_E^2.$$

These formulae may be obtained from the output shown in Box 12.3. Substituting in the appropriate components of variance, gives, for example:

PLANT2 expected mean square = 6 × 0.12450 + 2 × 0.05778 + 0.04218 = 0.9048 as in the ANOVA table (to 4 decimal places).

Appropriate denominators for the F-tests

The formulae for the expected mean squares illustrate how the *F*-ratios are constructed. There are two hypotheses being tested. The first is that $\sigma_L^2 = 0$, in which case the *F*-ratio $\dfrac{2\sigma_L^2 + \sigma_E^2}{\sigma_E^2}$ would be close to 1 (differing only for reasons of sampling error). The appropriate denominator is therefore the error mean square as before. The second null hypothesis however is that $\sigma_P^2 = 0$, in which case the *F*-ratio $\dfrac{6\sigma_P^2 + 2\sigma_L^2 + \sigma_E^2}{2\sigma_L^2 + \sigma_E^2}$ would again be close to 1. The appropriate denominator is now the expected mean square for LEAF2. The *F*-ratio is effectively looking at the variation between plants compared to the variation within plants, to ask if plant is a significant cause of variation.

These formulae have been used to calculate the *F*-ratios in the ANOVA table. The *F*-ratio for plant has been calculated using the expected mean square for leaf within plants as the denominator: 0.90478/0.15775 = 5.74, with 3 and 8 degrees of freedom. Note that the degrees of freedom for the *F*-ratio are those associated with the numerator and denominator respectively as always. The *F*-ratio for LEAF2 has been calculated using the error mean square, which represents disc to disc variation: 0.15775/0.04218 = 3.74, with 8 and 12 degrees of freedom.

Components of variance and conclusions

The disc to disc variation within a leaf is small (0.0422). Leaf to leaf variation within a plant is not much larger (0.0578). The plant component of variance is much greater (0.1245). So future experiments could be improved by using more plants. Whether this is the best way to improve your experimental design will depend upon not only the relative magnitudes of the components of

BOX 12.3 Two-way nested analysis of variance

General Linear Model

Word equation: CACONC2 = PLANT2 + LEAF2 (PLANT2)

PLANT2 and LEAF2 are random factors

Analysis of variance table for CACONC, using Adjusted SS for tests

Source	DF	Seq SS	Adj SS	Adj MS	F	P
PLANT2	3	2.71433	2.71433	0.90478	5.74	0.022
LEAF2 (PLANT2)	8	1.26200	1.26200	0.15775	3.74	0.020
Error	12	0.50621	0.50621	0.04218		
Total	23	4.48254				

Expected Mean Squares using Adjusted SS

Source	Expected Mean Square for each term
1 PLANT2	(3) + 2.0000(2) + 6.0000(1)
2 LEAF2 (PLANT2)	(3) + 2.0000(2)
3 Error	(3)

Error terms for tests using Adjusted SS

Source	Error DF	Error MS	Synthesis of Error MS
1 PLANT2	8.00	0.15775	(2)
2 LEAF2 (PLANT2)	12.00	0.04218	(3)

Variance components using Adjusted SS

Source	Estimated value
PLANT2	0.12450
LEAF2 (PLANT2)	0.05778
Error	0.04218

variance, but also the relative costs of increasing the number of individuals or the number of measurements per individual. In spite of the fact that there are only 3 and 8 degrees of freedom for the factor plant, both the variation between plants and the variation between leaves on the same plant, are significant ($p = 0.022$ and 0.020 respectively).

This example is of a nested experiment of three tiers. However, the chain of nesting can go on indefinitely. In each case, the appropriate denominator for the *F*-ratio at one level of the chain is provided by the level immediately below it. This is logical, because in moving from one level of nesting to the next, you are adding one component of variance, and you wish to test if that component is zero. You cannot construct an *F*-ratio for the bottom level of the chain (in this case, discs), as there is no replication within it.

12.5 Mixing random and fixed effects

This example illustrates how to mix random and fixed effects. It also demonstrates that biological individuals need not be random factors, and that the distinction between being a random or fixed factor very much depends upon the questions you are interested in.

A pig farmer is interested in comparing the genetic quality of his male pigs. She has five sires and ten dams, and allocates two dams to each sire. The weight of two piglets per litter is then monitored, and their weight gain over a two week period measured. The data are stored in the *dams and sires* dataset. The farmer is interested in these particular sires (presumably to select for stud or to set stud fees), and `SIRE` is therefore a fixed effect. However, the two dams are representative of female pigs in general, so `DAM` is a random factor. `DAM` is also nested within `SIRE`, because each dam was mated with only one sire. This will enable us to determine if the weight of the piglets is largely due to the male or female parent.

F-*ratios*

The analysis of this experiment is given in Box 12.4. As with the previous example, to calculate the *F*-ratio for the top level of nesting, the error term is the expected mean square of the level below. In this instance, the *F*-ratio for `SIRE` is obtained by comparing the variation between sires, with the variation between dams mated with the same sire (that is `DAM(SIRE)`). So:

$$F\text{-ratio for SIRE} = \frac{\text{Variation between sires}}{\text{Variation between dams mated with same sire}}$$

$$= \frac{0.25860}{0.25780} = 1.00.$$

The *F*-ratio for `DAM` however is obtained by comparing the variation between pigs from the same sire (but different dams) compared to the variation between pigs with the same dam and sire (error):

$$F\text{-ratio for DAM} = \frac{\text{Variation between pigs from same sire but different dams}}{\text{Variation between pigs from same dam and sire}}$$

$$= \frac{0.25780}{0.08266} = 3.12.$$

Conclusions

There is no evidence that the sires differ from each other ($F = 1$, df = 4,5, $p = 0.484$). The evidence that the weight gain of piglets is influenced by the mother is close to significance ($F = 3.12$, df = 5,10, $p = 0.059$).

BOX 12.4 An analysis mixing fixed and random factors

General Linear Model

Word equation: WGAIN = SIRE + DAM (SIRE)

SIRE is a fixed factor and DAM is a random factor

Analysis of variance table for WGAIN

Source	DF	SS	MS	F	P
SIRE	4	1.03439	0.25860	1.00	0.484
DAM (SIRE)	5	1.28898	0.25780	3.12	0.059
Error	10	0.82658	0.08266		
Total	19	3.14994			

Expected Mean Squares using Adjusted SS

Source	Expected Mean Square for each term
1 SIRE	(3) + 2.0000(2) + Q[1]
2 DAM (SIRE)	(3) + 2.0000(2)
3 Error	(3)

Error terms for tests using Adjusted SS

Source	Error DF	Error MS	Synthesis of Error MS
1 SIRE	5.00	0.25780	(2)
2 DAM (SIRE)	10.00	0.08266	(3)

Variance components using Adjusted SS

Source	Estimated value
DAM (SIRE)	0.08757
Error	0.08266

Variance components with fixed effects

As before, the error variance (σ_E^2) is estimated at the lowest level of nesting, namely between pigs of the same parentage. The expected mean square for dams contains variance due to error, and variance due to differences between dams when mated to the same sire (that is, DAM (SIRE), σ_D^2). Due to the fact that there are two dams per sire, this is multiplied by 2 giving:

$$\text{Expected mean square for DAM} = 2\sigma_D^2 + \sigma_E^2.$$

The formula linking variance components and expected mean squares for SIRE contains the term Q[1]. The variation between sires is due to: (1) variation between piglets from different sires; (2) variation between piglets from different dams but the same sire; (3) variation between piglets even when they have the same parents. The term Q[1] stands for the following:

$$\frac{\Sigma(\rho_i - \bar{\rho})^2}{a - 1}$$

where ρ_i = the true weight gain for piglets from the ith sire, $\bar{\rho}$ = true mean weight gain for all piglets, and a stands for the number of sires (and therefore $a - 1$ are the degrees of freedom for sires). Because SIRE is a fixed effect in this analysis, then we are estimating the fixed differences ρ_i. (This contrasts with the variance component σ^2_D, which is an estimate of how dams will vary from each other). Q[1] is then multiplied by 4 in the linking formula, because there are 4 piglets per sire.

SIRE was treated as a fixed effect because we supposed the farmer to be interested in these specific sires. This decision however was based on the questions being asked, rather than any intrinsic property of the factor. If the farmer was interested instead in whether sires in general influenced the quality of their offspring, and it was reasonable to assume that these five sires were a random selection of sires in general, then the analysis could be repeated with SIRE as a random effect.

Finally, there is an alternative way of asking the same question, if we are only interested in the differences between sires. The dams cannot be simply ignored, because piglets from the same dam do not provide independent evidence of the quality of that particular sire. However, the piglet weights may be averaged for each dam, so providing one data point for each sire-dam combination. The resulting analysis is then a model with one explanatory factor, SIRE, as shown in Box 12.5. Notice how the F-ratio and the degrees of freedom are exactly the same as for the nested and random effects analysis. This is because the two analyses are exactly equivalent. The only advantage we obtain from doing the more complex nested analysis is to gain information on the extent to which our dams vary. Many of the apparently sophisticated analyses that overcome non-independence in various ways turn out, as in this case, to be exactly equivalent to averaging over the non-independent datapoints and performing a straightforward GLM.

BOX 12.5 An alternative analysis to answer a simple question from a nested design

General Linear Model

Word equation: AVPIGLET = SIRE

SIRE is categorical

Analysis of variance table for AVPIGLET, using Adjusted SS for tests

Source	DF	Seq SS	Adj SS	Adj MS	F	P
SIRE	4	0.5177	0.5177	0.1294	1.00	0.484
Error	5	0.6446	0.6446	0.1289		
Total	9	1.1624				

12.6 Using mock analyses to plan an experiment

If you plan to do an experiment with random effects, it is very important to design it carefully. The design of an experiment determines the *X* variables and the length of the *Y* variable. Indeed one can view designing an experiment as designing the *X* variables. The only thing preventing the analysis being done at this stage is the absence of values for *Y*. Obviously, we can't know the results of the experiment before doing it. However, there are important properties of a design that can be explored, and which don't actually depend upon the values of *Y*. We can study these properties by constructing a null data set containing the *X* variables as you propose to use them and filling the *Y* variable with random numbers. This is because different experimental designs will produce different *F*-ratios, when nesting is involved. The degrees of freedom in your *F*-ratio test will depend upon the denominator, and the lower these degrees of freedom, the less powerful is the test. So different designs can be compared by running through dummy analyses, and discovering which gives the most powerful test for a particular variable. This will also warn you in advance of any designs for which you may have to resort to a synthetic denominator.

In this example, we use dummy analyses to pick out the best possible design for an experiment. The aim is to do a mate choice experiment using pheasants. The key question is whether melanic females are courted more vigorously by males than the normal female phenotype. One replicate of the trial involves placing a male and a female in separate cages, screened from each other, in the evening. The screen is removed during the hours of darkness. The male's behaviour is then recorded for an hour, starting 20 minutes after dawn. The 'vigour of courtship' is measured as the fraction of that hour in which the male's wattles are erect.

The main constraint is that only 36 measurements can be taken, owing to the number of days of the breeding season and the likely frequency of rain (which dampens ardour). So, given this constraint, how many males and females should be used, and what pattern of pairing? There are three possible extremes: (1) six males and six females used, with each possible combination of pairing; (2) six males and thirty-six females, with each male used six times and each female used only once; (3) six females and thirty-six males, with each female used six times and each male used only once. In all three plans, half of the females would be melanic and half would be the normal phenotype. Which of these three plans will best test our key question of the influence of melanism on the sexual attractiveness of female pheasants? The dummy data used in these analyses are stored in the *pheasants* dataset.

Plan 1: six males and six females, with each possible combination of pairing

Both the factors `MALE` and `FEMALE` are random. However, `FEMALE` is nested within the treatment variable `MELANIC` (as the females belong to two groups).

BOX 12.6 Designing an experiment with random effects: Plan 1

General Linear Model

Word equation: WATTLE = MELANIC + MALE + FEMALE (MELANIC)

MELANIC is categorical, MALE and FEMALE are random factors

Analysis of variance table for WATTLE, using Adjusted SS for tests

Source	DF	Seq SS	Adj SS	Adj MS	F	P
MELANIC	1	0.003762	0.003762	0.003762	0.25	0.645
MALE	5	0.052756	0.052756	0.010551	2.06	0.105
FEMALE (MELANIC)	4	0.060705	0.060705	0.015176	2.96	0.039
Error	25	0.128077	0.1281077	0.005123		
Total	35	0.245300				

Expected Mean Squares using Adjusted SS

Source	Expected Mean Square for each term
1 MELANIC	(4) + 6 (3) + Q[1]
2 MALE	(4) + 6 (2)
3 FEMALE (MELANIC)	(4) + 6 (3)
4 Error	(4)

Error terms for tests using Adjusted SS

Source	Error DF	Error MS	Synthesis of Error MS
1 MELANIC	4.00	0.015176	(3)
2 MALE	25.00	0.005123	(4)
3 FEMALE (MELANIC)	25.00	0.005123	(4)

Variance components using Adjusted SS

Source	Estimated value
MALE	0.00090
FEMALE (MELANIC)	0.00168
Error	0.00512

The analysis of this plan is given in Box 12.6. We are primarily interested in MELANIC, but need to account for variation between males and females. Thus the word equation is:

WATTLE = MELANIC + MALE + FEMALE (MELANIC).

Our main concerns are how powerful this analysis is. The *F*-ratio of MELANIC uses the denominator FEMALE (MELANIC), as it compares the variation in courtship between males and the two groups of females, compared to variation within one melanic group. FEMALE (MELANIC) has 4 degrees of freedom, and this is one important determinant of the power of the test.

Another student decided to be more ambitious, and fitted a model which also included the interaction between male and melanic, given that some males may prefer melanic females, and others may not. However, the output in Box 12.7 indicates that a synthetic denominator is needed to provide an *F*-ratio, and so this *F*-ratio is inexact.

To test for an interaction using an exact test, it would be necessary to have replicate measures of each male/melanic combination. If the three different females within each melanic group were assumed to be identical, then this

BOX 12.7 Designing an experiment with random effects: Plan 1 with an interaction

General Linear Model

Word equation: WATTLE = MELANIC | MALE + FEMALE (MELANIC)

MELANIC is categorical, MALE and FEMALE are random factors

Analysis of variance table for WATTLE, using Adjusted SS for tests

Source	DF	Seq SS	Adj SS	Adj MS	F	P
MELANIC	1	0.003762	0.003762	0.003762	0.23	0.657†
MALE	5	0.052756	0.052756	0.010551	1.73	0.282
MELANIC * MALE	5	0.030540	0.030530	0.006108	1.25	0.322
FEMALE (MELANIC)	4	0.060705	0.060705	0.015176	3.11	0.038
Error	20	0.097537	0.0976537	0.004877		
Total	35	0.245300				

†Not an exact *F*-test

Expected Mean Squares using Adjusted SS

Source	Expected Mean Square for each term
1 MELANIC	(5) + 6 (4) + 3 (3) + Q[1]
2 MALE	(5) + 3 (3) + 6 (2)
3 MELANIC * MALE	(5) + 3 (3)
4 FEMALE (MELANIC)	(5) + 6 (4)
5 Error	(5)

Error terms for tests using Adjusted SS

Source	Error DF	Error MS	Synthesis of Error MS
1 MELANIC	4.06	0.016407	(3) + (4) – (5)
2 MALE	5.00	0.006108	(3)
3 MELANIC * MALE	20.00	0.004877	(5)
4 FEMALE (MELANIC)	20.00	0.004877	(5)

Variance components using Adjusted SS

Source	Estimated Value
MALE	0.00074
MELANIC * MALE	0.00041
FEMALE (MELANIC)	0.00172
Error	0.00488

would be the case. However, female has been declared to be a random variable, so different females are not being treated as identical. This is why the *F*-ratio produced is an approximation.

Plan 2: six males and thirty six females, with each male used six times and each female used only once

In the second plan, each female is used only once, so female individuals are being treated as replicate units. We cannot estimate the variation between females, because this is now confounded with error. The word equation becomes WATTLE = MELANIC | MALE, with MALE being declared as random. This analysis is given in Box 12.8.

BOX 12.8 Designing an experiment with random effects: Plan 2

General Linear Model

Word equation: WATTLE = MELANIC | MALE

MELANIC is categorical, MALE is a random factor

Analysis of variance table for WATTLE, using Adjusted SS for tests

Source	DF	Seq SS	Adj SS	Adj MS	F	P
MELANIC	1	0.003762	0.003762	0.003762	0.62	0.468
MALE	5	0.052756	0.052756	0.010551	1.73	0.282
MELANIC * MALE	5	0.030540	0.030540	0.006108	0.93	0.481
Error	24	0.158242	0.158242	0.006593		
Total	35	0.245300				

Expected Mean Squares using Adjusted SS

Source	Expected Mean Square for each term
1 MELANIC	(4) + 3.0000(3) + Q[1]
2 MALE	(4) + 3.0000(3) + 6.0000(2)
3 MELANIC * MALE	(4) + 3.0000(3)
4 Error	(4)

Error terms for tests using Adjusted SS

Source	Error DF	Error MS	Synthesis of Error MS
1 MELANIC	5.00	0.006108	(3)
2 MALE	5.00	0.006108	(3)
3 MELANIC * MALE	24.00	0.006593	(4)

Variance components using Adjusted SS

Source	Estimated value
MALE	0.00074
MELANIC * MALE	−0.00016
Error	0.00659

The error term for MELANIC is now MALE | MELANIC, with 5 degrees of freedom. This is a more powerful test, and it is possible to test the interaction. The only information that has been lost is an estimate of the variance component for females. This was not however a primary aim of the experiment.

Plan 3: six females and thirty six males, with each female used six times and each male used only once

In the third plan, males are treated as the replicate unit, and so do not appear in the word equation. The word equation becomes WATTLE = MELANIC + FEMALE (MELANIC). See Box 12.9.

Again the error term for MELANIC is FEMALE (MELANIC) with 4 degrees of freedom. The factor MALE does not even appear in the model, so it is certainly not possible to test for the interaction between MALE and MELANIC.

BOX 12.9 Designing an experiment with random effects: Plan 3

General Linear Model

Word equation: WATTLE = MELANIC + FEMALE (MELANIC)

MELANIC is categorical, FEMALE is a random factor

Analysis of variance table for WATTLE, using Adjusted SS for tests

Source	DF	Seq SS	Adj SS	Adj MS	F	P
MELANIC	1	0.003762	0.003762	0.003762	0.25	0.645
FEMALE (MELANIC)	4	0.060705	0.060705	0.015176	2.52	0.062
Error	30	0.180833	0.180833	0.006028		
Total	35	0.245300				

Expected Mean Squares using Adjusted SS

Source	Expected Mean Square for each term
1 MELANIC	(3) + 6 (2) + Q[1]
2 FEMALE (MELANIC)	(3) + 6 (2)
3 Error	(3)

Error terms for tests using Adjusted SS

Source	Error DF	Error MS	Synthesis of Error MS
1 MELANIC	4.00	0.015176	(2)
2 FEMALE (MELANIC)	30.00	0.006028	(3)

Variance components using Adjusted SS

Source	Estimated value
FEMALE (MELANIC)	0.00152
Error	0.00603

So in summary, Plan 2 best suits the aim of testing the factor MELANIC for significance. It provides the *F*-ratio with the most powerful test (the denominator has 5 degrees of freedom), and it gives the added bonus of being able to test for the interaction (without having to approximate the *F*-ratio). This conclusion, however, was not obvious without doing the dummy analyses first. It is also an example of a dilemma that must be sorted out at the design stage. When random effects are involved, it is often hard to think through the consequences, so these kind of exercises are invaluable. The general principle however holds for all experiments—work out the model you aim to fit before doing the experiment.

12.7 Summary

- Two properties distinguish fixed and random factors: (1) what exactly needs to be repeated in the repetition of an experiment (the levels for a fixed effect, and the population for a random effect); (2) conclusions about random factors also apply to some wider population of which the experimental levels were taken to be a random sample.

- Typical examples of random factors are biological individuals (humans, starlings, plants etc.).

- Four new concepts are introduced relevant to the analysis of random factors: (1) components of variance; (2) formulae that link components of variance and expected mean squares; (3) nesting; (4) appropriate denominators for *F*-tests.

- In models containing a random effect, two variances must be estimated: the error variance, and the variance between individuals represented by the random factor (e.g. the variance between starlings).

- The expected mean square for a factor in a random effects model may contain a number of components of variance: e.g. variance due to error, due to differences between leaves on the same plant, due to differences between plants. Formulae may be derived which link these components of variance to the expected mean square of the ANOVA table.

- The concept of nesting was introduced, in which levels of a random factor fall into groups (e.g. leaves on different plants). The random factor may be nested within another random factor (e.g. leaves within plants) or within a fixed effect (female pheasants within the fixed effect melanic).

- In a model with random effects, the appropriate denominator for the *F*-ratio may not be the error variance. If there is nesting, the appropriate denominator for one factor is the mean square of the term in the level of nesting below it.

- Dummy analyses provide a useful tool to help in the design of experiments, particularly those with random effects.

12.8 Exercises

Examining microbial communities on leaf surfaces

An experiment was performed to determine the effect of two soil treatments on the density of bacteria found on the leaf surfaces of sugar beet. Six plants were grown in an environmental chamber, three in Soil type 1 and three in Soil type 2. Ten samples were taken from each of the six plants. For each sample, a fixed biomass of leaf was homogenised, and a fixed volume of a designated dilution spread on a plate. The bacterial density was then assessed by counting the number of colonies for each plate. The data were stored in the *bacteria2* dataset in variable TREATMNT (1 or 2) and PLANT (1 to 6), with the number of colonies being recorded in DENSITY.

Two sets of analyses were conducted on these data, to ask if the soil treatment influenced the density of bacteria on the leaf surface. The first analysis fitted the word equation DENSITY = TREATMNT and the second analysis fitted the word equation DENSITY = TREATMNT + PLANT (TREATMNT), declaring plant as a random factor. These are shown in Box 12.10.

(1) Interpret Analysis 1 on the assumption it is a reasonable analysis to perform.

(2) Interpret the second analysis, emphasising any differences in the conclusions. Explain why any such differences may have arisen and which analysis you prefer.

(3) What is the difference between random and fixed effects, and between crossed and nested factors?

(4) Which denominator was used to calculate the *F*-ratio of 2.58 in the second analysis?

(5) Which parts of the output would you take into account, and how, in deciding how to allocate sampling effort in a repeat of the experiment between more plants and more readings per plant?

BOX 12.10(a) Analysis 1

General Linear Model

Word equation: DENSITY = TREATMNT
TREATMNT is categorical

Analysis of variance table for DENSITY, using Adjusted SS for tests

Source	DF	Seq SS	Adj SS	Adj MS	F	P
TREATMNT	1	9543.2	9543.2	9543.2	12.50	0.001
Error	58	44265.9	44265.9	763.2		
Total	59	53809.1				

BOX 12.10(b) Analysis 2

General Linear Model

Word equation: DENSITY = TREATMNT + PLANT (TREATMNT)

TREATMNT is categorical, PLANT is random and nested within TREATMNT

Analysis of variance table for DENSITY, using Adjusted SS for tests

Source	DF	Seq SS	Adj SS	Adj MS	F	P
TREATMNT	1	9 543.2	9 543.2	9 543.2	2.58	0.183
PLANT (TREATMNT)	4	14 777.8	14 777.8	3 694.5	6.77	0.000
Error	54	29 488.1	29 488.1	546.1		
Total	59	53 809.1				

Expected Mean Squares using Adjusted SS

Source	Expected Mean Square for each term
1 TREATMNT	(3) + 10 (2) + Q[1]
2 PLANT (TREATMNT)	(3) + 10 (2)
3 Error	(3)

Error terms for tests using Adjusted SS

Source	Error DF	Error MS	Synthesis of Error MS
1 TREATMNT	4.00	3694.5	(2)
2 PLANT (TREATMNT)	54.00	546.1	(3)

Variance components using Adjusted SS

Source	Estimated value
PLANT (TREATMNT)	314.8
Error	546.1

How a nested analysis can solve problems of non-independence

Return to the *sheep* dataset first examined in Section 8.6 in which Dr Sharp has a hypothesis that male sheep look up more frequently while eating than female sheep. The data on six sheep (3 males and 3 females) are recorded in the *sheep* dataset as five columns: (i) DURATION of feeding time in minutes; (ii) NLOOKUPS, the number of lookups; (iii) SEX, coded as 1 for female and 2 for male; (iv) SHEEP, coded as 1 to 3 for each female and 1 to 3 for each male and (v) OBSPER, the number of the observation period from 1 to 20.

One solution to the non-independence of the data which is discussed in Chapter 8 is to summarise the measurements so that one sheep contributes one datapoint. However, now that we have discussed nested analyses, there is an alternative solution.

(1) Conduct a more appropriate analysis using a nested design.

(2) Compare the *F*-ratios and degrees of freedom for LUPRATE in the nested analysis with the single summary analysis.

13 Categorical data

In the previous chapters, the focus has been on data with a continuous Y variable. For categorical data, most analyses are now carried out with log-linear models which are beyond the scope of this text. However with general linear modelling facility, there are some analyses which can be done with categorical data. This chapter will outline these analyses, and compare them to another popular method of analysis—the contingency table.

13.1 Categorical data: the basics

Categorical data are usually counts of individual items, and so are discrete, rather than continuous. The data presented in Table 13.1, of male and female students classified into having blue eyes or not, are categorical. The units (a person) arc independent individuals. This example will be used to devise a contingency table analysis.

Contingency table analysis

The null hypothesis is that the proportion of blue eyed people is the same for males and females. Probability theory allows us to calculate the number we would expect in each cell of the table if this were true. The total proportion of blue-eyed people in the population (from which our table is a random sample) is: $\frac{88+70}{952}$. So we would expect $\frac{158}{952} \times 467$ males to be blue eyed

Table 13.1 A two-by-two table of categorical data

	Males	Females	Total
Blue eyed	88	70	*158*
Not blue eyed	379	415	*794*
Total	*467*	*485*	*952*

Table 13.2 Expected and observed values for a two-by-two contingency table

	Males	Females	Total
Blue eyed	88 77.5	70 80.5	*158*
Not blue eyed	379 389.5	415 404.5	*794*
Total	*467*	*485*	*952*

and $\frac{158}{952} \times 485$ females to be blue eyed. Similarly, the fraction of the population that is not blue eyed is $\frac{794}{952}$, so we would expect $\frac{794}{952} \times 467$ males not to be blue eyed and $\frac{794}{952} \times 485$ females not to be blue eyed.

This can be generalised into the formula

$$\text{Expected value for a cell} = \frac{\text{Row total} \times \text{Column total}}{\text{Grand Total}}.$$

The expected values are shown in Table 13.2 in bold, beneath the observed values.

The question is how close are the expected and observed values. Their 'closeness' is quantified by calculating a chi-squared value according to the following formula:

$$\chi^2 = \sum \frac{(O - E)^2}{E}$$

where O = observed values and E = expected values. The derivation of this formula is discussed in Section 13.3, so for the purpose of this example, we will accept it as a measure of how closely the observed values correspond with the values we would expect under the null hypothesis. In this case

$$\chi^2 = \frac{(88 - 77.5)^2}{77.5} + \frac{(70 - 80.5)^2}{80.5} + \frac{(379 - 389.5)^2}{389.5} + \frac{(415 - 404.5)^2}{404.5} = 3.35.$$

Any value of chi-squared has a number of degrees of freedom associated with it (the greater the number of cells, the greater will be the value of chi-squared, regardless of the closeness of the observed and expected values, so the size of the data set needs to be taken into consideration). In calculating the expected values, we are constrained by the marginal totals (that is, the total number of males and females, blue eyed and not blue eyed people are fixed). Given this, once one expected value has been calculated, then all other expecteds could have been calculated just by ensuring that the rows and columns added up correctly. So in this two-by-two example, there is only one degree of freedom. If we were analysing these data using GLMs, the equivalent question would be 'is the interaction between sex and colour of eyes significant?'. The interaction term would also have one degree of freedom. A general formula for calculating the number of degrees of freedom for contingency tables is:

$$\text{Degrees of freedom} = (\text{Number of rows} - 1) \times (\text{Number of columns} - 1).$$

For a chi-squared of 3.35 on 1 degree of freedom, $p = 0.0672$. The conclusion is that there is insufficient evidence to reject the null hypothesis that the proportion of males and females with blue eyes is the same.

When are data truly categorical?

It is not infrequent for categorical analyses to be used on inappropriate sets of data, so the first point to ascertain is whether or not the data are truly categorical. The data of Table 13.1 are categorical, so the contingency table analysis is valid. Consider however the following examples:

1. The 952 datapoints refer to 952 visits by honey bees to flowers classified by (a) size (small, large) and (b) colour (yellow, blue). The aim is to ask whether bees favour certain sizes or colours of flowers. While the number of visits could be regarded as a categorical variable, individual visits are extremely unlikely to be independent. An individual bee may be making return visits to the same flower, and be counted more than once. Bees also recruit others to visit the same flower. This probable lack of independence of visits means that it would be unwise to analyse these data by categorical methods.
2. The 952 datapoints refer to the number of holes made in four leaves, two of one plant species and two of another. Each of the four leaves were placed in a separate vial, with a caterpillar of species A or B. The aim is to see if the two caterpillar species have an equal propensity to eat the two different plant species. The holes in one leaf are not independent, but made by one caterpillar. To take an extreme but quite probable case, suppose one of the caterpillars died before eating. Then there would be no holes at all, showing that holes cannot be considered as independent items. So again, a contingency table analysis would not be appropriate.
3. Four animals take part in a training programme, and then tested to see how long they take to complete a task. There are two different training periods (B1 and B2) and two different test procedures (A1 and A2) for the two cross-classifications. The cell entries are then the number of minutes each animal took to complete the task. Here, the mistake has been to make a continuous variable artificially categorical. If we rounded to the nearest second, the number would be 60 times greater—but the number of independent items in the experiment would remain unchanged. It can be seen that the true sample size in this case is 4, not 952.

So in summary, for data to be truly categorical, they must be **independent counts of individuals items.** What distribution describes categorical data? When analysing continuous variables, GLM made the assumption that they were Normally distributed, and if not, they were transformed to be normally distributed. Categorical data, however, consist of counts, so the appropriate distribution will often be Poisson, binomial or hypergeometric. We shall pursue the **Poisson distribution** here, as theory shows that the other cases are equivalent to Poisson variables subject to linear constraints.

13.2 The Poisson distribution

Two properties of a Poisson process

The Poisson distribution is central to the analysis of categorical data. It is a discrete distribution used to describe randomness in space or time. Why should we assume that categorical data follow a Poisson distribution? Not all categorical data will, but it is a useful starting point. It is as if we are operating with the grand null hypothesis that nothing is influencing our data except chance. If this is true, then a Poisson distribution would describe the number of counts we would expect per unit time or per unit space, and the data would describe a Poisson distribution around the mean of the fitted values. The Poisson distribution is also discrete—suitable for describing count data.

Not all categorical data will follow a Poisson distribution. One example of a process which can be approximated to a Poisson process is a radioactive substance which emits alpha particles. There are a vast number of atoms, each of which has a tiny probability of decaying in the next second. This may be described by the binomial distribution, with parameters n (for number of atoms) and p (for the probability of any one atom decaying in the next second). In cases where n is very large however and p very small, the Poisson distribution provides a good approximation to describe the number of alpha particles emitted per second, described by the mean number of particles emitted per second. An example of a 'Poisson process in space' may be the distribution of parasites across hosts. It is known however that with some species, parasites tend to aggregate in a few hosts, in which case, the number of parasites per host follows a clumped rather than a Poisson distribution. How can these two different scenarios be distinguished?

If any process follows a Poisson process, it fulfils two criteria: **independence** and **homogeneity**. In distinguishing a Poisson process from a clumped or uniform one, it is useful to think in terms of 'items' and 'containers'. These abstract terms allow us to consider processes that occur in space or time under one umbrella. The criteria then become 'independence of items' and 'homogeneity of containers'. In the radioactive example, the items are the particles emitted from the radioactive substance—the emission of one alpha particle must not influence the emission of any other alpha particles. (This is not quite true, as one alpha particle may collide with a nucleus to make a heavier nucleus, which can then decay.) The containers, in this instance, then become units of time. The condition of homogeneity will be fulfilled when the probability that an alpha particle will be emitted remains constant per unit time. (This will also be true as long as the half life of the radioactive material is very much greater than the observation period.) For the hosts and parasites, the items become the parasites, and the containers are the hosts. Parasites will follow a Poisson distribution in space, if one parasite is quite independent of any other parasite (for example, there is no reproduction within hosts), and if the hosts are homogeneous (for example, equally susceptible and equally exposed). For some species of hosts and parasites these criteria may be fulfilled, while for others they are not.

By considering these two properties of the Poisson process, it is often possible to make an educated guess about whether or not a variable is likely to follow a Poisson distribution. There is a more precise way of determining the randomness of a distribution, and that is the dispersion test. This is described later in this section.

The mathematical description of a Poisson distribution

The formula for a Poisson distribution is:

$$\text{Probability that } y \text{ events occur in 1 time unit} = \frac{e^{-\lambda}\lambda^{y}}{y!}$$

where λ = the mean number of events per time unit, and the $y!$ is called y factorial, and equals y multiplied by every integer between y and 1: i.e. $y! = y \times (y-1) \times (y-2) \ldots \times 1$. This formula may be converted to describe Poisson distributions in space by referring to the probability of events occurring per unit space rather than per unit time. So what does this distribution look like?

The shape of the Poisson distribution changes, depending upon the mean (Fig. 13.1). If the mean is less than 1, the mode is zero. Given that it is not possible for counts to be less than zero, this leads to considerable right hand skew (Fig. 13.1(a)). As the mean increases, the mode moves along the x axis, and the distribution becomes more symmetrical. We can always construct a Normal distribution with the same mean and variance as a given Poisson distribution. When the mean is greater than 5, statisticians begin to feel it is acceptable to use this Normal distribution as an approximation to the Poisson. (This can only ever be an approximation, however, given that the Normal distribution is continuous and the Poisson distribution is discrete). This is a property that is made use of extensively in the analysis of categorical data, and we return to this point when we discuss the origins of the chi-squared test (Section 13.3).

You will notice that the formula for the Poisson distribution has only one parameter, namely λ. It follows that the variance must depend upon λ, and in

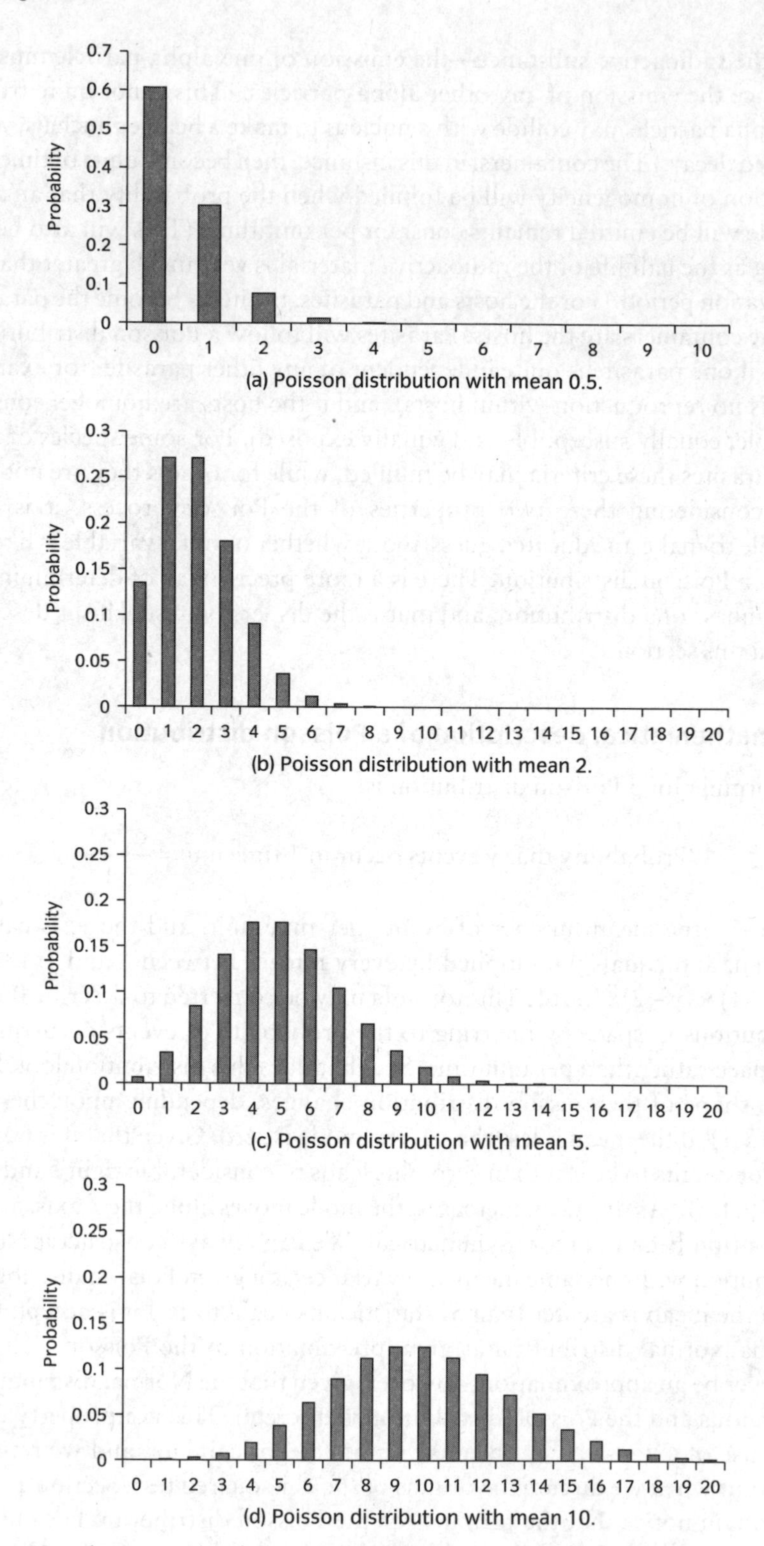

Fig. 13.1 How the Poisson distribution changes shape as the mean increases.

fact it can be shown that the variance equals λ, and so equals the mean. So only one parameter, the mean, is needed to define the distribution. This contrasts with the Normal distribution, where two parameters are required—the mean and the variance. This also illustrates why categorical data cannot be analysed using GLMs without transformation. As the mean increases, so does the variance, contravening the assumption of homogeneity of variance (Chapter 9). It is actually possible to transform data following a Poisson distribution to have homogenous variance by using the square root transformation (the weakest of the alternatives discussed in Chapter 9). When we look at GLM methods of analysing categorical data, we will make use of this.

The dispersion test

It is possible to investigate whether a set of data follow a Poisson distribution, by testing whether the mean equals the variance. The ratio of the variance to the mean should be 1, but is unlikely to be 1 exactly. The extent to which we might expect the ratio to deviate from 1 will depend upon the size of the data set—smaller data sets could vary quite considerably from 1, yet still be drawn from a Poisson distribution. This is the essence of a **dispersion test**. Under the null hypothesis of Poissonness, the ratio

$$(n-1)\frac{s^2}{m}$$

will follow approximately a chi-squared distribution with $n - 1$ degrees of freedom. (s^2 = estimate of the variance, m = estimate of the mean, n = number of counts in the data set). (This is only approximate, as the chi-square distribution is a continuous distribution, yet the ratio above is calculated from discrete data, and so can only take a finite number of values.) If the variance:mean ratio is significantly less than 1, then the data contain very little variation at all. This kind of distribution may also be referred to as **underdispersed**, because there is less variation than expected under the Poisson hypothesis. The data points are too similar, and this may be because they lack independence. Alternatively, if the variance:mean ratio is significantly greater than 1, then the probability distribution is aggregated (clumped) over space or time. This alternative is called **overdispersed**, because there is more variation than expected. Overdispersion may occur because important explanatory variables have been omitted. Once the variation they explain is eliminated, then this could reduce the variance:mean ratio to a value more appropriate for a Poisson distribution. For example, if the distribution of parasites across males followed a Poisson distribution with $\lambda = 1.2$, but the distribution across females was Poisson with $\lambda = 3.6$, then the population as a whole would be overdispersed. However, once the variation due to gender has been taken into account, the residual variation would follow a Poisson distribution.

The distribution of helminth numbers in two bird species was studied, by dissecting 119 individuals of each species *post mortem*. The number of helminths in each bird was recorded and stored in two variables SPA and SPB in the *parasites* dataset. The descriptive statistics for the two frequency distributions are given in Box 13.1, with two histograms in Fig. 13.2, to give a visual description of the shape of the two distributions.

BOX 13.1 Descriptive statistics of the parasite data. (Q1 = lower quartile; Q3 = upper quartile)

Descriptive statistics

Variable	N	Mean	Median	Mode	StDev	Min	Max	Q1	Q3
SPA	119	2.126	2.000	2.000	1.232	0.000	5.000	1.000	3.000
SPB	119	2.739	1.000	0.000	4.045	0.000	22.000	0.000	4.000

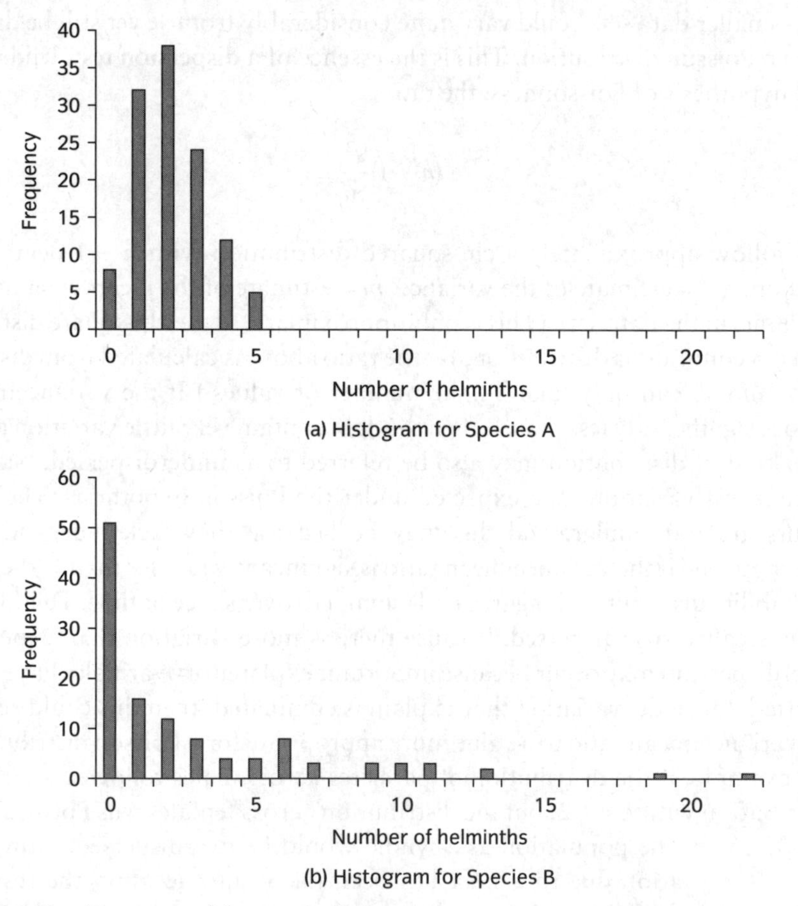

Fig. 13.2 Two histograms of the parasite data.

BOX 13.2 A dispersion test for the helminth parasites in Bird species A

Test

$$\chi^2_{118} = \frac{118 * (Variance(SpA))}{Mean(SpA)} = 84.2$$

Result

The two-tailed probability of obtaining this value of χ^2 or something more extreme under the null hypothesis is 0.0162.

The first thing to notice from the descriptive statistics is that while the means are similar (2.13 and 2.74 to 2 decimal places), the variances are very different (the squares of the standard deviations are 1.52 and 16.4). In one case it is less than the mean, and in the other it is greater. Are either of these differences significant? The frequencies of the parasites in the two bird species are plotted on the same axes in Fig. 13.2. The frequency has right skew in species B, but is more symmetrical for species A.

In Box 13.2, a dispersion test is carried out for both species. The test is carried out in three steps: (1) The chi-squared value is obtained from the variance:mean ratio multiplied by $n - 1$ (which is 118 in this case); (2) The chi-squared value is converted into a p-value, by finding the probability of getting that particular chi-squared value, or anything less than that; (3) The p-value is then converted into the relevant p-value for the test.

Step 1

The χ^2 value is calculated from the variance:mean ratio and degrees of freedom for Species A. Under the null hypothesis these data are from a Poisson distribution, so this calculated quantity should come from the chi-squared distribution with 118 degrees of freedom.

Step 2

The next step is to calculate the probability of getting the calculated value of χ^2, or anything less than this, if it comes from a χ^2distribution with 118 degrees of freedom. In other words, what is the area of the probability distribution of χ^2_{118} to the left of the calculated value (84.2). This probability may be found in tables, or directly from your statistical package, and is 0.0081. This corresponds to the probability of being to the left of 84.2 in Fig. 13.3, which depicts the χ^2 distribution with 118 degrees of freedom.

Step 3

Is this the final probability for our test? No—the dispersion test is effectively a two-tailed test, because we are interested in whether the variance is greater or

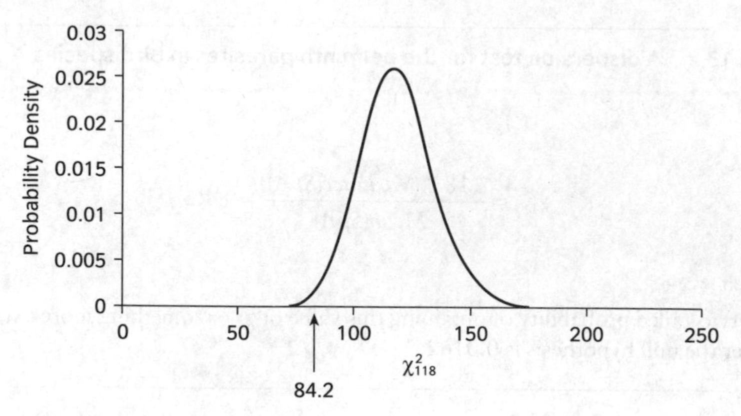

Fig. 13.3 A chi-squared distribution with 118 degrees of freedom, with the test statistic from the first dispersion test for parasites in Species A marked with an arrow.

less than the mean. To convert this p-value into the appropriate p-value for our test, it should be multiplied by 2. So the final answer for the dispersion test is a p-value of 0.0162. In this case the variance is significantly less than the mean, and we conclude that the distribution is underdispersed.

The dispersion test can then be repeated for the helminth parasites in bird species B, as shown in Box 13.3.

In this case, the value of χ^2 is much higher at 704.8. This is so high as to be off the scale of Fig. 13.3, and the probability of being to the left of this value is 1. To convert this to a p-value relevant to our test, we need to subtract it from 1 (as it is in the right hand half of our distribution and we are interested in the tails) and multiply by 2 (as it is a two tailed test), giving a p-value of effectively zero. In this example, the variance is very much greater than the mean, and we would conclude that the distribution of the parasites in Bird species B is overdispersed, or aggregated.

So the parasites in Bird species A are underdispersed. Underdispersion must arise from non-independence of items (in this case of parasites, through spacing

---

BOX 13.3 **A dispersion test for the helminth parasites in Bird species B**

Test

$$\chi^2_{118} = \frac{118 * (Variance(SpB))}{Mean(SpB)} = 704.8$$

Result

The two-tailed probability of obtaining this value of χ^2 or something more extreme under the null hypothesis is <0.001.

---

out via competition for example, or in this case perhaps from an immune response). In contrast, the parasites in Bird species B were overdispersed. This can also arise from non-independence (for example, by the parasites reproducing inside the host, being attracted to each other, or perhaps the presence of parasites weakening the host and so making it more vulnerable to further parasitism). However, overdispersion may also arise from heterogeneity of containers (in this case hosts; for example males and females differing in susceptibility, or hosts of different ages having being exposed to parasites for different periods of time, or different regions of habitat having different levels of infestation).

This example has illustrated a situation in which the chi-squared test is two tailed. Chi-squared values for contingency table tests are one tailed, and so it is often the one tailed value which is provided in tables. For this reason it is worth giving an explicit rule for converting a one tailed p-value to a two tailed p-value. If the one tailed p-value is denoted *onep*, and the two tailed p-value is *twop* then:

If $onep < 0.5$, then $twop = 2 \times onep$.

If $onep > 0.5$, then $twop = 2 \times (1\text{-}onep)$.

13.3 The chi-squared test in contingency tables

Derivation of the chi-squared formula

Analysing categorical data using the Poisson distribution

Categorical data with low means may come from a Poisson distribution which is very right skew (for example Fig. 13.1a). These kinds of data sets can be analysed by using the Poisson distribution directly. For example, a newspaper may report that one town has two cases of a rare form of childhood leukaemia, and claim that this particular village is suffering from a blight. In towns of roughly the same size, the national average is 0.5 children. How can we determine if 2 is significantly greater than 0.5? The null hypothesis is that the incidence of this disease occurs at random throughout the UK, that is, incidence follows a Poisson distribution with mean 0.5. What, then, is the probability of observing 2 or something more extreme? This is equivalent to observing 1 – (zero or one child with the disease). Using the formula given in the previous section, this is $1 - 0.910 = 0.09$. So only 9% of the time would you expect to come across a town with 2 children suffering from this disease. For a two tailed significance value, this should be doubled to 18%, which is greater than the 5% threshold. We would conclude that there is no evidence from these data that the town is unusual.

This example illustrates how the Poisson distribution can be used directly to obtain probability values. However, when the mean is much larger, for example 150, this technique could be arduous. Fortunately, when the mean is greater than 5, we can use the Normal approximation of the Poisson distribution, as illustrated in the following example.

Analysing categorical data using the normal approximation to the Poisson distribution

There was concern that the number of applicants to read biology at a hypothetical university was in decline. In 1994, 204 people applied, but in 1995 only 180. Is this drop significant? First of all, how do we calculate the expected number of applicants? On average, the number of applicants has been $\frac{204+180}{2} = 192$. So how does each year vary from the expected mean? We can make use of the fact that the data are both categorical and assume they follow the Poisson distribution, and also the mean is sufficiently large to use the Normal approximation. To see how close the data are from their expected values, we calculate a z value for each year:

$$z = \frac{\text{observed value} - \text{mean}}{\text{standard deviation}}.$$

However, if these data follow a Poisson distribution, then it is also true that the mean is equal to the variance. Consequently, this formula can be rewritten as:

$$z = \frac{\text{observed} - \text{expected}}{\sqrt{\text{expected}}}$$

where the expected value is the mean. In Table 13.3 these z values are calculated.

So the z value tells us how many standard deviations the observed value is from the mean. In Normal distributions, 96% of all values are within two standard deviations of the mean (see the revision section), so the fact that our two observed values are within 1 standard deviation suggests that the decline in applicant number is not significant.

Table 13.3 z values for biology applicants

	1994	1995
Observed values	204	180
Mean value	*192*	*192*
z value	$\frac{204-192}{\sqrt{192}} = 0.866$	$\frac{180-192}{\sqrt{192}} = -0.866$

We have come to this conclusion however by looking at the two z values separately. Is there any way we can combine the information into one test statistic? The two values will always sum to zero, but if squared and then summed, will provide a measure of the closeness of the observed values to the mean. In general terms:

$$\sum z^2 = \sum \left[\frac{(\text{Observed} - \text{Expected})^2}{\text{Expected}} \right].$$

This formula is the formula for chi-squared—and so we can see that the chi-squared distribution with 1 degree of freedom is equal to a squared normal distribution. By analysing the data in this way, we are assuming that the data are both truly categorical (and so follow a Poisson distribution with the variance equal to the mean), but also can be approximated by the Normal distribution (as long as the mean is large enough). To ensure the approximation to Normality is close enough, conditions are often made as to the sample size within each cell. Stringent conditions require that all expected values are ≥5, while less stringent conditions state that all should be ≥1 and at least 80% ≥5.

Each cell makes a contribution to the overall chi-squared value. By summing the cell contributions, we are using all available information in one test statistic to test one overall hypothesis. This is an effective way to use the data, and avoids such problems as multiplicity of p-values (see Chapter 11). The square root of each chi-squared contribution also provides information on whether one particular cell deviates significantly from its expected value. This can be useful in interpreting the results of a chi-squared test, as shown in the next example.

Inspecting the residuals

Charles Darwin believed that it was better for plants if they were fertilised by other individuals rather than be self-fertilised. Selfing is more of a risk for a plant if it is large, because successive pollinator visits are more likely to be between flowers on the same plant. To reduce this risk, Darwin hypothesised, larger plants are more likely to be dioecious (that is, each individual plant has either male flowers or female flowers) rather than hermaphrodite (where each flower has both male and female parts). He expected monoecy (where each plant has separate male and female flowers) to be intermediate. A chi-squared analysis of New Zealand herbs, shrubs and trees (data of Table 13.4 supplied to Darwin in 1856 by J. D. Hooker and stored in the *Darwin* dataset, in variables `STRATEGY`, `PLSIZE` and `NOSP`) appears to support Darwin's theory (Box 13.4).

The chi-squared statistic is highly significant ($p < 0.0005$). Under Darwin's hypothesis, the residuals on the diagonal should be positive. This would be because more herb species than expected are hermaphrodite, and more tree species than expected are dioecious. Consequently, residuals off the diagonal

Table 13.4 Plant species of New Zealand classified by plant type and mating system (J.D. Hooker)

	Hermaphrodite	Monoecious	Dioecious
Herbs	379	102	19
Shrubs	88	30	31
Trees	56	40	12

BOX 13.4 Chi-squared analysis of Darwin's hypothesis

Contingency table analysis

	Hermaphrodite	Monoecious	Dioecious	ALL
Herbs	379 345.44 1.81	102 113.61 −1.09	19 40.95 −3.43	500
Shrubs	88 102.94 −1.47	30 33.85 −0.66	31 12.20 5.38	149
Trees	56 74.62 −2.16	40 24.54 3.12	12 8.85 1.06	108
ALL	523	172	62	757

Chi-sq = 63.282, DF = 4, p-value = 0.000

Cell contents:
Count
Exp freq
St resid

should be negative. By inspecting the residuals, we can see if the significance is in the direction expected.

While the residuals for hermaphrodite herbs and dioecious trees are in the direction expected, there are some notable exceptions. Dioecious shrubs and monoecious trees have high positive residuals, which was not predicted by the hypothesis. It is now accepted that performing such analyses on counts of species is wholly unacceptable. One assumption of the Poisson distribution may have been contravened—that of independence. If the plant species belong to the same phylogenetic group, the same genus for example, then the mating system may have evolved only once for that whole group. One evolutionary event should be used as one count in the contingency table. Species within the same genus are unlikely to be independent.

So by inspecting the residuals, we can see if the pattern of significance is in the direction expected. If a chi-squared analysis is very significant, but there is no meaningful pattern in the residuals, then one possibility is that the assumption of independence has been contravened. Non-independence is equivalent to giving each count more weight than it deserves. For example, if the mating system of 10 species of plants arose from just one evolutionary event in Table 13.4, this is the same as multiplying each observed value by 10. This will increase the numerator of the χ^2 statistic 100 fold, but the denominator by only 10 fold as illustrated below:

$$\frac{(\text{Observed} \times 10 - \text{Expected} \times 10)^2}{\text{Expected} \times 10} = \frac{100 \times (\text{Observed} - \text{Expected})^2}{10 \times \text{Expected}} = 10 \times \chi^2.$$

So the chi-squared value will be increased by a factor of 10, without altering the degrees of freedom. Non-independence artificially inflates the chi-squared value, and so can cause spurious significance.

13.4 General linear models and categorical data

Using contingency tables to illustrate orthogonality

Contingency table analysis tests for an association between two explanatory variables, as has been illustrated by the analysis of blue-eyed males and females and the *Darwin* dataset. Here, we will use this test to illustrate the relationship between two orthogonal variables, rather than to actually analyse data. Recalling the analysis of the *wheat* dataset of Chapter 7, the variables VARIETY and SOWRATE are orthogonal. We discussed the meaning of orthogonality in Chapter 5, where we defined two variables as being orthogonal when knowledge of one variable provides no knowledge of the other. In the light of contingency table analysis, an equivalent definition would be that two variables have no association. In Box 13.5, a contingency table analysis between VARIETY and SOWRATE gives a chi-squared value of exactly zero. This is another manifestation of their orthogonality, as knowing which sowrate is given to a plot provides no information about which variety is grown in that plot.

In this case, the design was not only orthogonal, but also balanced. However, balance is not necessary for orthogonality, as discussed in Chapter 5. In Box 13.6, a different design has been analysed, taken from Fig. 5.7 Plan C, in which the plots are allocated proportionately across blocks and treatments. Again the chi-squared value is exactly zero, illustrating no association between the two explanatory variables. This also illustrates exactly what the contingency

BOX 13.5 Contingency table analysis between two orthogonal variables in a balanced design

Chi-square test:

Rows: SOWRATE Columns: VARIETY

	1	2	All
1	3	3	6
	3	3	6
2	3	3	6
	3	3	6
3	3	3	6
	3	3	6
4	3	3	6
	3	3	6
All	12	12	24
	12	12	24

Chi-sq = 0.000, DF = 3, *p*-value = 1.000

Cell contents:

Number of plots in that treatment combination

BOX 13.6 Contingency table analysis between two orthogonal variables in an unbalanced design

Chi-square test:

Rows: BLOCKS Columns: TREATMENTS

	T1	T2	T3	T4	Total
1	2	4	4	6	16
	2.00	4.00	4.00	6.00	
2	1	2	2	3	8
	1.00	2.00	2.00	3.00	
3	2	4	4	6	16
	2.00	4.00	4.00	6.00	
4	3	6	6	9	24
	3.00	6.00	6.00	9.00	
Total	8	16	16	24	64

Chi-sq = 0.000, DF = 9, *p*-value = 1.000

Cell contents:

Number of plots in that treatment combination

table analysis is designed to detect: whether the proportionate allocation of counts between the rows changes as you move across the columns.

This section has used one method of analysis as a tool to illustrate what is meant by orthogonality. It complements our illustrations of orthogonality between continuous and categorical variables in Section 6.3. There we conducted a GLM between a continuous and categorical variable, showing that the categorical variable could provide no information about the continuous one. For the rest of this chapter we return to the analysis of data.

Analysing by contingency table and GLMs

There are a limited number of analyses on categorical data that may be carried out in the GLM framework. The following example is first analysed by contingency table analysis, and then by GLM.

A field trial investigated the resistance in barley to fungal attack by powdery mildew. Five barley strains were compared (STRAIN), under four watering regimes (WATER). The number of mildew spots (SPOTS) was counted for six plants in each treatment combination (data stored in the *powdery mildew* dataset). For the frequency of spots to be considered a Poisson process, the occurrence of spots needs to be independent, and the plants need to be equally susceptible. Initially the assumption will be made that this is the case.

A contingency table analysis of this data set is presented in Box 13.7. The chi-squared value is not significant ($p = 0.798$). The conclusion is that the proportion of spots in the different water treatments is the same across the different strains. Our conclusion is not that the number of spots in the different water treatments are the same, nor that the number of spots is the same for the different strains. The chi-squared statistic of a contingency table analysis is testing the **interaction** between the two explanatory variables.

It is also possible to analyse this data set using GLM methods. In a first attempt at doing this in Box 13.8, the researcher has fitted the model

$$\text{SPOTS} = \text{WATER} + \text{STRAIN}.$$

The plot of the residuals against the fitted values in Fig. 13.4 however illustrates that there is clearly a problem with heterogeneity of variance. This is to be expected if the data follow a Poisson distribution, as the variance increases with the mean. The answer is to transform the data. In fact, if the data really do follow a Poisson distribution, the problem should be solved exactly by the square root transformation. The analysis of the square root of spots is given in Box 13.9. The residual graphs of Fig. 13.5 shows some improvement—but is this sufficient? In this case there is a further method by which we can test the null hypothesis that the data are Poisson. After the square root transformation, and the statistical elimination of important explanatory variables, if the data are drawn from a Poisson distribution, the error mean square should on average equal 0.25. (It is not necessary to understand why this is the case, to make use of this fact.)

BOX 13.7 Contingency table analysis for the resistance of barley to powdery mildew

Contingency table analysis

Rows: WATER Columns: STRAIN

	1	2	3	4	5	ALL
1	6 5.27 0.32	5 6.02 –0.42	5 6.32 –0.53	6 5.12 0.39	6 5.27 0.32	28
2	17 15.62 0.35	14 17.85 –0.91	17 18.74 –0.40	16 15.17 0.21	19 15.62 0.86	83
3	5 5.27 –0.12	7 6.02 0.40	6 6.32 –0.13	5 5.12 –0.05	5 5.27 –0.12	28
4	7 8.84 –0.62	14 10.11 1.22	14 10.61 1.04	7 8.59 –0.54	5 8.84 –1.29	47
ALL	35	40	42	34	35	186

Chi-sq = 7.832, DF = 12, p-value = 0.798

Cell contents:
Count
Exp freq
St resid

BOX 13.8 GLM analysis of the number of powdery mildew spots on barley

General Linear Model

Word equation: SPOTS = WATER + STRAIN

WATER and STRAIN are categorical

Analysis of variance table for SPOTS, using Adjusted SS for tests

Source	DF	Seq SS	Adj SS	Adj MS	F	P
WATER	3	403.400	403.400	134.467	20.66	0.000
STRAIN	4	12.700	12.700	3.175	0.49	0.745
Error	12	78.100	78.100	6.508		
Total	19	494.200				

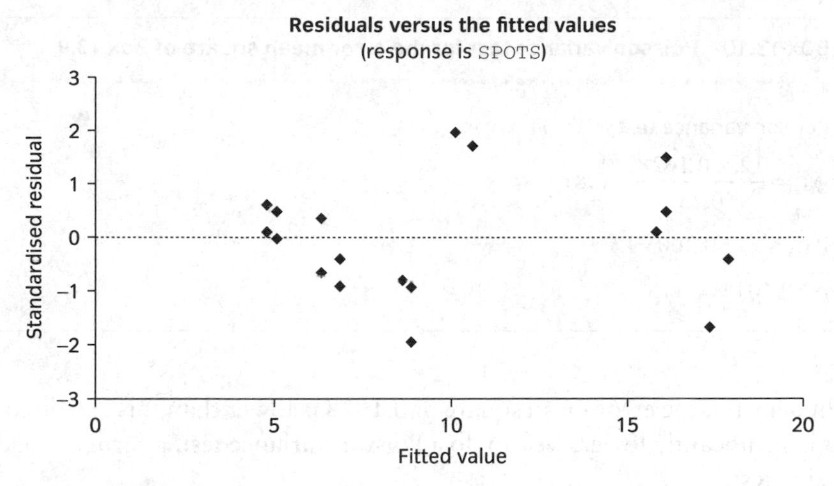

Fig. 13.4 Residual plot for SPOTS.

BOX 13.9 GLM analysis of the square root of the number of powdery mildew spots on barley

General Linear Model

Word equation: SQRTSP = WATER + STRAIN
WATER and STRAIN are categorical

Analysis of variance table for SQRTSP, using Adjusted SS for tests

Source	DF	Seq SS	Adj SS	Adj MS	F	P
WATER	3	9.7311	9.7311	3.2437	19.92	0.000
STRAIN	4	0.3728	0.3728	0.0932	0.57	0.688
Error	12	1.9541	1.9541	0.1628		
Total	19	12.0579				

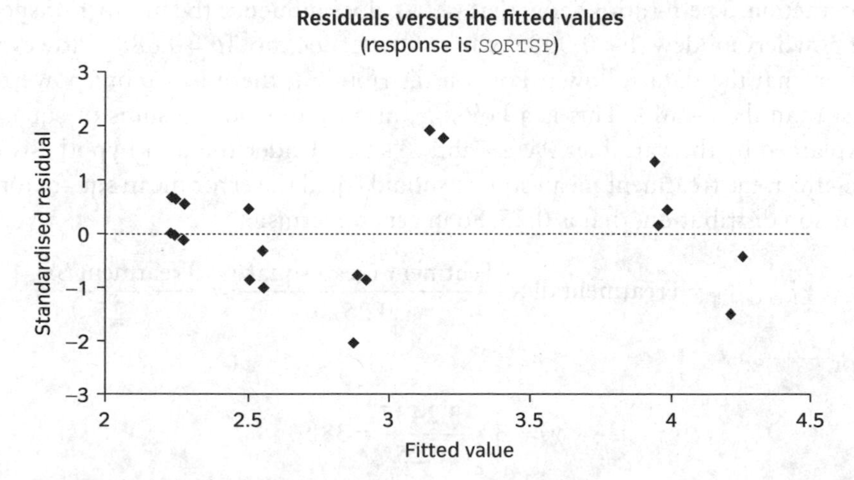

Fig. 13.5 Residual plot for SQRTSP.

BOX 13.10 Poisson variance test for the error mean square of Box 13.9

Poisson variance test

$$\chi^1_{12} = \frac{12 \times 0.1628}{0.25} = 7.81$$

$P(X \leq x) = 0.2005$

$p = 0.401$

In our case, the error mean square of 0.1628 is lower than this. To discover if it is significantly lower, we can do a **Poisson variance test**, a form of the dispersion test.

$$\chi^2_{\text{Error df}} = \text{Error df} \times \frac{\text{Error Mean Square}}{0.25}.$$

The null hypothesis is that the error distribution is Poisson.

The Poisson variance test is carried out in Box 13.10 for the analysis of `SQRTSP` from Box 13.9. Once again, this is a two tailed test, because we are investigating whether the error mean square differs significantly from 0.25. If it was significantly greater, then this would indicate overdispersion, and if significantly less, then the data would be underdispersed. The chi-squared value of 7.81 gives a cumulative probability of 0.2005. To convert this to the p-value relevant to our test, this should be multiplied by 2 to give 0.401. We do not reject the null hypothesis that the data derive from a Poisson distribution.

So we have finally reached the position where we can accept the GLM analysis of `SQRTSP` as being a valid analysis of Poisson data. What conclusions do we draw from this analysis? The **GLM analysis** investigates the **main effects** of `WATER` and `STRAIN`, in contrast to the **contingency table**, which tested the **interaction**. The F-ratios show that `WATER` does influence the number of spots of powdery mildew ($p < 0.0005$) while `STRAIN` does not ($p = 0.688$). However, given that the data follow a Poisson distribution, there is a more powerful test than the F-ratio. This is a Poisson variance test on the sums of squares explained by the variables `WATER` and `STRAIN`. Under the null hypothesis of no effect, the treatment mean square should equal the error mean square for a Poisson distribution, that is 0.25. So in general terms:

$$\chi^2_{\text{Treatment df}} = \text{Treatment df} \times \frac{\text{Teatment mean square}}{0.25} = \frac{\text{Treatment SS}}{0.25}.$$

For `WATER`:

$$\chi^2_3 = 3 \times \frac{3.2437}{0.25} = 38.9.$$

For STRAIN:

$$\chi^2_4 = 4 \times \frac{0.0932}{0.25} = 1.49.$$

These chi-squared statistics return p-values of 0.000 and 0.828 respectively for a one tailed test (we are only interested in the chi-square being significantly greater than the hypothesised value). These results agree with our earlier conclusions. These tests are always approximate, but are the appropriate tests to do when we wish to rely on the assumption of Poisson variability. In this circumstance they are more powerful than the F-tests, because there is no need to use the error mean square to estimate the extent of error variation. In this case, the Poisson variance test suggests that the assumption of Poissonness is reasonable.

Finally, would it be possible to test the main effects using contingency table analysis? This could be done by collapsing the table to look at one explanatory variable at a time. Table 13.5 gives the appropriate contingency table for WATER—the data are the marginal totals from the larger contingency table. Under the null hypothesis, the number of spots should now be equally divided between the four categories, but otherwise the chi-squared statistic is calculated as before.

This gives $\chi^2_3 = 43.4$ which corresponds to $p = 0.0000$. A similar analysis can be done on the marginal totals of STRAIN, given in Table 13.6.

In this case $\chi^2_4 = 1.37$, which corresponds to $p = 0.850$. The conclusions from the two methods of analysis are consistent.

Table 13.5 Contingency table to investigate the main effect of WATER on powdery mildew

	WATER				
	1	2	3	4	
Observed	28	83	28	47	*186*
Expected	46.5	46.5	46.5	46.5	

Table 13.6 Contingency table to investigate the main effect of STRAIN on powdery mildew

	STRAIN					
	1	2	3	4	5	
Observed	35	40	42	34	35	*186*
Expected	37.2	37.2	37.2	37.2	37.2	

BOX 13.11 Analysing the square root of seedling numbers by treatment

General Linear Model

Word equation: SQRTSLD = TRT1

TRT1 is categorical

Analysis of variance table for SQRTSLD, using Adjusted SS for tests

Source	DF	Seq SS	Adj SS	Adj MS	F	P
TRT1	1	5.5586	5.5586	5.5586	9.49	0.005
Error	28	16.4063	16.4063	0.5859		
Total	29	21.9650				

This leaves the question of which method of analysis is to be preferred. The contingency table analysis does not provide a single analysis in which the main effects and interactions may be looked at together, nor is there a natural way to examine main effects in the presence of overdispersion. We are reaching the limits of GLMs here, and the next step would be to use log-linear models to analyse these sorts of data (see Chapter 14).

Omitting important variables

If categorical data are analysed by GLM methods, and the Poisson variance test suggests overdispersion, one cause may be the omission of an important explanatory variable. A field experiment was carried out to investigate the germination of seedlings under two different treatments (TRT1). The response variable was the number of seeds which germinated, so a GLM analysis was carried out on the square root of the seedling number. The data are taken from the *seedling germination* dataset. The output is given in Box 13.11.

The Poisson variance dispersion test gives:

$$\chi^2_{28} = 28 \times \frac{0.5859}{0.25} = 65.6$$

which gives a final p-value of <0.001, suggesting that the data are overdispersed. This may be caused by heterogeneous data. In fact, the experiment was carried out in three blocks, corresponding to three different greenhouses, so the data were reanalysed, including BLOCK1 as a factor (Box 13.12).

The dispersion test now gives:

$$\chi^2_{26} = 26 \times \frac{0.1561}{0.25} = 16.2$$

which gives a final p-value of 0.137. The null hypothesis of Poissonness is no longer rejected.

BOX 13.12 Analysing the square root of seedling numbers by block and treatment

General Linear Model

Word equation: SQRTSLD = BLOCK1 + TRT1

BLOCK1 and TRT1 are categorical

Analysis of variance table for SQRTSLD, using Adjusted SS for tests

Source	DF	Seq SS	Adj SS	Adj MS	F	P
BLOCK1	2	12.3474	12.3474	6.1737	39.55	0.000
TRT1	1	5.5586	5.5586	5.5586	35.61	0.000
Error	26	4.0590	4.0590	0.1561		
Total	29	21.9650				

Analysing uniformity

In contrast with the previous experiment, another student decided to collect data on the germination of Brassica seedlings in a set of vegetable beds. The six beds acted as blocks, and there were five treatments giving different levels of protection against herbivores. The number of cotyledons that emerged after one week were counted, and analysed using the square root transformation (Box 13.13). The data are stared in the *Brassica* dataset.

In this case, both BLOCK and TREAT are highly significant. In fact, once the variance due to these factors has been eliminated, there is very little variance left. The dispersion test for Poissonness gives

$$\chi^2_{20} = 20 \times \frac{0.0895}{0.25} = 7.16.$$

BOX 13.13 Analysing the square root of cotyledon numbers by block and treatment

General Linear Model

Word equation: SRNCOT = BLOCK + TREAT

BLOCK and TREAT are categorical

Analysis of variance table for SRNCOT, using Adjusted SS for tests

Source	DF	Seq SS	Adj SS	Adj MS	F	P
BLOCK	5	23.9098	23.9098	4.7820	53.45	0.000
TREAT	4	20.5340	20.5340	5.1335	57.37	0.000
Error	20	1.7895	1.7895	0.0895		
Total	29	46.2333				

This gives a final p-value of 0.00774. So the null hypothesis of Poissonness is rejected. Once these two explanatory variables have been taken into account, there is little variation in cotyledon number left—even less than would be expected by chance. In these situations, we fall back on the F-ratio tests, rather than performing the more powerful dispersion tests.

What kinds of factors may cause sub-Poisson variance? There are two obvious possibilities:

1. The fraction of seeds that germinated and were counted was not small. A fixed number of seeds were sown, and so the number of seedlings counted represent a fraction of a fixed total. The distribution which describes proportions is the binomial distribution, with variance npq (where n = total number of seeds, p = proportion which germinated, and $q = 1 - p$). If p is small, then q is close to 1, and the binomial variance can be approximated by the Poisson variance, np. However, if p is not small, then np will provide an overestimate of the variance observed.
2. Competition could produce a very even distribution of seedlings. If one seedling shades another, or secretes substances which inhibit the growth of other seedlings nearby, then the germination of each seed is no longer independent. The breaking of this assumption of the Poisson distribution will lead to sub-Poisson variation.

It is important to stress that the dispersion test is an approximate test. It is reasonably valid for degrees of freedom up to 100 when the mean is greater than 15. It is truly appalling when the mean is less than 5. This is one of the reasons why generalised linear models are more satisfactory for dealing with categorical data.

13.5 Summary

- For data to be truly categorical, they must be independent counts of individual items.
- Categorical data that are cross classified may be analysed by contingency table analysis.
- The distribution underlying the analysis of categorical data is the Poisson distribution. This is a one parameter distribution, in which the mean equals the variance.
- At low means (<5), the distribution has a strong right skew. Above a mean of five, the distribution is more symmetrical, and the normal approximation is close enough for use in the chi-square test.
- Two properties of a Poisson process are independence of items and homogeneity of containers. Independence refers to the occurrence of one event being independent

of all other events. Homogeneity refers to the probability of any event occurring remaining constant over specified intervals of time or space.

- The dispersion test may be used to test the equality of the mean and variance of a distribution. If the variance is significantly lower than the mean, the data are underdispersed (uniform). If the variance is significantly greater than the mean the data are overdispersed (aggregated). The validity of the dispersion test depends upon the mean and sample size.

- The chi-squared formula relies on the dual assumptions of the data deriving from a Poisson distribution, and the mean being large enough for the normal approximation to be valid.

- Inspecting the residuals in a contingency table analysis provides information about the direction of the effect. Highly significant χ^2 values with no apparent pattern amongst the residuals may suggest that the counts are not independent, and the significance is spurious.

- Categorical data may be analysed in the GLM framework by using the square root transformation. Models may then be fitted to test the main effects of contingency tables, while the $r \times c$ contingency table analysis tests the interaction.

- In the GLM analysis, the error mean square may be tested for its closeness to 0.25 by the Poisson variance test. If the data do derive from a Poisson distribution, the treatment effects may be examined using chi-square tests rather than F-ratio tests.

13.6 Exercises

Soya beans revisited

Twenty-four plots of Soya beans were divided into three equal groups in a completely randomised design, and each group received a different formulation of a selective weed-killer. We first looked at this *soya bean* dataset in the first exercise of Chapter 9 where we decided that the square root transformation stabilised the variance most successfully. Here, the square root transform of the Y variable, SQRTDAM, is analysed, and we examine the output more closely. See Box 13.14.

(1) If DAMAGE were a count of independent items, then a square root transformation should result in an error mean square of 0.25. Do you think it is plausible that DAMAGE is a count of independent items?

(2) Perform a formal test of the hypothesis via a Poisson variance test. What is the p-value for rejecting the Poisson hypothesis?

(3) Accepting the Poisson hypothesis, test for an effect of WDKLR using a chi-square test. What is the p-value?

(4) Do you prefer this chi-square test or the GLM F-ratio test? Why?

BOX 13.14 Soya bean data revisited

General Linear Model

Word equation: SQRTDAM = WDKLR
WDKLR is categorical

Analysis of variance table for SQRTDAM, using Adjusted SS for tests

Source	DF	Seq SS	Adj SS	Adj MS	F	P
WDKLR	2	46.180	46.180	23.090	83.18	0.000
Error	21	5.829	5.829	0.278		
Total	23	52.010				

Term	Coef	SECoef	T	P
Constant	5.3268	0.1075	49.53	0.000
WDKLR				
1	−1.6860	0.1521	−11.09	0.000
2	−0.0256	0.1521	−0.17	0.868
3	*1.7116*			

Fig trees in Costa Rica

The spatial distribution of a species of fig tree was studied in Costa Rica. The number of individuals was counted in each of 100 quadrats at each of three sites. The questions of interest were whether the sites varied in the density of individuals, and whether individuals were randomly distributed within sites. Two variables are stored in the *fig tree* dataset: SITE (1 to 3) and NINDIVS (the number of individuals in a quadrat).

(1) Under what assumptions could the data from site 1 be viewed as a sample of size 100 from a Poisson distribution?

(2) Using your package to obtain information on means and variances, what informal conclusions can you draw about the distribution of fig trees within and between sites?

(3) Perform formal tests of the hypotheses that fig trees are distributed randomly within and between sites, using dispersion tests.

(4) Perform a GLM analysis to answer the same questions.

(5) Suggest some potential reasons why a species may be more variable in its numbers in quadrats than is consistent with a Poisson distribution.

(6) Suggest some potential reasons why a species may be less variable in its numbers in quadrats than is consistent with a Poisson distribution.

14 What lies beyond?

The *General* Linear Model is indeed General, and the methods you will have learnt from this book will suffice for an important range of statistical problems. We can have as many variables as we like in a model (subject to the availability of degrees of freedom), each can be continuous or categorical, and we can include interactions at will (subject to considerations of marginality). Hypothesis testing and estimation are routinely available. But there are limits to what the General Linear Model can do for us. Even though it represents a substantial goal for the statistics component of an undergraduate biology course, there are many situations in which it is not enough. This chapter aims to indicate some of the types of problem for which further methods are required, and to cast a glance at potential solutions.

14.1 Generalised Linear Models

The assumptions of GLM do not apply to categorical data. In Chapter 13, we therefore looked at other methods, and also looked at ways of transforming categorical data to make GLMs more reasonable. Let us take a step back from those solutions, and ask what exactly goes wrong if we apply GLM methods to categorical data? Table 14.1 lists the differences between data with Normal error, suitable for GLM, and categorical data with Poisson error.

Table 14.1 Comparing data with Normal and Poisson errors

Normal/GLM	Poisson/discrete
Symmetric and continuous error	Asymmetric and discrete error
Variance independent of mean (=homogeneity of variance)	Variance dependent on the mean (=heterogeneity of variance)
Variance unknown	Variance known (=mean)
Adding is natural	Multiplying is natural

A class of models called *Generalised Linear Models* extends the range of General Linear Models to cover both categorical data, and other situations. It does this by allowing three kinds of flexibility, involving specifying three elements, as follows

1. *Variance function.* This function relates how the error variance depends on the fitted value. Thus, for an ordinary GLM, the variance function just equals one for all possible fitted values, as the scatter is assumed to be the same across the whole model. For Poisson error, the variance function equals the fitted value, because the variance of a Poisson distribution equals its mean. In general, any positive function can be specified.
2. *Link function.* The right hand side of fitted value equations are of exactly the same form in Generalised Linear Models as they take in General Linear Models. They have categorical terms with conditional brackets, continuous terms with slopes, and all the kinds of interaction. But instead of making this equal the fitted value directly, the expression is called the Linear Predictor (LP). Then we have to specify how the Linear Predictor is turned into the Fitted Value (FV). The Identity Link, in which we just make Linear Predictor equal to the Fitted Value, is obviously the link function for an ordinary GLM. The log link, which specifies LP = log(FV), in other words FV = exp(LP), essentially causes the effects of the x variables on fitted values to multiply instead of add. (Link functions are named for the function that transforms FV to LP rather than the more obvious way round.) Thus the log link is suitable for categorical data, in which the natural null hypotheses are multiplicative. Other links are suitable for other purposes, and in principle any monotonic function can be used.
3. *Whether the error variance is known or needs to be estimated.* The error variance is taken equal to a scale factor times the variance function. Sometimes this scale factor has to be estimated, as in the ordinary GLM, where we estimate s^2. Sometimes, as with Poisson errors, we know that variance equals the mean, so the scale factor is known in advance to equal 1. Over- or underdispersion in categorical data is handled by allowing the scale factor to be estimated. Thus the final element is to decide that the scale factor should be estimated from the data, or to specify its numerical value.

Apart from these three extra elements, Generalised Linear Models are fundamentally the same as General Linear Models.

The method for categorical data thus uses the *log* link function, and a *linear* variance–mean relationship; and this is why the Generalised Linear Model suitable for categorical data is called the log–linear model. It will handle all kinds of analyses that chi-square methods and square root transformations will not. For example, three-way contingency tables are simply log–linear models with three categorical x variables. One particularly important way that three-way tables arise is through combining a number of two-way tables.

Continuous variables can be included as explanatory variables in log–linear models. If you have serious analyses of categorical data to perform, then log–linear models are the right method.

Another frequent situation in biology is to have a binary response variable, such as dead or alive, male or female. These data are likely to follow a binomial distribution rather than a Normal one. For these cases the Generalised Linear Model with the logistic link, and binomial error, provides what is known as the Logistic Regression. Again there can be categorical and/or continuous explanatory variables, and the possibility of interactions between them in their influence on the *y* variable.

The last use we will mention here of Generalised Linear Models concerns the issue we discussed in Chapter 9, when we saw that transformations were very useful in combatting broken assumptions in a GLM. However, we also saw the possibility that a transformation that brought about homogeneity of variance might not correct non-linearity. With Generalised Linear Models, we have much more scope to find an analysis that meets the assumptions, because the generalisation allows a variance function to be specified, and a link function. Problems of heterogeneity and non-linearity are therefore much more easily soluble.

It is important to understand how much is familiar in the Generalised Linear Model. There are model formulae, parameters with coefficients and standard errors, main effects and interactions, and nesting. Adjusted means are calculated from the coefficients, using a fitted value equation. You can still do polynomial regressions, and marginality must still be respected. There are fitted values, residuals and model criticism, and outliers and influential datapoints. Variation is partitioned and statistical elimination is an important concept. We often have to make choices of model, and have the dubious assistance available of stepwise methods. Computers understand model formulae for Generalised Linear Models, just as humans do, so you can issue commands to perform analyses in the same language in which you discuss them. Learning about the General Linear Model in this book is therefore an excellent preparation for this larger family of more advanced techniques.

14.2 Multiple *y* variables, repeated measures and within-subject factors

All but one of the examples in this book have been restricted to a single *y* variable at a time, and severe warnings have been made about independence of datapoints and the problems of repeated measures. Repeated measures include time-series, and analyses that involve what are called in social sciences 'within-subject factors', and are involved in many important biological applications. Readers should be aware that many biologists and statisticians would regard the views we put forward in Chapter 8 (namely, that repeated measures on the same individual should not be allowed to occupy more than one datapoint, and if required multiple measurements can be encoded in multiple *y* variables leading to a MANOVA) as too severe. Methods and packages are available for dealing with repeated measures that do permit many datapoints for each individual. At the level of professional acceptability, the important thing is to use methods that are well regarded within your particular field. Good practice evolves, and it would not surprise us if there was a trend towards the principle of one datapoint per individual, as MANOVA methods become more widely understood. There is a moral quality to the extent of replication, and we believe in what might loosely be termed the 'deserved degrees of freedom'.

14.3 Conclusions

Generalised Linear Models are therefore an extremely powerful tool that extends what we know of General Linear Models, allowing us to tackle an even wider range of statistical problems. The availability of Generalised Linear Model facilities on computer platforms is unfortunately not yet quite as good as that of General Linear Models. But they are the way of the future, and reading this book will prepare you for learning about them.

15 Answers to exercises

Chapter 1

Melons

(1) The null hypothesis is that there is no difference in the mean yield between the varieties of melon.

(2) The null hypothesis is rejected (with $p < 0.0005$). We would conclude that there are significant differences in the mean yield of melons between the varieties. We estimate that variety 2 has the highest mean yield, and varieties 1 and 3 the lowest mean yields.

(3) The model produces an estimate of 15.6 for the unexplained variance with 18 degrees of freedom.

(4) The standard error of the mean is calculated by $\frac{s}{\sqrt{n}}$ where $s = \sqrt{15.6} = 3.95$.

This gives a standard error of 1.612 for varieties 1, 2 and 4, and a standard error of 1.975 for Variety 3.

(5) This information could be presented as means and their associated confidence intervals. The formula for a confidence interval is:

$$\text{Mean} \pm t_{\text{crit}}\text{SE}_{\text{mean}}.$$

In this case, the critical t value could be for a 95% confidence interval and must have 18 degrees of freedom, giving 2.10. This gives the intervals presented in Table 15.1.

Table 15.1 Confidence intervals

Mean	95% Confidence interval
20.49	(17.11, 23.88)
37.40	(34.02, 40.79)
20.46	(16.32, 24.61)
29.90	(26.51, 33.28)

BOX 15.1 Analysis for dioecious trees

Word Equation: FLOWERS = SEX

SEX is categorical

Analysis of variance table for FLOWERS

Source	DF	SS	MS	F	P
SEX	1	171841	171841	1.18	0.284
Error	48	7017255	146193		
Total	49	7189097			

Dioecious trees

(1) SEX is a categorical variable with two levels. To test the null hypothesis that male and female trees produce the same flowers, we need to fit the word equation

FLOWERS = SEX.

This would give the ANOVA table of Box 15.1.

We would therefore conclude that male and female trees do not have significantly different numbers of flowers.

(2) The data could be illustrated graphically in a number of ways. Here is a boxplot, in which the rectangle represents the middle 50% of the data, the line across the box being the median (middle value), and the tails stretching between the upper quartile and the maximum value, and between the lower quartile and the minimum. See Fig. 15.1.

This illustrates that while the medians are very close, female trees have much greater variability in the number of flowers than males.

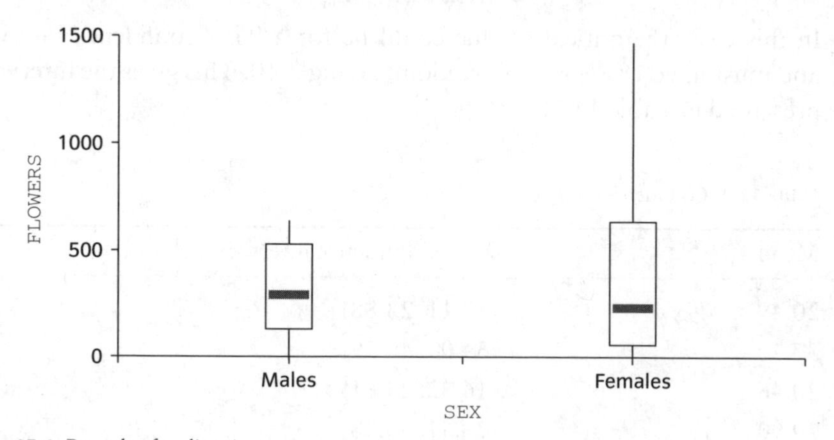

Fig. 15.1 Box plot for dioecious trees.

Chapter 2

Does weight mean fat?

(1) The fitted model is given by the equation FAT = 26.9 + 0.0207 WEIGHT.

(2) The proportion of explained variance is $\frac{1.33}{218.42} = 0.006$.

(3) The slope is estimated to be 0.02069, with the standard error of this estimate as 0.06414. This information could also be presented as a confidence interval, with the critical t value having 17 degrees of freedom. This would give (−0.115, +0.156) as the 95% confidence interval for the slope.

(4) The slope is not significantly different from zero ($p = 0.751$).

(5) A zero slope would imply that WEIGHT provides no information about FAT.

Dioecious trees

(1) Plotting flowers against DBH gives the graph of Fig. 15.2.

(2) A regression analysis would use the word equation, FLOWERS = DBH, and would provide the output shown in Box 15.2.

This gives the fitted values equation as: FLOWERS = −481.16 + 4.5128 DBH.

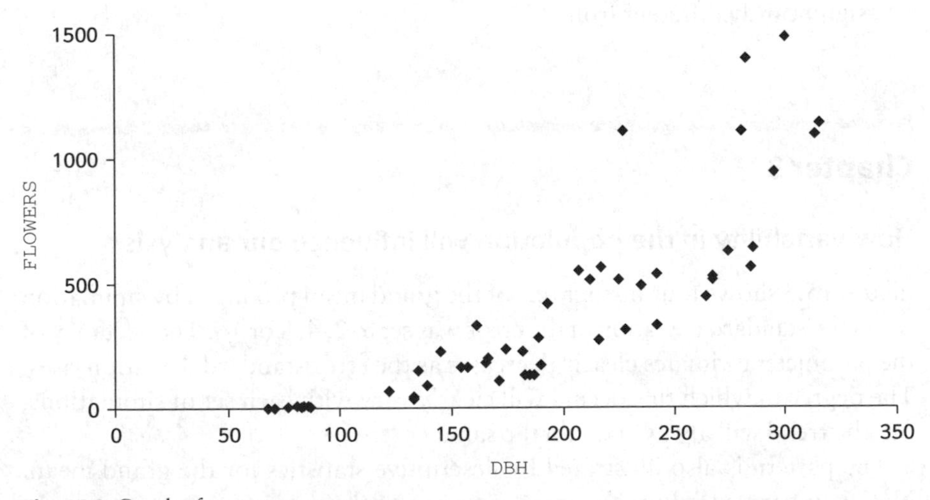

Fig. 15.2 Graph of FLOWERS versus DBH.

BOX 15.2 Analysis for dioecious trees

Regression analysis

Word Equation: FLOWERS = DBH

DBH is continuous

Analysis of variance table for FLOWERS

Source	DF	SS	MS	F	P
Regression	1	5060723	5060723	114.13	0.000
Residual Error	48	2128374	44341		
Total	49	7189097			

Coefficients table

Predictor	Coef	SECoef	T	P
Constant	−481.16	86.24	−5.58	0.000
DBH	4.5128	0.4224	10.68	0.000

(3) To test the null hypothesis that the slope is not significantly different from 4, we would calculate the test statistic as:

$$t_s = \frac{4.5128 - 4}{0.4224} = 1.21$$

for which $p = 0.232$. Therefore we conclude that the slope is not significantly different from 4.

Chapter 3

How variability in the population will influence our analysis

Figure 15.3 shows four histograms of the grand mean produced by simulation when the standard deviation of the error was set to 2, 4, 8 or 16. The accuracy of the parameter estimates clearly decreases as the error standard deviation rises. The degree to which this occurs will clearly vary with each set of simulations, but the trend will always remain the same.

The pattern is also illustrated by descriptive statistics for the grand mean. Notice in particular how the standard error of the mean (an inverse measure of the accuracy of our parameter estimates) increases with σ. Box 15.3 gives details. The standard error of the mean is given in the StDev column as it is the standard deviation of the sampling distribution.

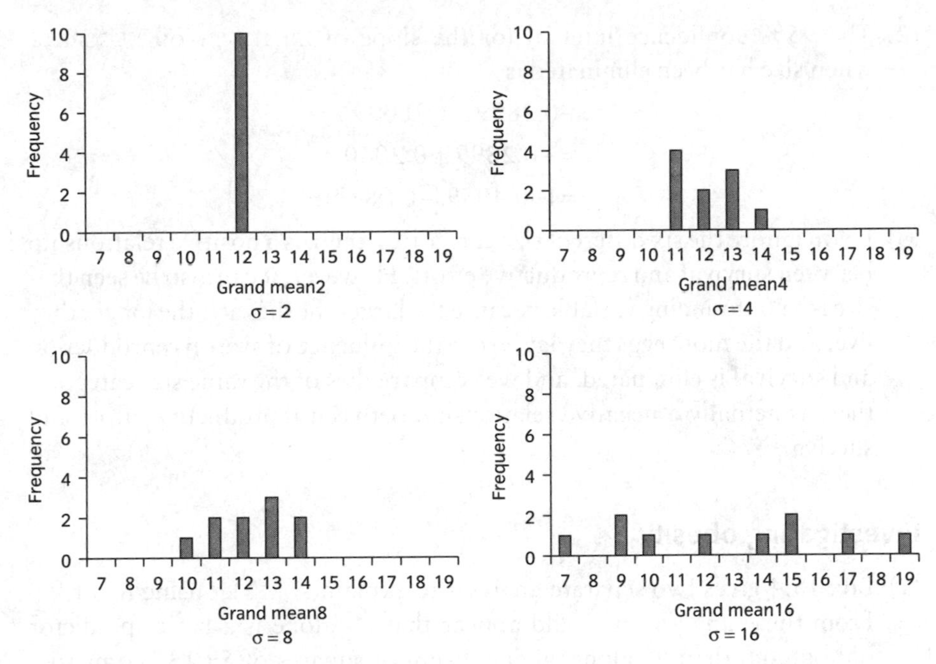

Fig. 15.3 Variation of parameter estimates.

BOX 15.3 Accuracy of grand mean

Descriptive statistics for the grand mean when σ is varied

Sigma (σ)	N	Mean	Median	StDev
2	10	12.114	12.154	0.296
4	10	12.217	11.998	1.198
8	10	12.285	12.636	1.241
16	10	12.71	13.05	3.94

Chapter 4

The cost of reproduction

(1) The 95% confidence interval for the slope of LLONGVTY on LEGGRATE is

$$\begin{aligned} &0.2813 \pm t_{24}0.1165 \\ &= 0.2813 \pm 0.2405 \\ &= (0.0408, 0.5218) \end{aligned}$$

(2) The 95% confidence interval for the slope of LLONGVTY on LEGGRATE when size has been eliminated is

$$-0.2899 \pm t_{23}0.0996$$
$$= -0.2899 \pm 0.2060$$
$$= (-0.4959, -0.0839)$$

(3) If you ignore the six different size categories, there is a positive relationship between survival and reproductive effort. However, it can also be seen that size is a confounding variable because the larger the flies are, the longer they live, and the more eggs they lay. Once the influence of size on reproduction and survival is eliminated, and we compare flies of the same size category, there is actually a negative relationship between reproductive effort and survival.

Investigating obesity

(1) Box 15.4 gives two separate analyses to explain FOREARM using HT or WT. From these analyses it would appear that WT alone is a better predictor of FOREARM than HT alone, giving a sum of squares of 59.137 compared to 0.944. That someone's weight can act as a predictor of obesity, but their height cannot, seems intuitively sensible.

(2) Box 15.5 shows the analysis using both explanatory variables together. From this it can be seen that the *F*-ratios (based on the adjusted sums of squares) for both WT and HT have increased—so together they increase each other's informativeness. In fact, HT is now significant ($p = 0.009$). This is because the combination of someone's height and weight provides much better predictive power for obesity than knowing one or other of these pieces of information.

BOX 15.4(a) First analysis of FOREARM

General Linear Model

Word equation: FOREARM = HT

HT is continuous

Analysis of variance table for FOREARM, using Adjusted SS for tests

Source	DF	Seq SS	Adj SS	Adj MS	F	P
HT	1	0.944	0.944	0.944	0.18	0.678
Error	37	199.094	199.094	5.381		
Total	38	200.038				

BOX 15.4(b) Second analysis of FOREARM

General Linear Model

Word equation: FOREARM = WT
WT is continuous.

Analysis of variance table for FOREARM, using Adjusted SS for tests

Source	DF	Seq SS	Adj SS	Adj MS	F	P
WT	1	59.137	59.137	59.137	15.53	0.000
Error	37	140.901	140.901	3.808		
Total	38	200.038				

BOX 15.5 Analysis of FOREARM using both explanatory varaiables

General Linear Model

Word equation: FOREARM = HT + WT
HT and WT are continuous

Analysis of variance table for FOREARM, using Adjusted SS for tests

Source	DF	Seq SS	Adj SS	Adj MS	F	P
HT	1	0.944	24.777	24.777	7.68	0.009
WT	1	82.970	82.970	82.970	25.72	0.000
Error	36	116.124	116.124	3.226		
Total	38	200.038				

Term	Coef	SE Coef	T	P
Constant	17.452	8.274	2.11	0.042
HT	−0.17173	0.06196	−2.77	0.009
WT	0.23317	0.04598	5.07	0.000

(4) These patterns and conclusions could be predicted from the third analysis alone by comparing the sequential and adjusted sums of squares for the two explanatory variables. There is relatively little difference for the variable WT, but a substantial difference for the variable HT. The low sequential sum of squares for height indicates that alone in the model it has poor explanatory power. The high adjusted sum of squares however indicates improved explanatory power with WT in the model.

Chapter 5

Growing carnations

(1) Yes—the sequential and adjusted sums of squares are identical, indicating that this data set is orthogonal.

(2) Yes—it was worthwhile blocking for BED, as this explained a significant amount of variation. Without BED as an explanatory variable, the variation explained by BED has been left as error variation, so reducing the *F*-ratios for the two treatments (and reducing the precision of all parameter estimates).

(3) The sequential and adjusted sums of squares are no longer exactly the same in Box 5.8, owing to loss of orthogonality. However, the differences are only slight, and do not alter our conclusions about which variables are significant.

(4) The two graphs are displayed in Fig. 15.4.

The dorsal crest of the male smooth newt

(1) The analysis of Box 15.6 shows that LSVL is a significant predictor of LCREST giving $p < 0.0005$.

(2) It is a good idea to include POND as local conditions may well influence the relationship between LCREST and LSVL. In this particular study however POND is insignificant, so its inclusion does not matter in this case as $p = 0.881$, see Box 15.7.

(3) If the data were collected over the breeding season, it may well have been the case that the crest was growing during the course of the study—and

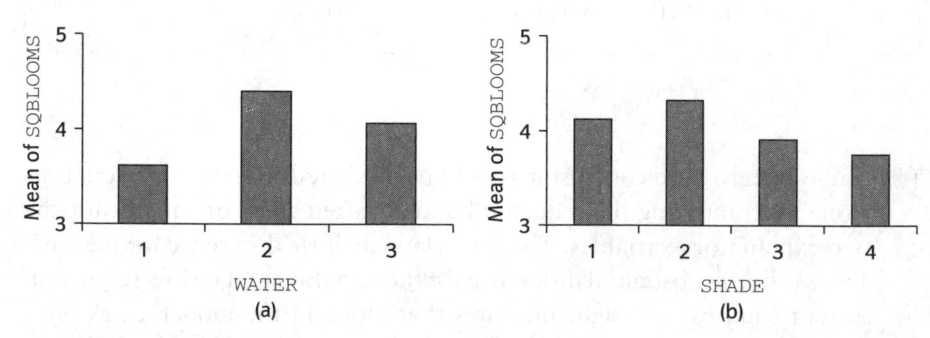

Fig. 15.4 Barcharts for SQBLOOMS.

BOX 15.6 Dorsal crest analysis

General Linear Model

Word equation: LCREST = LSVL

LSVL is continuous

Analysis of variance table for LCREST, using Adjusted SS for tests

Source	DF	Seq SS	Adj SS	Adj MS	F	P
LSVL	1	2.3894	2.3894	2.3894	45.81	0.000
Error	85	4.4337	4.4337	0.0522		
Total	86	6.8231				

Coefficients table

Term	Coef	SECoef	T	P
Constant	−9.381	1.501	−6.25	0.000
LSVL	5.0870	0.7516	6.77	0.000

BOX 15.7 Further dorsal crest analysis

General Linear Model

Word equation: LCREST = POND + LSVL

POND is categorical, LSVL is continuous

Analysis of variance table for LCREST, using Adjusted SS for tests

Source	DF	Seq SS	Adj SS	Adj MS	F	P
POND	9	0.32519	0.24063	0.02674	0.48	0.881
LSVL	1	2.30483	2.30483	2.30483	41.78	0.000
Error	76	4.19310	4.19310	0.05517		
Total	86	6.82312				

therefore the relationship between LCREST and LSVL would almost certainly have changed as the data were collected. In this case, inclusion of DATE could eliminate any such seasonal effects, and allow the relationship between LCREST and LSVL to be investigated over and above any seasonal changes. To detect these circumstances, DATE could be included as a continuous variable in the model: if significant then seasonal effects would be important.

Chapter 6

Conservation and its influence on biomass

(1) BIOMASS $= 2.21156 - 0.02443 - 0.002907 \times 200 + 0.10574 = 1.711$ (to 3 decimal places).

(2) BIOMASS $= 2.21156 + 0.02443 - 0.002907 \times 300 - 0.12526 = 1.239$ (to 3 decimal places).

(3) The evidence that the biomass of vegetation depends upon being in a conservation area is weak ($p = 0.101$, which is not significant). Biomass of vegetation in the sample was lower in conservation areas.

(4) There is strong evidence that soil type affects biomass ($p < 0.0005$), biomass being highest on chalk and lowest on loam.

(5) The slope of altitude on biomass represents the effect of an additional metre on biomass, a 95% confidence interval being given by:

$$\begin{aligned} &-0.002907 \pm t_{45} \times 0.00013 \\ &= -0.002907 \pm 2.0141 \times 0.00013 \\ &= (-0.00317, -0.00265). \end{aligned}$$

(6) The adjusted sum of squares for CONS is considerably lower than the sequential. This is due to a sharing of information between CONS, ALT and SOIL. This suggests that conservation areas tend to be more common at particular altitudes and/or on particular soil types. The very strong effect of CONS, based on Seq SS, shows that it has a high correlation with BIOMASS. However, the low Adj SS suggests that this could be accounted for by a correlation with ALT or SOIL.

(7) Randomised experiments allow us to infer causation from correlation, while observational studies do not.

Determinants of Grade Point Average

(1) An analysis of the *grades* data set is given in the Box 15.8. There is no evidence that either YEAR or MATH predict GPA ($p = 0.094$ and 0.124 respectively). The scores in the sample were higher for Year 1, and the trend in the sample was for higher MATH scores to be associated with higher GPA—but this trend is very slight, so the evidence that this is the case for the population is weak. The evidence that VERBAL predicts GPA is strong ($p < 0.0005$), with higher VERBAL scores being associated with higher GPA.

(2) For the first year with a VERBAL score of 700 and a MATH score of 600:

GPA $= 0.6582 + 0.06521 + 700 \times 0.002288 + 600 \times 0.000937 = 2.887$.

For the second year with a VERBAL score of 600 and a MATH score of 700:

GPA $= 0.6582 + (-0.06521) + 600 \times 0.002288 + 700 \times 0.000937 = 2.622$.

BOX 15.8 **Analysis of the grades dataset**

General Linear Model:

Word equation: GPA = YEAR + VERBAL + MATH

YEAR is categorical, VERBAL and MATH are continuous

Analysis of variance table for GPA, using Adjusted SS for tests

Source	DF	Seq SS	Adj SS	Adj MS	F	P
YEAR	1	1.1552	0.8460	0.8460	2.84	0.094
VERBAL	1	6.7595	5.1600	5.1600	17.32	0.000
MATH	1	0.7092	0.7092	0.7092	2.38	0.124
Error	196	58.3961	58.3961	0.2979		
Total	199	67.0200				

Coefficients table

Term	Coef	SE Coef	T	P
Constant	0.6582	0.4404	1.49	0.137
YEAR				
1	0.06521	0.03870	1.69	0.094
2	*−0.06521*			
VERBAL	0.002288	0.000550	4.16	0.000
MATH	0.000937	0.000608	1.54	0.124

Chapter 7

Antidotes

(1) The coefficients for the full model are given in Table 15.2. The coefficients give the interaction diagram of Fig. 15.5.

(2) From the ANOVA table, we conclude that the interaction ANTIDOTE × DOSE is significant ($p = 0.015$). In other words, the effectiveness of the two antidotes changed depending upon dose of toxin administered, to different degrees. The interaction diagram suggests that the effectiveness of Antidote 2 changed little with toxin dose, but with Antidote 1 there was a very marked change.

(3) Given that the interaction is significant, then the 'one complicated story' illustrated by the interaction diagram is a good way to present these results. An alternative would be to present the two by four table of means and their associated standard errors as in Table 15.3.

Table 15.2 Coefficients for antidote analysis

ANTIDOTE	DOSE 5	10	15	20
1	0.897	8.657	21.997	33.757
2	0.127	0.500	1.593	2.053

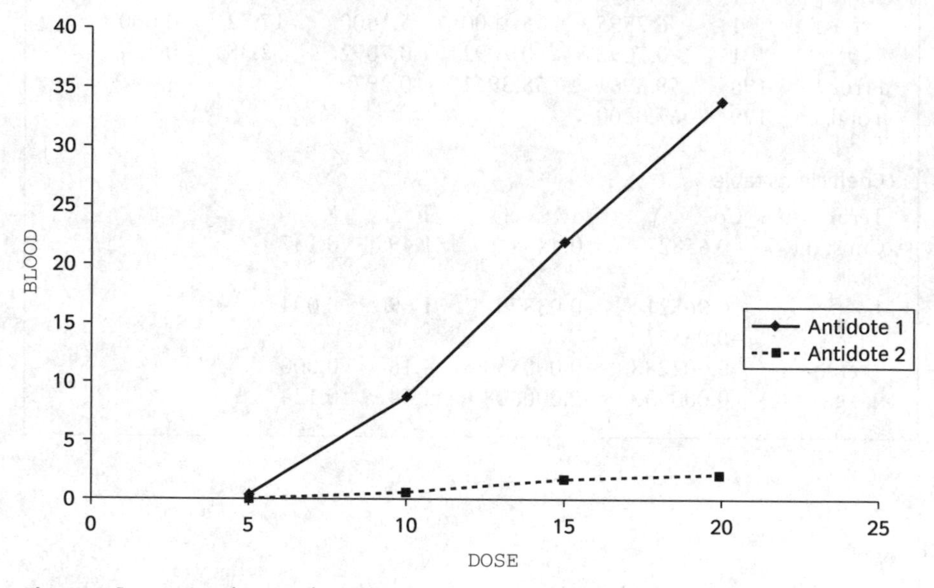

Fig. 15.5 Interaction diagram for antidote analysis.

Table 15.3 Means and their standard errors

ANTIDOTE × DOSE		Mean	SE Mean
1	5	0.8967	4.434
1	10	8.6567	4.434
1	15	21.9967	4.434
1	20	33.7567	4.434
2	5	0.1267	4.434
2	10	0.5000	4.434
2	15	1.5933	4.434
2	20	2.0533	4.434

There is another interesting point that we should notice about this table, and that is that the standard error of the mean is greater than the mean itself in many cases. The standard deviation (s) must be even greater. If s is a good estimate of the error standard deviation across the whole dataset, this implies negative concentrations, which are clearly nonsensical. The more obvious conclusion is that s is not a good estimate of error SD for all treatment combinations, and that the variance is heterogeneous. In Chapter 9 we will discuss how to deal with this.

Weight, fat and sex

The analysis in Box 15.9 fits the male and female relationships in the same analysis by use of the interaction term.

(1) This give the equation $\texttt{FAT} = 11.571 + 0.1855 \times \texttt{WEIGHT}$ for males.

(2) This give the equation $\texttt{FAT} = 5.239 + 0.4029 \times \texttt{WEIGHT}$ for females.

(3) There is evidence that the slopes differ, the interaction term having a p-value of 0.035.

BOX 15.9 Analysis for weight, fat and sex

General Linear Model

Word equation: FAT = SEX + WEIGHT + SEX * WEIGHT

SEX is categorical and WEIGHT is continuous

Analysis of variance table for FAT, using Adjusted SS for tests

Source	DF	Seq SS	Adj SS	Adj MS	F	P
SEX	1	90.321	2.108	2.108	1.05	0.322
WEIGHT	1	87.105	79.542	79.542	39.59	0.000
SEX * WEIGHT	1	10.857	10.857	10.857	5.40	0.035
Error	15	30.138	30.138	2.009		
Total	18	218.421				

Coefficients table

Term	Coef	SE Coef	T	P
Constant	8.405	3.091	2.72	0.016
SEX				
1	−3.166	3.091	−1.02	0.322
2	*3.166*			
WEIGHT	0.29420	0.04676	6.29	0.000
WEIGHT * SEX				
1	0.10869	0.04676	2.32	0.035
2	*−0.10869*			

Chapter 8

How non-independence can inflate sample size enormously

(1) The data points are not independent. There should be only one data point per sheep. If a sheep has a tendency to look up frequently (for example, it has a nervous disposition), then this will be the case during each of the 20 observation periods, yet this will have been taken as 20 independent pieces of information on look up rate.

(2) The average duration of feeding and the average number of look ups should be calculated for each sheep, and used to calculate an average look up rate (AVLUPRATE), giving a dataset of size 6, which is shown and analysed in Box 15.10. In contrast to Box 8.7, SEX is no longer significant ($p = 0.251$).

(3) The undergraduate has spent far too much time in scrutinising a very limited number of sheep. So far as this analysis is concerned he would be far better off recording data from a greater number of sheep. There could be more intricate analyses of sheep behaviour that would justify this intensity of observation.

Combining data from different experiments

(1) He does not realise that they may be analysed in one General Linear Model, as long as the data from the different years are blocked for year.

BOX 15.10 Sheep analysis

SEX	SHEEP	AVLUPRATE
1	1	0.132729
1	2	0.234845
1	3	0.167530
2	4	0.194874
2	5	0.217035
2	6	0.322807

General Linear Model

Word equation: AVLUPRAT = SEX

SEX is categorical

Analysis of variance table for AVLUPRAT, using Adjusted SS for tests

Source	DF	Seq SS	Adj SS	Adj MS	F	P
SEX	1	0.006641	0.006641	0.006641	1.80	0.251
Error	4	0.014739	0.014739	0.003685		
Total	5	0.021379				

BOX 15.11 **Bird data combined**

General Linear Model

Word equation: YOUNG = YEAR + SONGDAY

YEAR is categorical, SONGDAY is continuous

Analysis of variance table for YOUNG, using Adjusted SS for tests

Source	DF	Seq SS	Adj SS	Adj MS	F	P
YEAR	4	10.698	19.782	4.945	4.06	0.008
SONGDAY	1	12.202	12.202	12.202	10.02	0.003
Error	38	46.282	46.282	1.218		
Total	43	69.182				

Coefficients table

Term	Coef	SECoef	T	P
Constant	5.4922	0.8637	6.36	0.000
YEAR				
1	0.1776	0.3746	0.47	0.638
2	1.4089	0.3772	3.73	0.001
3	−0.4916	0.4761	−1.03	0.308
4	0.4453	0.3998	1.11	0.272
5	*−1.5402*			
SONGDAY	−0.10913	0.03448	−3.17	0.003

(2) I would combine all data into three variables: YEAR, SONGDAY and YOUNG, and then fit the word equation: YOUNG = YEAR + SONGDAY. This asks the question whether the number of young may be explained by the number of days spent singing, once the differences between the years have been taken into account.

(3) As the analysis in Box 15.11 illustrates, Dr Glaikit's hypothesis is supported by the data. The earlier a blackbird starts to sing, the more young are fledged ($p = 0.003$), and the number of young produced also do vary significantly from year to year ($p = 0.008$).

Chapter 9

Stabilising the variance

(1) The square root transformation provides the most stable variance within groups.

Stabilising the variance in a blocked experiment

(1) The response variable is a count—and therefore we might expect it to follow a Poisson distribution. This would require a square root transformation to stabilise the variance.

(2) Working on this assumption, the output shown in Box 15.12 was produced.

(3) See Fig. 15.6 and Box 15.13.

BOX 15.12 Analysis of blocked experiment

General Linear Model

Word equation: SQRTNCOT = BLOCK + TRTMNT

BLOCK and TRTMNT are categorical

Analysis of variance table for SQRTNCOT, using Adjusted SS for tests

Source	DF	Seq SS	Adj SS	Adj MS	F	P
BLOCK	5	23.9098	23.9098	4.7820	53.45	0.000
TRTMNT	4	20.5340	20.5340	5.1335	57.37	0.000
Error	20	1.7895	1.7895	0.0895		
Total	29	46.2333				

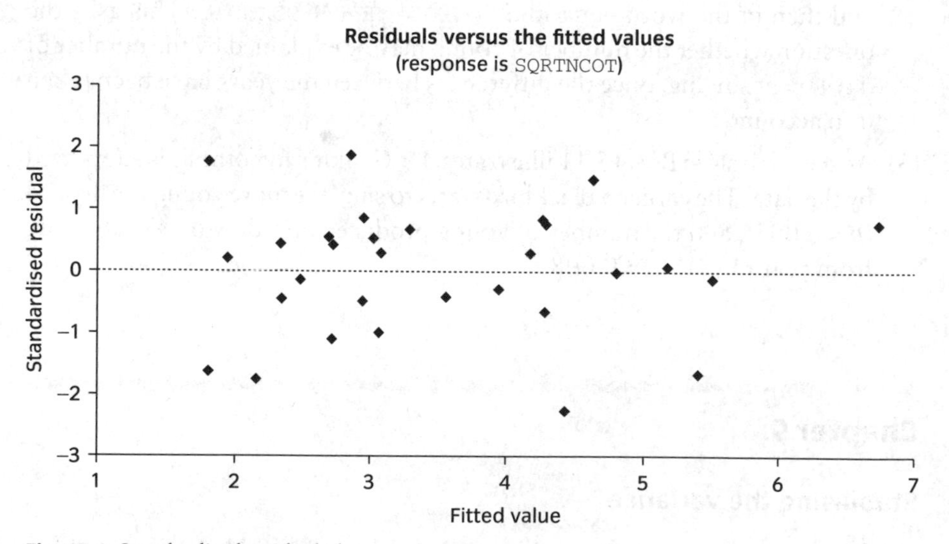

Fig. 15.6 Standardised residual plot.

BOX 15.13 Examining residuals of the square root transform analysis

Descriptive statistics for the standardised residuals by TRTMNT

TRTMNT	N	Mean	Median	StDev
1	6	−0.000	0.297	0.936
2	6	−0.000	−0.108	1.091
3	6	0.000	−0.075	0.425
4	6	−0.000	0.376	1.402
5	6	0.000	0.319	1.336

BOX 15.14 Analysis for log transform

General Linear Model

Word equation: LOGNCOT = BLOCK + TRTMNT

BLOCK and TRTMNT are categorical

Analysis of variance table for LOGNCOT, using Adjusted SS for tests

Source	DF	Seq SS	Adj SS	Adj MS	F	P
BLOCK	5	7.1245	7.1245	1.4249	21.54	0.000
TRTMNT	4	7.1755	7.1755	1.7939	27.12	0.000
Error	20	1.3230	1.3230	0.0662		
Total	29	15.6231				

(4) The other two transformations are given in Boxes 15.14–15.17 and Figs 15.7 and 15.8.

(5) As first expected, the square root transformation is best. This is confirmed by both the appearance of the residual plot and the similarly of the standard deviations of the residuals across treatments.

BOX 15.15 Examining residuals of the log transform analysis

Descriptive statistics for standardised residuals by TRTMNT

TRTMNT	N	Mean	Median	StDev
1	6	0.000	0.003	0.773
2	6	0.000	−0.155	0.870
3	6	0.000	−0.056	0.412
4	6	0.000	0.543	1.841
5	6	0.000	−0.264	1.042

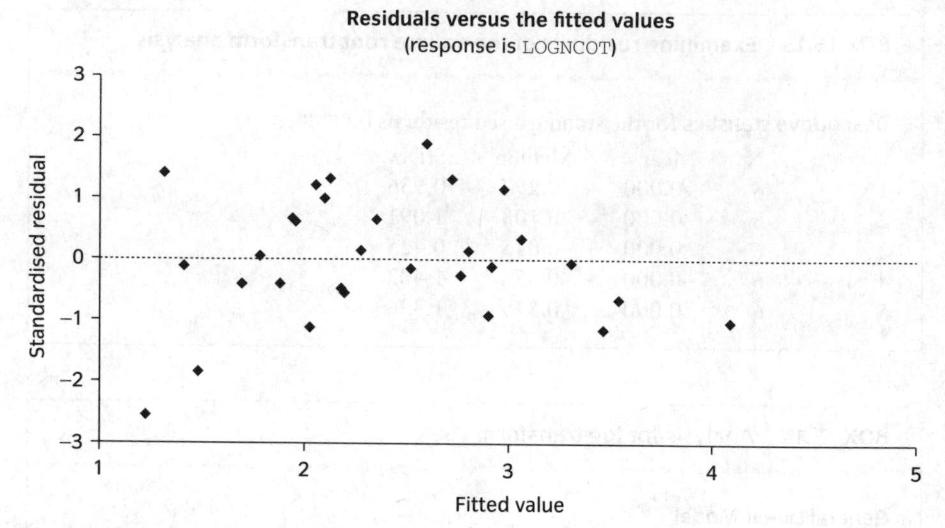

Fig. 15.7 Residual plot for log transform.

BOX 15.16 Analysis for inverse transform

General Linear Model

Word equation: INVNCOT = BLOCK + TRTMNT

BLOCK and TRTMNT are categorical

Analysis of variance table for INVNCOT, using Adjusted SS for tests

Source	DF	Seq SS	Adj SS	Adj MS	F	P
BLOCK	5	0.081283	0.081283	0.016257	3.74	0.015
TRTMNT	4	0.129070	0.129070	0.032267	7.43	0.001
Error	20	0.086895	0.086895	0.004345		
Total	29	0.297247				

BOX 15.17 Examining residuals of the inverse transform analysis

Descriptive statistics for standardised residuals by TRTMNT

TRTMNT	N	Mean	Median	StDev
1	6	0.000	0.176	0.658
2	6	0.000	0.176	0.624
3	6	0.000	−0.007	0.313
4	6	0.000	−0.612	2.084
5	6	0.000	−0.023	0.859

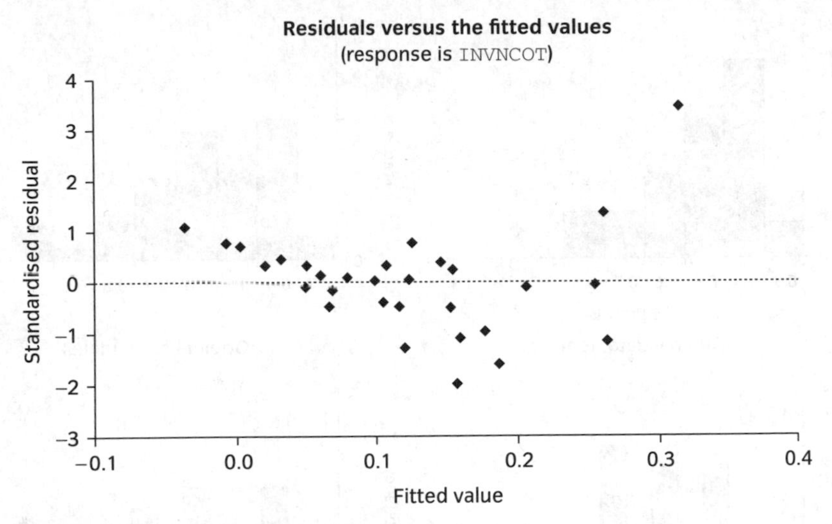

Fig. 15.8 Residual plot for inverse transform.

Lizard skulls

(1) The residual plot suggests both nonlinearity and inhomogeneity of variance.

(2) The data are allometric measurements—so we might expect to need a transformation—logarithms are often used in such cases (they compress larger values disproportionately more than smaller ones). Logs can also solve problems of nonlinearity.

(3) I prefer the second model—the residual plot is better, and the inclusion of site as a factor in the model allows any differences between the sites to be accounted for.

(4) The second residual plot does not show any evidence of nonlinearity or inhomogeneity.

(5) Site is not significant ($p = 0.992$). This does not necessarily mean that the lizards are of a similar size in the different sites. It does mean that the datapoints from the different sites all lie around the same straight line. Points from some sites may be further along the line than others.

(6) $t = \dfrac{1.9904 - 3}{0.2747} = -3.68$. This gives a p-value of 0.0004 for a t distribution with 70 degrees of freedom. From this, we could conclude that the slope is significantly different from 3.

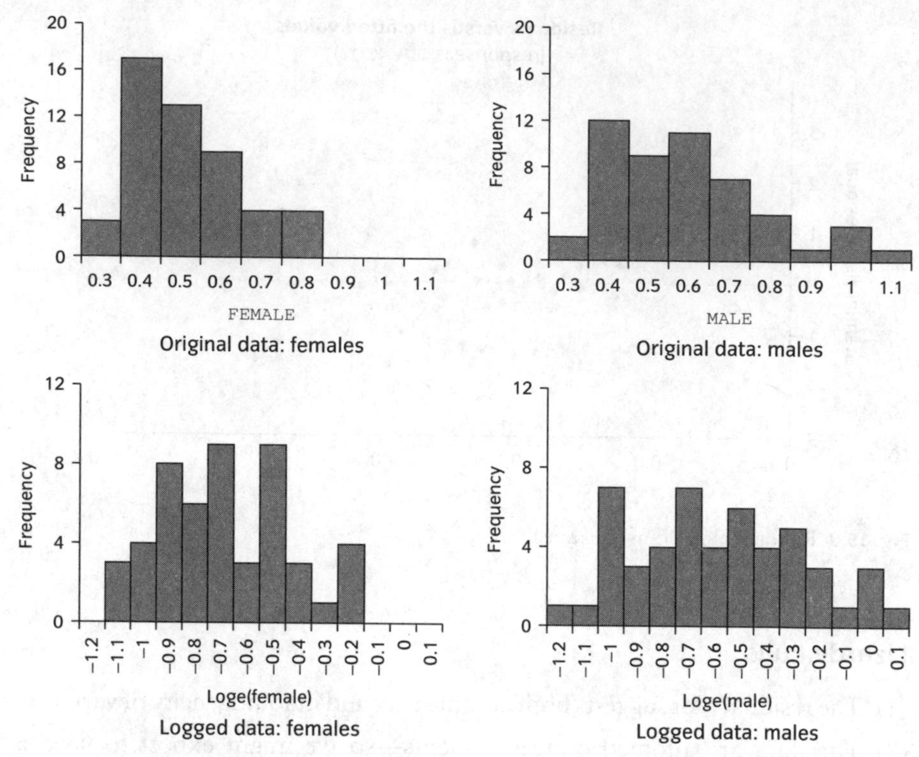

Fig. 15.9 Histograms for 'perfect' model.

Checking the 'perfect' model

(1) The original data and logged data produce the histograms in Fig. 15.9.

Both the original datasets appear to have right skew, while the logged data appear more symmetrical.

(2) The original data and the logged data produce the Normal probability plots illustrated in Fig. 15.10.

For both males and females, the original data deviate from the straight line, while the logged data are closer to the line. Many packages may provide a p-value to test the straightness of the Normal probability plot. Here, the p-value is sufficiently low for us to reject the hypothesis of Normality for the raw data, but not for the logged data ($p = 0.025$ and $p = 0.033$ for the raw data and $p = 0.411$ and $p = 0.775$ for logged data).

(3) Means and standard deviations for females and males are given in Table 15.4.

(4) Examples of some of the graphs produced from using mean and standard deviation of the female squirrel weights are given in Fig. 15.11. In spite of the fact that we know that these data were drawn at random from a

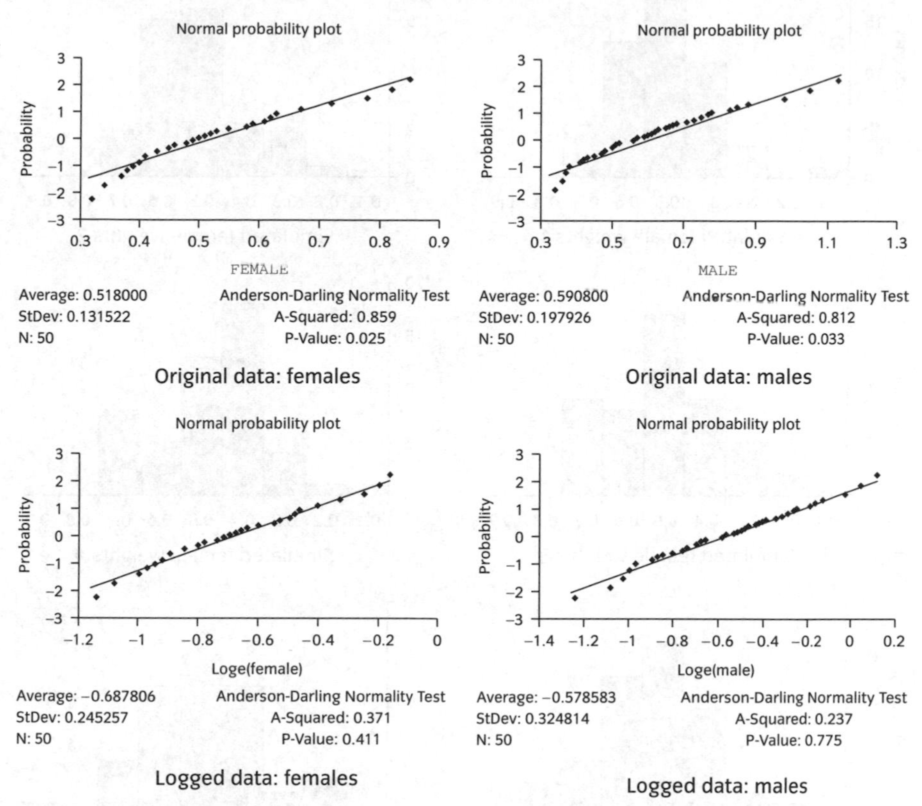

Fig. 15.10 Normal probability plots for the *squirrel* dataset.

Table 15.4 Means and standard deviations for the *squirrel* dataset

	Mean	Standard deviation
Females	0.5180	0.1315
Males	0.5908	0.1979

Normal distribution, with the respectable sample size of 50, the distributions still take a variety of shapes.

(5) Examples are given in Fig. 15.12. In these ten examples, none departs significantly from Normality according to the Anderson-Darling Normality Test, though one does come close at $p = 0.072$ (In fact we would expect 1 in 20 Normal datasets to depart from the straight line with a p-value of ≤ 0.05, as with all significance tests).

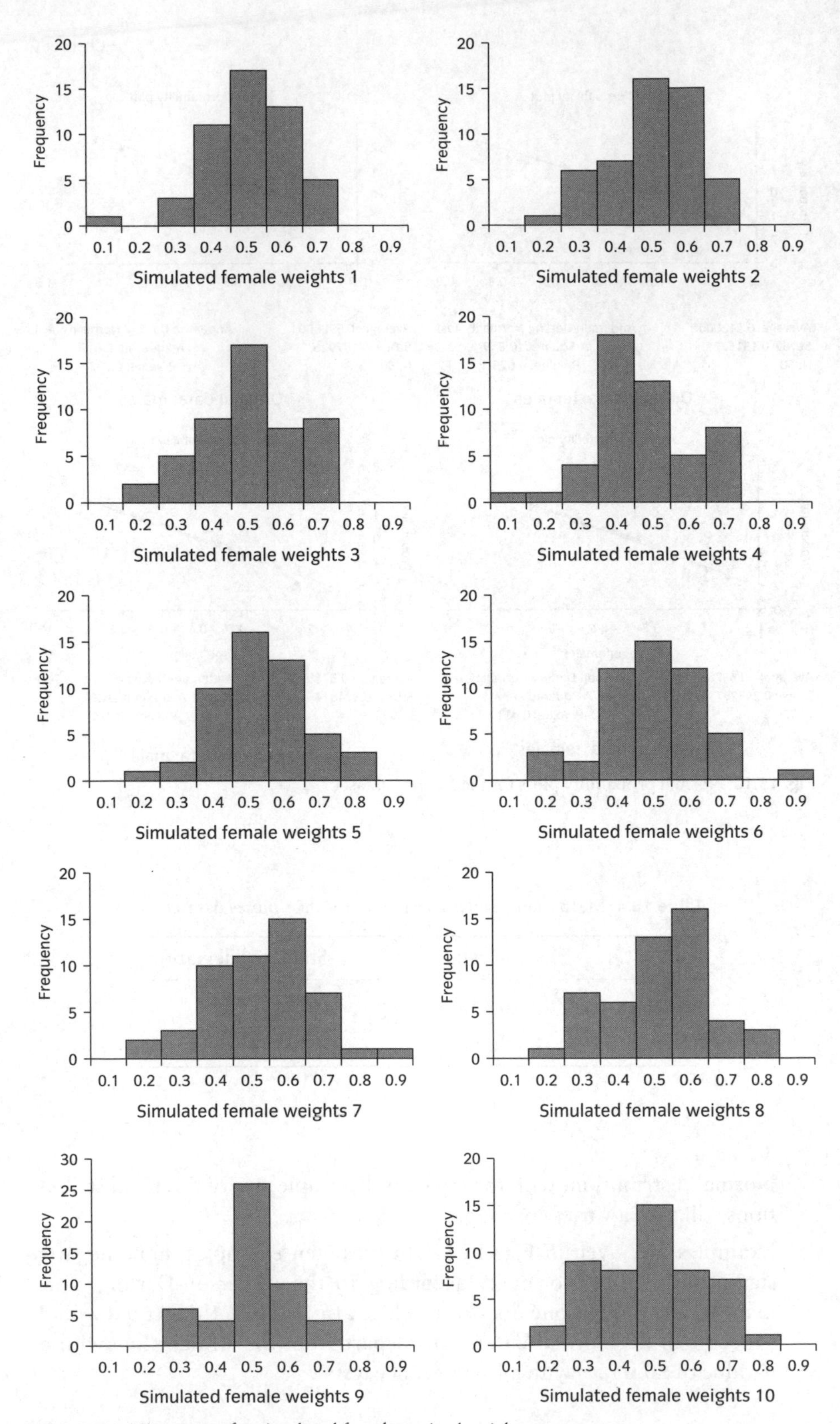

Fig. 15.11 Histograms for simulated female squirrel weights.

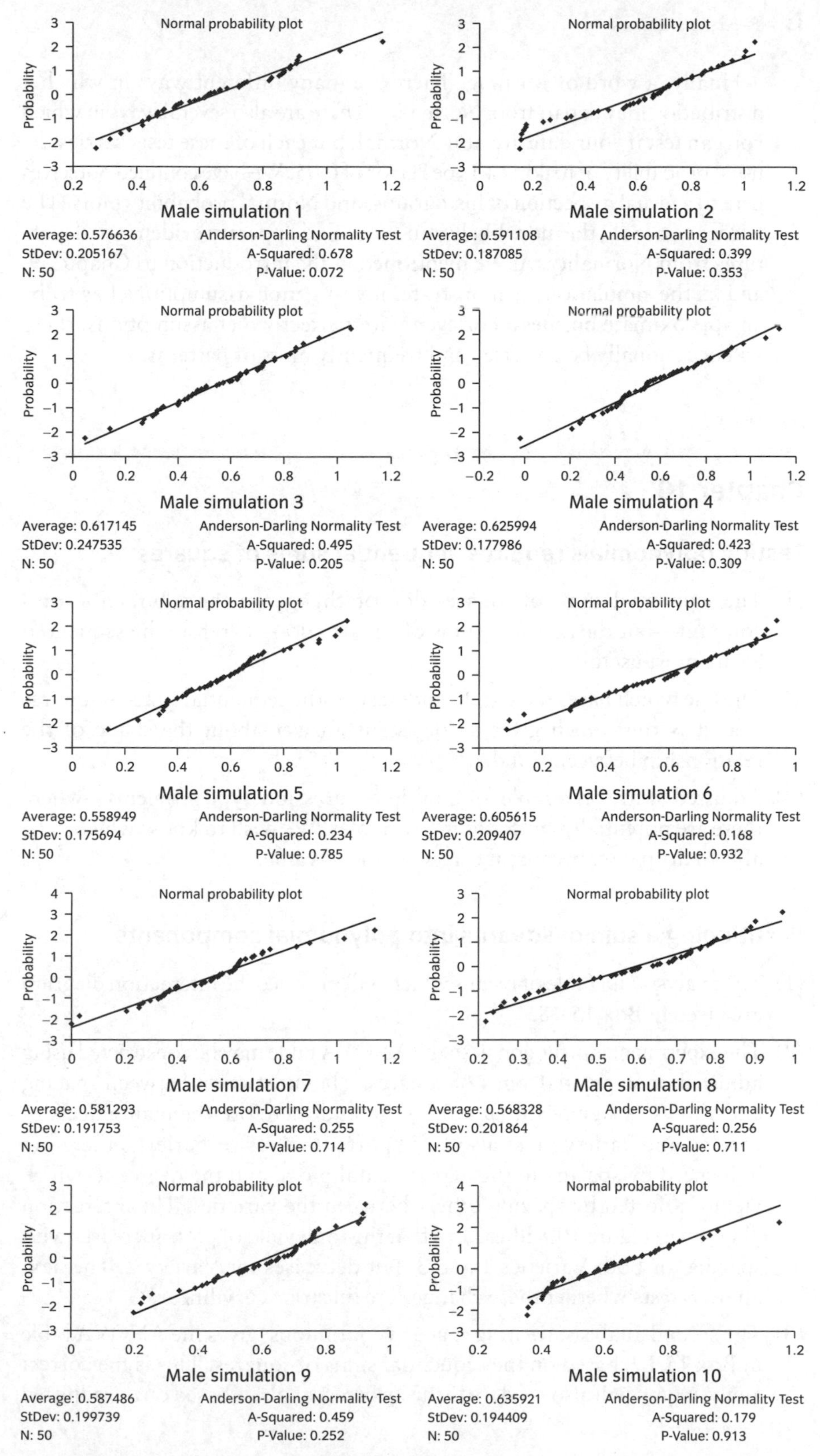

Fig. 15.12 Normal probability plots for simulated male squirrels.

Finally, a word of warning. There are many different ways in which a distribution may depart from Normality. There are also several ways in which you can test if your data are non-Normal, but each of these tests carries the usual probability of making a Type I error of 0.05. We have confined ourselves here to a visual inspection of histograms, and Normal probability plots. The harder you look, the more likely you are to find apparent evidence of departures from Normality. As we mentioned in the introduction to Chapter 9, and as the simulations demonstrate, investigating assumptions has to be an approximate business. For even with perfectly met assumptions, there will occasionally be patterns, and frequently hints of patterns.

Chapter 10

Testing polynomials requires sequential sums of squares

(1) The sequential sums of squares do not change on the subtraction of a constant—but the adjusted sums of squares do—therefore the sequential SS are most useful.

(2) The one which bases the significance test on the sequential sums of squares —as it is this which gives a consistent answer about the shape of the relationship between X and Y.

(3) Adjusted SS are preferable for 'multiple regression' type problems—where there are potentially confounding variables. We wish to know which variable to drop next because it offers no added value.

Partitioning a sum of squares into polynomial components

(1) The analysis and table of means which will produce the interaction diagram are given in Box 15.18.

(2) The experiment is orthogonal, so the ANOVA table may be constructed using adjusted or sequential sums of squares. The interaction between spacing and variety is significant ($p < 0.0005$), and therefore the main effects of spacing and variety must also be important. Yield of barley is therefore influenced by spacing in the experimental plots, and the degree to which yield is affected by spacing differs between the varieties. The interaction diagram in Figure 10.8 illustrates that the total yield of plots increases with spacing for both Varieties 1 and 3, but decreases for Variety 2. The next analysis tests whether these changes are linear or curvilinear.

(3) The second analysis, treating space as continuous, gives the ANOVA table in Box 15.19, based on the sequential sums of squares. This is the correct analysis (you will also need to do the incorrect analysis based on the adjusted

BOX 15.18 Analysis for barley yield

General Linear Model

Word equation: BYIELD = BBLOCK + BSPACE | BVARIETY

BBLOCK, BSPACE and BVARIETY are categorical

Analysis of variance table for BYIELD, using Adjusted SS for tests

Source	DF	Seq SS	Adj SS	Adj MS	F	P
BBLOCK	3	255.64	255.64	85.21	4.82	0.009
BSPACE	2	155.06	155.06	77.53	4.39	0.024
BVARIETY	2	1027.39	1027.39	513.69	29.07	0.000
BSPACE * BVARIETY	4	765.44	765.44	191.36	10.83	0.000
Error	24	424.11	424.11	17.67		
Total	35	2627.64				

Least squares means for BYIELD

BSPACE		Mean	SECoef
1		55.25	1.214
2		57.83	1.214
3		60.33	1.214
BVARIETY			
1		51.33	1.214
2		57.67	1.214
3		64.42	1.214
BSPACE * BVARIETY			
1	1	47.50	2.102
1	2	62.25	2.102
1	3	56.00	2.102
2	1	50.75	2.102
2	2	58.50	2.102
2	3	64.25	2.102
3	1	55.75	2.102
3	2	52.25	2.102
3	3	73.00	2.102

sums of squares, to complete Table 15.5). (The column of sequential sums of squares suffers from rounding errors as $155.05 \neq 155.06$).

(4) The sequential sums of squares for the linear and quadratic components of BSPACE sum to the sum of squares (adjusted or sequential) for BSPACE as a categorical variable, as do the degrees of freedom.

(5) The linear component of the polynomial decomposition is significant ($p = 0.007$), but the quadratic component is not ($p = 0.978$). Therefore there is no evidence that the lines are curved, but strong evidence that the slopes are not equal to zero.

BOX 15.19 Second analysis for barley yield

General Linear Model

Word equation: BYIELD = BBLOCK + BVARIETY | BSPACE | BSPACE

Analysis of variance table for BYIELD, using Sequential SS for tests

Source	DF	Seq SS	Adj SS	Seq MS	F	P
BBLOCK	3	255.64	255.64	85.21	4.82	0.009
BVARIETY	2	1027.39	38.17	513.69	29.07	0.000
BSPACE	1	155.04	3.59	155.04	8.77	0.007
BVARIETY * BSPACE	2	759.08	5.65	379.54	21.48	0.000
BSPACE * BSPACE	1	0.01	0.01	0.01	0.00	0.978
BVARIETY * BSPACE * BSPACE	2	6.36	6.36	3.18	0.18	0.836
Error	24	424.11	424.11	17.67		
Total	35	2627.64				

Table 15.5 Table of SS for barley yield

	Adjusted sum of squares	Sequential sum of squares	Degrees of freedom
Linear term for BSPACE in the second analysis	3.59	155.04	1
Quadratic term BSPACE * BSPACE in the second analysis	0.01	0.01	1
Sum of the above	3.60	155.05	2
Categorical variable BSPACE in the first analysis	155.06	155.06	2

(6) The conclusions differ in that we can describe the shape of the relationship between BYIELD and BSPACE as linear, but with differing slopes for different varieties.

Chapter 11

Finding the best treatment for cat fleas

(1) Analysis 1 suggests that there is no significant difference between the two treatments in their efficacy against cat fleas (p for TRTMT is 0.199).

(2) $R^2 = 0.297$ for Analysis 2, and $R^2 = 0.371$ for Analysis 3.

BOX 15.20 Analysis for cat fleas

General Linear Model

Word equation: LOGFLEAS = TRTMT + NCATS + CARPET
TRTMT and CARPET are categorical, NCATS is continuous

Analysis of variance table for LOGFLEAS, using Adjusted SS for tests

Source	DF	Seq SS	Adj SS	Adj MS	F	P
TRTMT	1	1.612	5.750	5.750	9.09	0.003
NCATS	1	23.730	22.962	22.962	36.31	0.000
CARPET	1	6.103	6.103	6.103	9.65	0.003
Error	85	53.759	53.759	0.632		
Total	88	85.203				

(3) I would expect model criticism techniques to illustrate that the third model is best, as it explains a higher proportion of the variance and the *p*-values are more significant—both consistent with an improvement in the specification of the model.

(4) The extra variables have clearly helped in the treatment comparison by reducing the error mean square—with TRTMT being significant in the third analysis ($p = 0.004$). However, the variable HAIRL is insignificant ($p = 0.613$)—so the final model could exclude this variable, as shown in Box 15.20.

For the final model:

$$R^2 = 0.369$$
$$R^2 \text{ adj} = 0.347.$$

This compares with the following values for analysis 3:

$$R^2 = 0.371$$
$$R^2 \text{ adj} = 0.341.$$

The raw R^2 for Analysis 3 is higher—because it contains an extra variable (even though this variable is insignificant). The adjusted R^2, on the other hand, is highest for the final model—which only includes significant variables.

Multiplicity of *p* values

Table 15.6 represents the results we obtained from 10 analyses of datasets produced as described in the question.

Table 15.6 Multiple regression with unrelated variables

Analysis	p value of whole regression	p value of most significant variable	R^2	R^2 adj
1	0.853	0.154	0.216	0
2	0.106	0.028	0.503	0.241
3	0.792	0.117	0.241	0
4	0.182	0.029	0.457	0.171
5	0.101	0.053	0.506	0.246
6	0.919	0.130	0.181	0
7	0.454	0.055	0.352	0.011
8	0.380	0.060	0.377	0.049
9	0.954	0.388	0.156	0
10	0.323	0.037	0.397	0.079

(1) For our analyses, the p value for the whole regression was never significant (however, we would expect significance 5% of the time on average).

(2) In contrast, the analysis did produce a significant variable in 3 out of 10 cases (i.e. about 30% of the time). The probability of finding one significant variable out of 11 possibilities is much higher than concluding that the regression as a whole is significant, in our null data sets.

(3) The value for R^2 adj is always lower than that of R^2, and in four cases was so low as to not give a value (to 3 significant figures). Given that our data set contains unconnected variables, this is providing us with more useful information about the explanatory power of our model.

(4) As the number of random X variables, used as explanatory variables, increases, we would expect to find that the most significant X variable has a p value below 0.05 more frequently. In other words, the probability of making a Type I error (concluding significance when there is none) would increase. The p value for the regression model as a whole, however, would only be significant 5% of the time on average, as before. The R^2 value would rise with the number of explanatory variables, as the proportion of variance explained will increase with the number of explanatory variables through chance alone. The R^2 adj however, would not increase, as this is an indication of the explanatory power of the regression per explanatory variable in the model.

Chapter 12

Examining microbial communities on leaf surfaces

(1) There is a significant difference in bacterial density between the two treatments ($p = 0.001$).

(2) In the second analysis, there is no significant difference between the treatments ($p = 0.183$). There is, however, a significant difference between the plants within treatments ($p < 0.0005$). The second analysis is to be preferred, because it treats plants as the independent units in the comparison of treatments, whereas the first analysis contravened independence by treating readings from the same plant as independent.

(3) A factor is random, if it is not the levels of that factor, but the population from which the sample is taken, that can be repeated in future experiments. The individuals represent a random sample of all possible individuals, and as such are representative of the wider population. Because of this property, conclusions drawn about the individuals may also be applied to the wider population. Fixed factors, in contrast, have levels that can be repeated exactly in future experiments. The conclusions drawn about those levels apply only to those specific levels. For crossed factors, all combinations of the two factors appear in the experimental design. Nested factors do not have all combinations in the experimental design; if B is nested in A, then each level of B occurs at only one level of A, but more than one level of B occurs at each level of A.

(4) The mean square for `PLANT (TREATMNT)` was used to calculate the *F*-ratio for `TREATMNT` in the second analysis. This is effectively comparing variation between plants in different treatments with variation between plants in the same treatment.

(5) There are two parts of the output which would assist with this decision: the degrees of freedom for the denominator of the *F*-ratio of `TREATMNT` (which are 4), and the variance components for `PLANT` and error (314.8 and 546.1 respectively). The variance component for error is considerably higher than for `PLANT`. The DF are also low for an *F*-ratio test. If I was primarily interested in a comparison between `TREATMNT`, then I would increase the number of plants. If I was primarily interested in the differences between plants given the same treatment (less likely), I would increase the number of readings per plant.

How a nested analysis can solve problems of non-independence

(1) A nested analysis would give the output shown in Box 15.21.

(2) The degrees of freedom and the *F*-ratio are identical.

BOX 15.21 Nested analysis

General Linear Model

Word equation: LUPRATE = SEX + SHEEP (SEX)
SEX is categorical, SHEEP is random

Analysis of variance table for LUPRATE, using Adjusted SS for tests

Source	DF	Seq SS	Adj SS	Adj MS	F	P
SEX	1	0.132816	0.132816	0.132816	1.80	0.251
SHEEP (SEX)	4	0.294774	0.294774	0.073693	26.89	0.000
Error	114	0.312387	0.312387	0.002740		
Total	119	0.739977				

Expected Mean Squares using Adjusted SS

Source	Expected Mean Square for Each Term
1 SEX	(3) + 20.0000(2) + Q[1]
2 SHEEP (SEX)	(3) + 20.0000(2)
3 Error	(3)

Error terms for tests using Adjusted SS

Source	Error DF	Error MS	Synthesis of Error MS
1 SEX	4.00	0.073693	(2)
2 SHEEP (SEX)	114.00	0.002740	(3)

Variance components using Adjusted SS

Source	Estimated value
SHEEP (SEX)	0.00355
Error	0.00274

Chapter 13

Soya beans revisited

(1) Possible—if (i) all plants are equally susceptible to herbicide damage, and equally exposed and (ii) the number of plants per plot is large and (iii) the chance of any one plant being damaged is small. If, for example, there was only one plant per plot, then it would either be damaged or not—limiting the count to either be 0 or 1 (which is called a Bernoulli variable).

(2) $$\chi^2_{21} = \frac{21 \times 0.278}{0.25} = 23.35.$$

From tables we find that $P(X \leq 23.35) = 0.326$. Therefore the p-value for our test is 0.652. So we cannot reject the null hypothesis that DAMAGE follows a Poisson distribution. (A Poisson distribution describes a count of independent items).

(3) $\chi_2^2 = \dfrac{2 \times 23.09}{0.25} = 184.7.$

From tables we find that $P(X \leq 184.7) = 1.0000$, so the p-value for this test <0.0001—highly significant. So we reject the null hypothesis that WDKLR has no effect on damage.

(4) The chi-square test is the correct one under the assumption of Poisson variance (which was upheld in this instance). If correct, it is also usually the more powerful under these circumstances.

Fig trees in Costa Rica

(1) If the number of individuals per quadrat satisfy the conditions of independence and homogeneity, then they may be viewed as a sample from a Poisson distribution. Independence, in this context, means that the presence of one individual is quite independent of the presence of another. Homogeneity means that the quadrats should be equally suitable for fig tree growth.

The means and variances within and between sites are given in Table 15.7.

(2) The means and variances within sites are very similar—suggesting that a Poisson distribution may be a good description of the distribution of fig trees within sites. Overall, however, the variance is nearly 10 times the mean—suggesting that the distribution of fig trees across sites is aggregated. This may result because some sites are generally more suitable for fig trees than others and indeed the means for the sites are quite different.

(3) The four dispersion tests are shown in Table 15.8. These tests confirm our conclusions drawn above. Within sites, the variance and mean are not significantly different. Between sites however the variance is significantly greater than the mean.

(4) The variable NINDIVS was transformed by taking the square root before analysis, to correct for the heterogeneity of variance expected from a Poisson distribution. The word equation SQRTN = SITE was then fitted to test for differences between sites. See Box 15.22.

Table 15.7 Means and variances for fig tree sites

Variable	N	Mean	Variance
Site 1	100	9.51	10.35
Site 2	100	18.93	19.44
Site 3	100	41.43	39.20
Overall	300	23.29	202.76

Table 15.8 Dispersion tests

Site 1	$\chi^2 = \dfrac{99 \times 10.35}{9.51} = 107.7$	$p = 0.258$
Site 2	$\chi^2 = \dfrac{99 \times 19.44}{18.93} = 101.7$	$p = 0.406$
Site 3	$\chi^2 = \dfrac{99 \times 39.2}{41.43} = 98.8$	$p = 0.487$
Overall	$\chi^2 = 299 \times \dfrac{202.76}{23.29} = 2603.1$	$p < 0.001$

BOX 15.22

General Linear Model

Word equation: SQRTN = SITE

SITE is categorical

Analysis of variance table for SQRTN, using Adjusted SS for tests

Source	DF	Seq SS	Adj SS	Adj MS	F	P
SITE	2	582.63	582.63	291.31	1091.99	0.000
Error	297	79.23	79.23	0.27		
Total	299	661.86				

If the data within sites are not significantly different from the Poisson distribution, then the error mean square should not be significantly different from 0.25. This is tested below:

$$\chi^2 = \frac{297 \times 0.270}{0.25} = 320.76.$$

With 297 degrees of freedom, $p = 0.164$. Thus we conclude, as above, that within sites the data are distributed according to the Poisson distribution.

(5) Aggregated distributions may arise when seedlings grow close to their parents, or when a few trees are the source of most viable offspring. Patchy habitats may also produce aggregation.

(6) Spatially uniform distributions may arise from competition creating spaces of a certain size between individuals—or alternatively from human intervention (cultivating trees in plantations), if *all* quadrats fell within the cultivated area.

REVISION SECTION

The basics

This revision section reviews the concepts with which you should already be familiar at the start of this course. If this is not the case, all topics are comprehensively covered in first level texts, an excellent example of which is Myra Samuel's textbook *Statistics for the life sciences*, published by Maxwell–Macmillan.

R1.1 Populations and samples

It is rarely possible to get an exact answer to a question. Normally, we have to make an estimate, and this may vary from a rough estimate to a more precise one. The first concept in statistics is to state just this in more formal terms. For example, we may be interested in the average height of men aged between 25 and 35 in the United Kingdom. Unless, by some supreme effort, we manage to measure every single man of that age, we will never know the exact answer for sure. Instead, we content ourselves with taking a sample, calculating the average, and hoping that it is representative of the whole population. The crucial point here is that the **sample** is a **random** selection of the whole **population**. For this to be the case, it is also important that the population has been precisely defined; for example, our sampling strategy would differ if our population was defined as being restricted to Oxfordshire.

Having taken our sample, we compute the mean by summing and dividing by the number of values:

$$\bar{y} = \frac{\sum y_i}{n}.$$

This is our estimate of the true population mean, μ. This concept of sampling to obtain an estimate also applies to conducting experiments. When we apply an experimental treatment and record data there is inevitably some error, and we are effectively sampling from all possible data readings that could occur when we apply that treatment. The resulting treatment mean can therefore be thought of as an estimate of the true mean, μ_A, which could only be obtained if there was no error involved in the experiment.

R1.2 Three types of variability: of the sample, the population and the estimate

Variability of the sample

Having computed the mean of the sample, from the same n data points we can obtain information about how variable the sample is. This is of interest, because it will give us an idea of whether our estimate is likely to be a rough one, or a more precise one. Two samples may have the same mean, but vary greatly in variability. For example, the range of scores in an undergraduate maths test extended from 40 to 72 for 30 participants—but the same group scored between 44 and 64 in an English test. In both cases, however, the mean score was 55 (Fig. R1.1).

The points in the first data set tend to be at a greater distance from the mean than in the second. If we define deviations as

$$\text{deviation} = \text{datapoint} - \text{mean}$$

then those data points greater than the mean have positive deviations, and those less than the mean have negative deviations. The deviations in the first data set will tend to be of greater absolute value than the second, but in both cases the deviations will sum to zero, because the definition of the mean is the central point. However, if squared before summing, then the resulting sum will give a measure of spread of the points around the mean; as shown in Table R1.1.

Clearly, there is much greater variability in math ability than English. This measure, referred to as **sums of squares**, has one major drawback—it is dependent upon sample size. In this example, a comparison of the sum of squares is valid, as the two data sets are of the same size. In general, however, the larger the data set, the greater

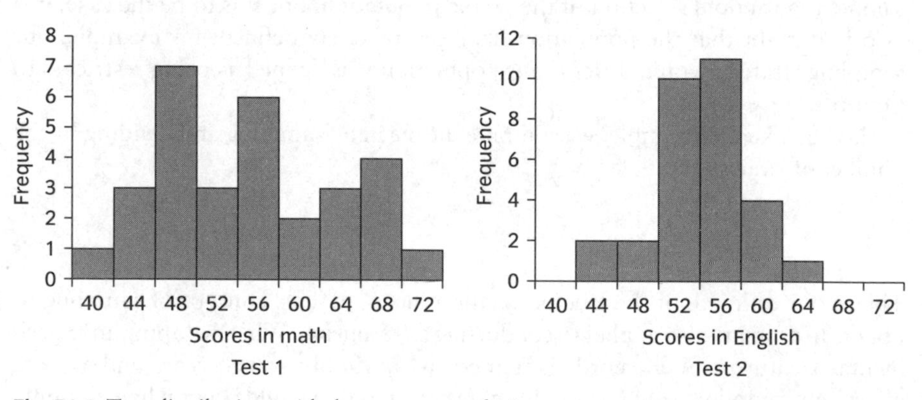

Fig. R1.1 Two distributions with the same mean of 55.

Table R1.1

For math scores	$\Sigma(y_i - \bar{y})^2 = 2100.6$
For English scores	$\Sigma(y_i - \bar{y})^2 = 591.2$

its sum of squares, so for valid comparisons between unequally sized data sets, a measure of variability that is independent of the number of datapoints is required. This can be simply remedied by taking account of sample size. We used n datapoints to define the mean, and then the same n datapoints (and the mean itself) to define how variable the sample is around the mean. But from the way in which the mean is calculated, the deviations must sum to zero, so we have only $n - 1$ independent pieces of information about how the sample varies about its mean. Thus our final measure of variability, the **variance**, may be calculated as:

$$s^2 = \frac{\Sigma(y_i - \bar{y})^2}{n - 1}.$$

The number of independent pieces of information that contribute to the calculation of a statistic, is called the **degrees of freedom.** To convert this measure back to the same units as the mean, we just take the square root. This gives us s, the **standard deviation.**

We have described here in intuitive terms why the denominator should be $n - 1$ rather than n, by referring to the number of independent pieces of information about the variance in the sample. However, in Appendix 2, we give a more formal proof of why $n - 1$ is the correct denominator.

Variability of the population

Just as we can never know the true population mean, so we can never know the true variance. Nevertheless it is useful to define it, as we will need to refer to the concept of the population variance. It is frequently referred to as σ^2, and our best estimate of this is the variance of our sample, s^2. It may be defined as the expected squared deviation around the true mean for all individuals in the population. The population is often defined as infinite, so the formula for the population variance involves notation that we have not yet covered.

Variability of the estimate

Having obtained our estimate of the mean ($\bar{y}$), we need to know how accurate this is. To answer this question, we will briefly digress to discuss Normal distributions.

The Standard Normal Distribution

Many continuous variables follow a distribution which is bell-shaped and symmetrical about its mean (see Fig. R1.2). Such distributions can often be approximated by a Normal distribution—which requires two parameters to define it; a mean and variance. A Standard Normal Distribution is simply a Normal distribution with mean 0 and standard deviation 1, and is often referred to as the z **distribution.** Any Normal distribution can be 'converted' to a Standard Normal Distribution by performing two operations. If a variable Y follows a Normal distribution with mean 5 and standard deviation 2 (Fig. R1.2), then if every value has 5 subtracted from it, Y will follow a distribution with mean 0. (It is as if we have shifted the whole distribution along the X axis). Then, if every value is divided by 2, the standard deviation of Y will become 1 (as if we have squashed the distribution closer

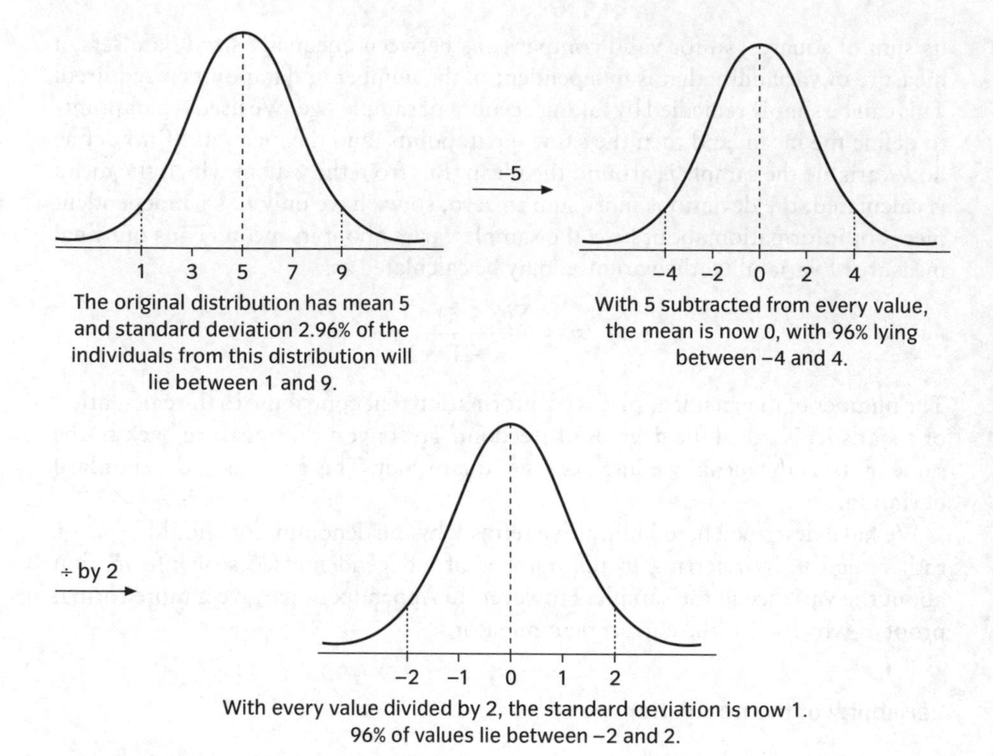

Fig. R1.2 Converting a Normal distribution to the Standard Normal Distribution.

around the mean). This process of converting a normal distribution to the standard normal distribution by shifting the mean and then altering the variance is called **standardising.**

In summary, to convert any Y variable to the z distribution, perform the following operation on Y:

$$z = \frac{Y - \mu}{\sigma}.$$

Why should we wish to do this? The Standard Normal Distribution has the useful property that 96% of its values lie within 2 standard deviations of the mean (and 68% within 1 standard deviation). This property is fundamental to much of the statistics we will do.

Accuracy of the estimate

Now we will return to quantifying the confidence we have in our estimate of the population mean. The value of $\bar{y}$ we have calculated is itself only one of many possible $\bar{y}$s we could have estimated from the same population, if we had drawn a slightly different sample by chance. So our mean can itself be thought of as a random sample of one from an infinite population of all possible means. If we imagine drawing another sample and calculating another $\bar{y}$, and repeating this process many

Table R1.2 Three types of variability

Source	Mean	Variance
Population	μ	σ^2
Sample	$\bar{y}$	s^2
Estimate of population mean ($\bar{y}$)	μ	$\frac{\sigma^2}{n}$

times, we could draw up a distribution of the way in which these $\bar{y}$s varied—the distribution of $\bar{Y}$. What would this distribution look like?

On average, it would be hoped that the mean of all these estimates would be μ, the population mean itself. In fact, all the $\bar{y}$s should be normally distributed around this mean (if our sample size is large enough—see the **central limit theorem**). Is it possible to say anything about the variance of this distribution? If we are drawing our samples (all size n) from a very variable population (that is to say σ^2 is high), then our estimates of the mean are likely to vary considerably around μ—and vice versa if σ^2 is relatively low. It is also true that if the samples taken are large, then our estimates are likely to be closer to μ than if our samples are small. Combining these two, the variance of the distribution of our estimates is $\frac{\sigma^2}{n}$, giving a standard deviation, or as it is sometimes called, **standard error of the mean** as $\frac{\sigma}{\sqrt{n}}$.

In summary, in this section we have discussed three types of variability as seen in Table R1.2.

Populations and samples have distributions that are easy to visualise, and from these samples we estimate parameters, of which the mean is the first and simplest. The estimates of these parameters have sampling distributions, which describe the values the estimate is likely to take. The sampling distributions cannot be plotted in the usual run of things, because we have only one estimate—to see a sampling distribution we need to perform a simulation in which an experiment is repeated many times. These sampling distributions are harder to envisage. The means of these distributions are the true values of the parameters in the population. The variance of these distributions will depend upon the parameter in question. In this case it is easy to rationalise why the variance of the estimates of $\bar{y}$ should be $\frac{\sigma^2}{n}$, but as the parameters we estimate from our sample become more complex, then so will the formulae for these variances. It is not necessary to know in detail how these variances are derived. If you are interested however Appendix 2 does go through the formal proof for the variance of the distribution of sample means. The key principles to understand are: (i) each parameter comes from a theoretical distribution (a sampling distribution) for which the standard error can be calculated (which your statistical package will do for you); (ii) the concept behind this standard error (it represents how accurate the estimate is); (iii) how you can make use of it.

R1.3 Confidence intervals: a way of precisely representing uncertainty

One thing we know almost for certain about $\bar{y}$ is that it is not exactly equal to μ—but can we make any statement about how close to μ it is likely to be? So far, we know that our estimate $\bar{y}$ comes from the distribution of all possible $\bar{y}$s that are Normally distributed around μ, with a variance of $\frac{\sigma^2}{n}$. This section outlines how to obtain a **confidence interval**: that is, the range of those parameter values that cannot be rejected from the data at the 5% level. (For a little more discussion about the definition of a confidence interval, then see Appendix 1). Parameters that have been estimated with great confidence will have narrow intervals associated with them—whilst those parameters about which we have less information will have wide confidence intervals.

From the properties of the Standard Normal Distribution, we know that 96% of all such $\bar{y}$s will lie within two standard deviations of μ. See Fig. R1.3.

Combining all this information, we can make the following statement:

for 96% of the time:

$$\mu - 2\frac{\sigma}{\sqrt{n}} < \bar{y} < \mu + 2\frac{\sigma}{\sqrt{n}}.$$

Conventionally however we are interested in a confidence level of 95% rather than 96%—so the 2 is changed to 1.96. We are also interested in stating a confidence interval for μ in terms of $\bar{y}$, so this expression is rearranged to give

$$\bar{y} - 1.96\frac{\sigma}{\sqrt{n}} < \mu < \bar{y} + 1.96\frac{\sigma}{\sqrt{n}}.$$

This is what we were after—whilst we can't say precisely what value μ is, we can make a statement about the range of values μ could take, and be consistent with the data at the 95% level.

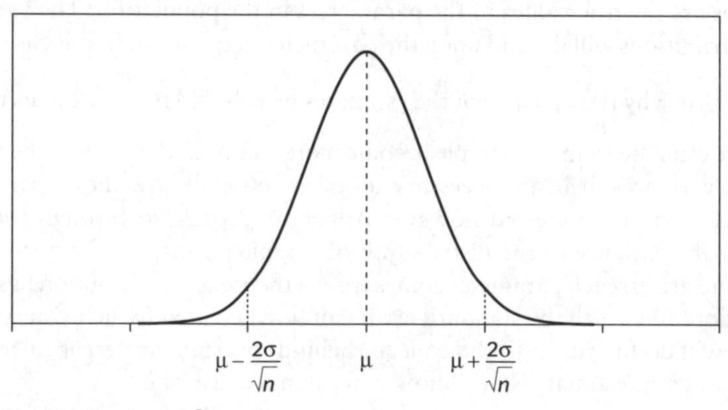

Fig. R1.3 The distribution of $\bar{Y}$.

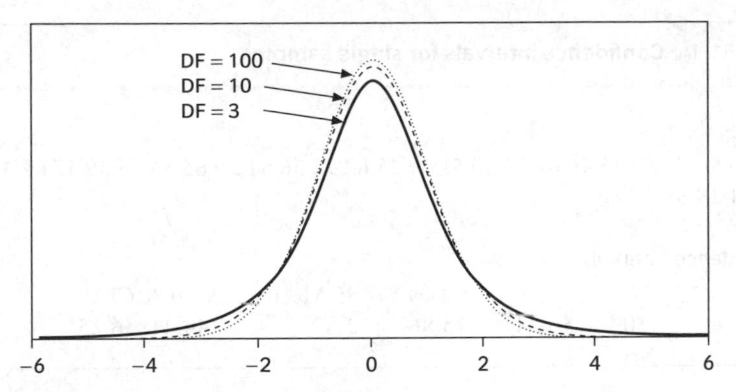

Fig. R1.4 *t* distributions for 3, 10 and infinite degrees of freedom.

When we come to put figures to this however there is a hitch. The expression includes σ, the population standard deviation—which can never be known. Instead, we only have an estimate of that parameter, namely *s*, the standard deviation of the sample. At the beginning of this century, we could have substituted *s* for σ, and been happy with that approximation. The smaller our sample however the less satisfactory this is, as *s* becomes a poorer estimate of σ (quite a different matter from the accompanying increase in the variance of $\bar{Y}$). This results in the distribution illustrated in Fig. R1.4 becoming flatter—it is no longer a Normal distribution, but a ***t* distribution**. This 'discovery' was made by Gossett in the 1900s—who then proceeded to construct tables to calculate exactly how much flatter than a Normal distribution the *t* distribution is.

The answer to this will depend upon how good our estimate of *s* is—in other words, how many independent pieces of information have been used to estimate *s*. These are the degrees of freedom for *s* (see Section R1.2) and will be $n - 1$. Figure R1.4 shows three *t* distributions, for degrees of freedom 3, 10 and ∞. As the degrees of freedom decrease, the *t* distribution becomes flatter and wider—so we need to move further away from the mean to encompass 95% of the distribution. As the degrees of freedom increase, the *t* distribution converges on the Normal distribution, until with large samples, it is hard to distinguish between the two.

Returning to our original problem, we need to ascertain exactly how many standard deviations from the mean are to encompass 95% of the population. This is done by looking up the appropriate **critical *t*-value** in *t* tables (denoted t_{crit} below). For example, when $n - 1 = 10$, then we need to move 2.228 standard deviations away from the mean. So our final formula becomes:

$$\bar{y} - t_{crit}\frac{s}{\sqrt{n}} < \mu < \bar{y} + t_{crit}\frac{s}{\sqrt{n}},$$

or in more general terms:

$$\text{estimate} \pm t_{crit} \times \text{standard error of the estimate.}$$

This is referred to as a **95% confidence interval**, because this is the range of values that μ could take, and still be consistent with the data at the 95% level. This confidence interval has been constructed for a mean—but similar intervals can be

BOX R1.1 Confidence intervals for single samples

Values EXAMRES

53 83 66 71 59 45 46 67 34 50 51 49 25 62 28 36 61 29 65 56 65 39 47 67 41 18 50 51 28 68

Confidence intervals

Variable	N	Mean	StDev	SE Mean	95.0 % CI
EXAMRES	30	50.33	15.86	2.89	(44.41, 56.25)

constructed for other parameters, and other levels of confidence, as shall be done in the main text.

As an example, the exam results of 30 students are given in Box R1.1, (stored in the variable EXAMRES), followed by a basic statistical description of the data, including the confidence interval.

R1.4 The null hypothesis—taking the conservative approach

This section outlines the principle behind a *t*-test, for the purpose of revising the concept of a null hypothesis and the method of testing such a hypothesis. The sample mean and variance have been calculated, and this information has been used to construct a confidence interval. This same information can be used to test a hypothesis. If our sample was a set of 30 differences between two groups (e.g. the difference in math scores before and after taking an undergraduate course), then if there was no improvement in ability over the duration of the course, the mean difference should be zero. If the differences are defined as DIFF = score after – score before, then it should be hoped that the mean difference is positive. If the course however had actually confused the students then it is possible that it is negative.

The first step is the construction of a **null hypothesis** (or H_0). Usually this errs on the side of caution, stating that no effect would be expected, but it is equally valid to hypothesise that the true mean takes some nonzero value. In this case:

H_0 : There is no difference between the scores, $\mu = 0$.

The alternative is that there is a difference—but it is rare that the direction of the difference would be stated (see the section on one and two-tailed tests). So the alternative hypothesis is phrased as:

H_A: $\mu \neq 0$.

The main principle behind the test is that we will not reject the null hypothesis unless there is convincing evidence that it is not true. Note that the null and alternative

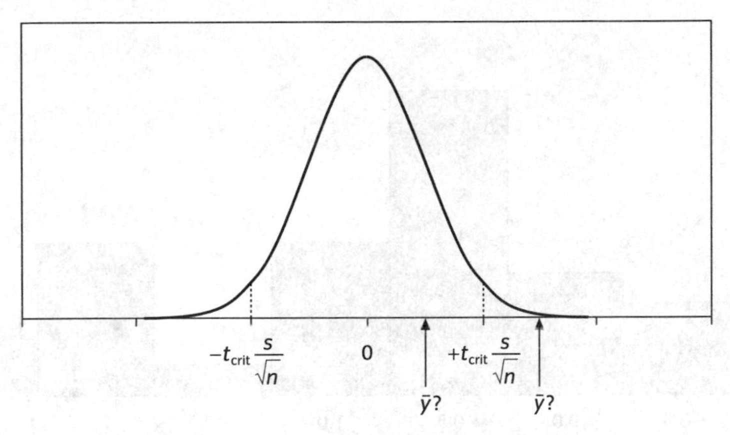

Fig. R1.5 The distribution of $\bar{Y}$ under the null hypothesis.

hypotheses are phrased in terms of population parameters—these encapsulate the 'great truths' we are after. It is almost certain that the sample mean, $\bar{y}$, will not be zero—and even if it was, this would not necessarily mean that μ was zero. So we assume that the null hypothesis is true, and then calculate the probability of having this dataset (or something more extreme) if this were the case. Convention has it that this probability must drop as low as 0.05 before we reject our assumption of H_0 being correct. This means that if the null hypothesis is true, there is a probability of 0.05 that we will reject it. This is referred to as a **Type I error**.

The *t*-test can be visualised pictorially. The null hypothesis is that the distribution of $\bar{Y}$ is such that its mean is zero (Fig. R1.5).

Our value of $\bar{y}$ is one datapoint from this theoretical distribution, and the question is, is it likely that it has come from our hypothesised distribution? There are two possible broad outcomes. In the first, $\bar{y}$ is sufficiently close to 0 to conclude that it may well come from this distribution. In the second, it is so far from 0 that we conclude that it is unlikely to be from this distribution. We are not however interested in this distance in absolute terms, but in terms of 'number of standard deviations'. Thus this distance in terms of standard deviations (referred to as the ***t*-statistic**) is:

$$t_s = \frac{\bar{y} - 0}{s/\sqrt{n}}.$$

This needs to be compared with the appropriate critical t value in tables, to determine if $\bar{y}$ is within the 95% limits (Fig. R1.5), or outside it. If $t_s > t_{crit}$ or $t_s < -t_{crit}$, then we conclude that there is a probability of less than 0.05 that we would have obtained such a value of t_s if $\bar{y}$ originates from a distribution of $\bar{Y}$ which has a mean of 0. In other words, we have a ***p*-value** < 0.05, and the strength of the evidence is such that it leads us to **reject the null hypothesis.**

To remind you exactly what this *p*-value means; it is the probability of obtaining this value of t_s, or a more extreme value, if the null hypothesis were true.

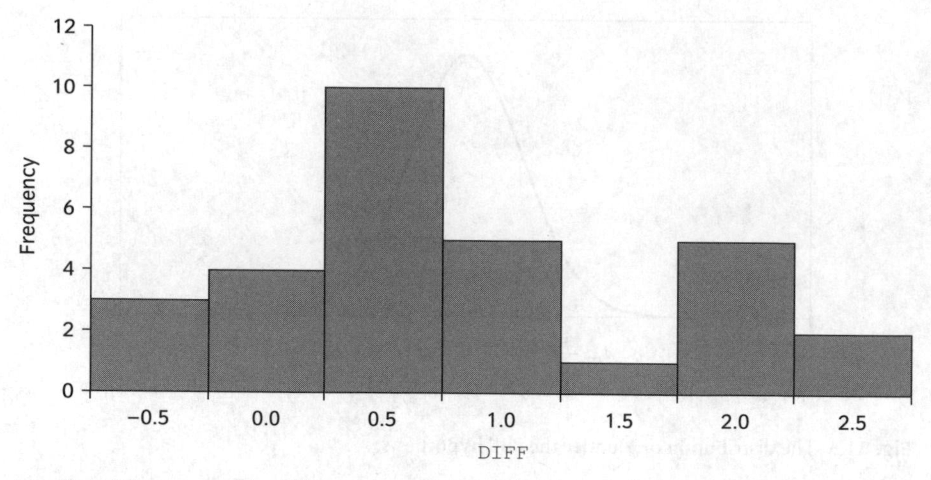

Fig. R1.6 Histogram of the difference in math scores before and after a course.

BOX R1.2 One sample *t*-test

***t*-test of the mean**

$H_0: \mu = 0$

$H_A: \mu \neq 0$

Variable	N	Mean	StDev	SE Mean	T	P
DIFF	30	0.862	0.838	0.153	5.64	0.0000

Turning back to the 30 unfortunate undergraduates sitting their maths tests, a histogram of their DIFF (= after − before) scores is shown in Fig. R1.6.

The histogram does indicate that some students did manage to perform worse in the after course test than before. We see in Box R1.2 that a one sample *t*-test however provides a low *p*-value (<0.00005), allowing us to conclude that in general, the undergraduates performed better after the course (note that we can infer the direction of the result from the fact that the mean difference is positive).

This method of hypothesis testing may be extended to include a hypothesised population mean of any value, or to test other parameters estimated from samples against hypothesised values of the equivalent population parameters. The whole process may be broken down into the following steps:

- Define the null and alternative hypotheses in terms of population parameters.
- Calculate the sample estimate of this population parameter.
- Calculate the standard error of this estimate.
- Calculate the *t*-statistic and compare with tables.

This method will now be applied to a different parameter estimate: the difference between two means of independent samples.

R1.5 Comparing two means

Two sample *t*-test

It is hypothesised that male and female squirrels differ in body mass. Fifty squirrels of each sex were measured, and the natural log of body mass recorded. Histograms of the data are plotted in Fig. R1.7, with females as LOG (FEM) and males as LOG (MALE) and stored in the *squirrels* dataset.

Under the null hypothesis, these 100 measurements would come from the same distribution. So:

H_0: $\mu_A = \mu_B$ or $\mu_A - \mu_B = 0$.

H_A: $\mu_A \neq \mu_B$.

From the data, we obtain samples means of $\bar{y}_A = -0.579$ and $\bar{y}_B = -0.688$ for males and females respectively. So $\bar{y}_A - \bar{y}_B = 0.109$. However, the test distribution is no longer the distribution of one sample mean $\bar{Y}$, but the distribution of the difference between two independent sample means $\bar{Y}_A - \bar{Y}_B$. The variance of this distribution is now:

$$s^2\left(\frac{1}{n_A} + \frac{1}{n_B}\right),$$

with the corresponding standard error being the square root of this. There are two methods of calculating this variance based on slightly different assumptions—the pooled and unpooled method. The pooled method has been used here. The *t*-statistic is then calculated as before:

$$t_s = \frac{(\bar{y}_A - \bar{y}_B) - 0}{\sqrt{s^2\left(\frac{1}{n_A} + \frac{1}{n_B}\right)}}.$$

In the case of the male and female squirrels, the standard error is 0.0575, giving a *t*-statistic of 1.90. This is less than the critical value of 1.98 for 98 degrees of freedom ($n_A + n_B - 2 = 98$), leading us to the conclusion that these two sets of data are not significantly different. See Box R1.6.

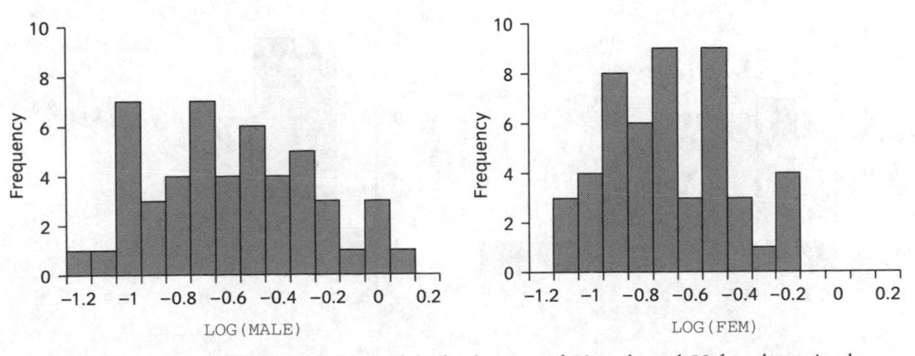

Fig. R1.7 Histograms of the natural log of the body mass of 50 male and 50 female squirrels.

BOX R1.6 Two sample *t*-test

Two sample *t*-test and confidence interval for LOG (MALE) vs LOG (FEM)

H_0: μ [LOG (MALE)] – μ [LOG (FEM)] = 0

	N	Mean	StDev	SE Mean
LOG (MALE)	50	−0.579	0.325	0.046
LOG (FEM)	50	−0.688	0.245	0.035

95% CI for Mean[LOG (MALE)] – Mean[LOG (FEM)]: (−0.005, 0.223)

t = 1.90 p = 0.061 DF = 98

Both use pooled StDev = 0.288

These formulae appear to be getting more complex—but in fact they both follow the pattern

$$t_s = \frac{\text{estimate} - \text{null hypothesis value}}{\text{standard error of estimate}}.$$

Alternative tests

In Fig. R1.7, the histograms of LOG (MALE) and LOG (FEM) for male and female squirrels showed distributions with considerable variance that were roughly symmetrical. However, Fig. R1.8 shows the same data plotted before the log transformation. What happens if a two sample *t*-test is conducted on these data?

The output in Box R1.7 indicates that there does appear to be a significant difference in the body mass of the two sexes.

So there appears to be a discrepancy between the transformed and untransformed analyses. Looking at the histograms of Fig. R1.8, both distributions are skew to the right—particularly male body masses. The parametric statistics we have used assume that the distributions concerned roughly follow a Normal distribution, and will be highly influenced by extreme points. The four points to the right of

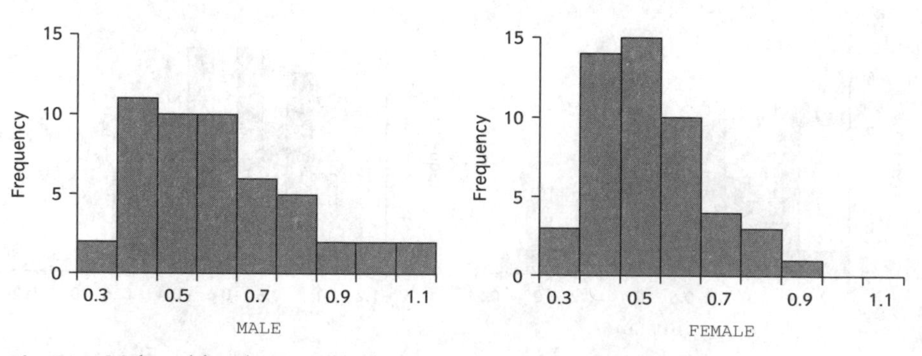

Fig. R1.8 Male and female squirrel body masses.

BOX R1.7 Two sample *t*-test of squirrel body masses

Two sample *t*-test and confidence interval for MALE vs FEMALE

H_0: μ (MALE) − μ (FEMALE) = 0

	N	Mean	StDev	SE Mean
MALE	50	0.591	0.198	0.028
FEMALE	50	0.518	0.132	0.019

95% CI for: μ (MALE) − μ (FEMALE): (0.006, 0.140)

t = 2.17 p = 0.033 DF = 98

Pooled standard deviation = 0.168

the male body mass plot will greatly increase the mean of the sample, and in doing so, suggest that the whole male body mass distribution is greater than it is in reality. When faced with such skew distributions, there are two options—to transform the data as was done earlier in this section, or to use tests which will not be so easily influenced by extreme values; namely nonparametric tests. This text will not be covering such tests—again many first level statistics books, including Samuels (1989) *Statistics for the life sciences* provide a very clear exposition on that topic.

This example, however, has highlighted that it is important to examine the data prior to analysis, and consider whether they follow a roughly symmetrical distribution. If not, then a variety of transformations are available to rectify this. If this advice is ignored, then an inappropriate test could produce a result of spurious significance, as has happened here. Chapters 8 and 9 examine the assumptions behind these parametric tests in more detail, how to test your data (and model) to see if these assumptions have been contravened, and how to solve any problems that arise.

One and two tailed tests

When comparing two samples, you can choose to conduct either a **one-tailed** or a **two-tailed** test. In the examples given, we have been opting for a two-tailed test, and the reason for this will become apparent as we describe these two alternatives.

Whether a test is one or two-tailed is actually defined by the alternative hypothesis (H_A) rather than by the null hypothesis (H_0). In comparing our male and female squirrels, H_A was that the logarithm of the body mass was different for the two sexes. Suppose however we were testing a particular hypothesis that males are on average heavier than females. H_0 would remain the same, as:

$$\mu \text{ of LOG (MALE)} - \mu \text{ of LOG (FEM)} = 0.$$

However, H_A would have direction and become:

$$\mu \text{ of LOG (MALE)} - \mu \text{ of LOG (FEM)} > 0.$$

In other words, we are looking for a difference between males and females in one particular direction only. If, on analysing the data, we found that females were very much heavier than males, we would not be able to reject the null hypothesis.

BOX R1.9 A one-tailed two sample *t*-test of the logarithms of squirrel body masses

Two sample *t*-test and confidence interval for LOG (MALE) vs LOG (FEM)

	N	Mean	StDev	SE Mean
LOG (MALE)	50	−0.579	0.325	0.046
LOG (FEM)	50	−0.688	0.245	0.035

95% CI for μ LOG (MALE) − μ LOG (FEM): (−0.005, 0.223)

t-test μ LOG (MALE) = μ LOG (FEM) (vs >): t = 1.90 p = 0.030 DF = 98
Both use pooled StDev = 0.288

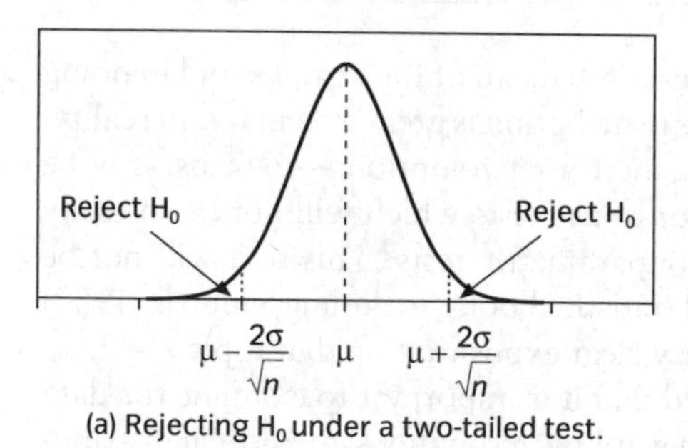

Fig. R1.9 Comparing one and two-tailed tests.

So how does this new directional alternative hypothesis affect how we conduct the *t*-test? In fact, the *t*-ratio is calculated in exactly the same way. The only difference comes into play when we convert the *t*-ratio into a *p*-value, as can be seen in Box R1.9.

In Box R1.6, we concluded that LOG (MALE) and LOG (FEM) were not significantly different. However, in Box R1.9, we would conclude that the variable LOG (MALE) is significantly greater than the variable LOG (FEM). How has this happened? In Fig. R1.9, the range of *t*-ratios which would lead us to reject the null hypothesis are shown: Fig. R1.9(a) illustrates this for a two-tailed test, and Fig. R1.9(b) for a one-tailed test. In both cases, the probability of making a Type I error (rejecting the null hypothesis when true) is 0.05. For two-tailed tests this is divided between two tails, whereas for the one-tailed test it is all in one tail.

In Box R1.9, we can see that with a one-tailed test, we have a significant *p*-value ($p = 0.03$), whereas with a two-tailed test, the *p*-value of 0.061 was not significant. Tempting though it is, under such circumstances, to choose the one-tailed test, we should consider the problems involved in doing this.

What would happen if we gathered our data, calculated our *t*-ratio, and only then decided on doing a one tailed test? We already know the direction of the difference, so we may well use this information in choosing the direction of our alternative hypothesis. However, this means that in reality our *t*-ratio could fall into the larger region illustrated in Fig. R1.10, and lead to us rejecting our null hypothesis more frequently. In fact, if we always chose the direction of H_A after observing the direction in our data set, then we would increase the probability of making a Type I error to 10%.

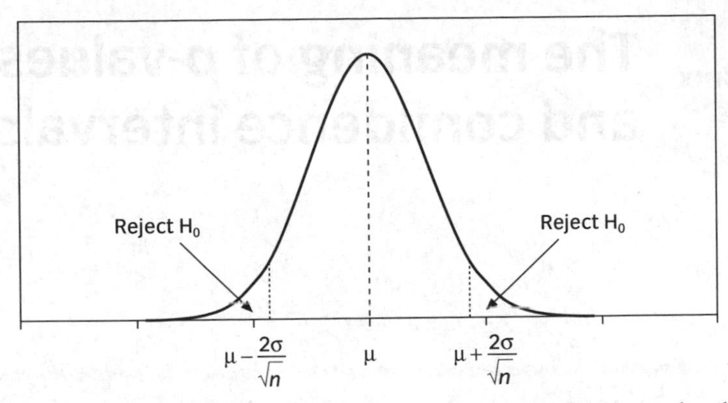

Fig. R1.10 Choosing the direction of H_A after observing the direction of the sample will lead to the probability of making a Type I error 10% of the time.

So it is advisable to use two-tailed tests, unless there are exceptional circumstances which dictate that a one-tailed test is most appropriate. Even if you are totally rigorous in fixing H_A before gathering your data, you may still find it difficult to convince others that you have been so, especially if a two-tailed test would not lead to the same conclusion!

R1.6 Conclusion

This revision section has presented a brief overview of the theory behind the simplest parametric tests. The concepts mentioned include:

- populations and samples; their parameters and the link between them
- mean and variance
- degrees of freedom
- standardising a Normal distribution
- confidence intervals
- the null hypothesis
- *t*-distribution and *t*-tests
- One and two-tailed tests
- the *p*-value

Another concept which has not been discussed, but with which you should be familiar before proceeding is the central limit theorem.

Some of these concepts will be discussed in more detail as we move through the early chapters. In particular, we will constantly be constructing null hypothesis, and the meaning of the *p*-value and degrees of freedom are discussed in more depth in the first two chapters and in Appendix 1.

APPENDIX ONE

The meaning of *p*-values and confidence intervals

What is a *p*-value?

A *p*-value lies between 0 and 1, and is an inverse measure of strength of evidence. So if $p = 0.001$, there is very strong evidence, and if $p = 0.2$ or 0.9 for example, we would conclude that the evidence is so weak that no conclusion can be drawn. The conventional threshold in science between concluding that the evidence is strong enough or not is 0.05.

Evidence of what exactly? Every statistical test performed has a null hypothesis associated with it, and the *p*-value measures the strength of evidence against the null hypothesis. Examples of null hypotheses are:

- fertiliser has no effect on yield
- a drug has no effect on patient recovery time
- men and women are equally intelligent.

A *p*-value measures strength of evidence, not magnitude of effect. If Fertiliser A has a moderately beneficial effect on yield compared to Fertiliser B, then in very small experiments there will usually be too little evidence to detect any difference. In moderately sized experiments, the effect will sometimes be demonstrated with $p \leq 0.05$, while in large experiments, the difference will often be established with$p \leq 0.001$. The magnitude of the difference, will, however, be the same in all these trials.

When we conclude that we have a significant result, with a *p*-value of 0.03, what exactly is this probability measuring? Here are a list of **incorrect** answers to that question:

1. The null hypothesis is true with probability 0.03.
2. There is a probability of 0.03 that there are no treatment differences.
3. When we conclude that our *F*-ratio is significant, 5% of the time the null hypothesis is in fact true, and there really are no treatment differences.

The *p*-value is, in fact, a conditional probability. It tells us the probability of obtaining that test statistic by chance *given that the null hypothesis is true*. In reality, we will never know if the null hypothesis is true or false—all we can do is to gather evidence against the null hypothesis. When the evidence is such that it is extremely unlikely that the null hypothesis is true, then we reject it. This system seems to be heavily biased in favour of the null hypothesis. While this seems to be the conservative attitude (i.e. assume that there is no effect of treatment until persuaded that this is

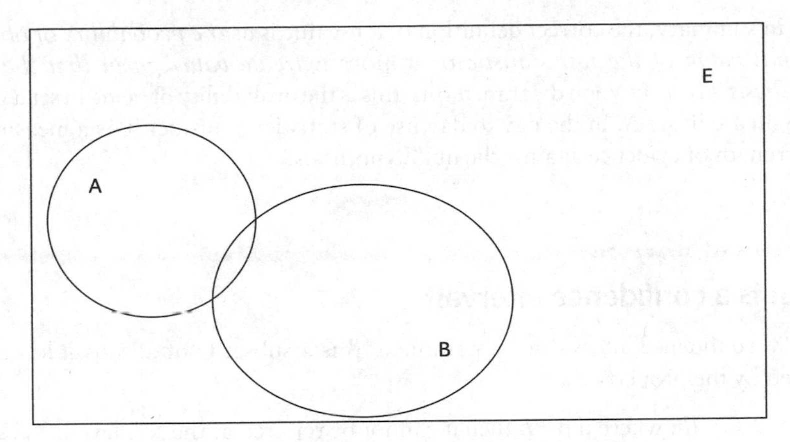

Fig. A1.1 The p-value Venn diagram. A = the event that the null hypothesis is true. B = the event that the null hypothesis is rejected.

very unlikely), it is not the only possible approach in statistics. Baysian statistics, for example, ascribes relative probabilities to different possible outcomes (see Hilborn and Mangel *The ecological detective*, Princeton University Press).

Conditional probabilities can be represented by Venn diagrams. In Fig. A1.1, we have two intersecting sets within the universal set.

Here E is the universal set, A is the set of experiments in which the null hypothesis is really true (and therefore the size or extent of A is unknowable), and B is the set of experiments in which the value of the test statistic leads us to reject the null hypothesis. In an ideal world in which experiments always lead us to the right conclusions, A and B would not overlap. However, we actually set the amount of overlap by setting the threshold p-value at 0.05—that is the probability that we fall into set B (reject the null hypothesis) if we are also in set A (the null hypothesis is true). This kind of mistake is a Type I error, and we accept that this will happen 5% of the time.

When we do an experiment, we don't actually know if we lie in set A (and the null hypothesis is true), or if we lie outside it (the null hypothesis is false). Thus we cannot quantify A or B. The smaller our p-value however the greater is the strength of the evidence against the null hypothesis.

Returning to our incorrect answers, we can now relate them to the Venn diagram.

1. The null hypothesis is true with probability 0.03.
2. There is a probability of 0.03 that there are no treatment differences.

Both of these are saying that the probability of the null hypothesis is true is 0.03—that is, the size of A is 0.03. However, this is incorrect and the size of A is unknowable.

3. When we conclude that our F-ratio is significant, 5% of the time the null hypothesis is in fact true, and there really are no treatment differences.

Set B represents the set of experiments in which we reject the null hypothesis, so this third statement is really phrased the wrong way round. It says, *given that we are in set B*, the probability of being in set A is 0.05. This would only be true if sets A and B are of exactly the same size—and there is no reason why they should be.

So in summary, the correct definition of a *p*-value is as *the probability of obtaining that value of the test statistic or a more extreme value given that the null hypothesis is true*. In Venn diagram terms, this is the probability of being in set B given that you are in set A. In the day to day use of statistics, consider it as a measure of the strength of evidence against the null hypothesis.

What is a confidence interval?

A 95% confidence interval for a parameter β is a subset C of all possible values, defined by the property that

$$C = \{b\text{: where if } \beta = b \text{ then it cannot be rejected at the 5\% level}\}.$$

A verbal description of this might be that the interval is the set of values which cannot be distinguished from the parameter estimate at the 95% level. A formulation that is often found in textbooks is that if C is a 95% confidence interval, then there is a 95% chance that β is contained in C. This statement, strictly speaking, defines the 'fiducial interval' of Fisher. In some situations, the fiducial and confidence intervals can be shown to exist and to be numerically equivalent. These situations include estimating the mean of a Normal distribution from a sample, but do not include estimating a binomial probability from a sample.

APPENDIX TWO

Analytical results about variances of sample means

The mathematical arguments behind many of the statistical truths discussed in this book are very technical. However, some of these arguments turn out to be quite simple if phrased in the language of **expectation algebra**. This appendix introduces that language, and uses it to present the following truths:

- If we take a random sample of size n from a population, and repeat this many times, the variance of those samples is the population variance divided by n.
- The variance of a sample calculated with $n - 1$ in the denominator is an unbiased estimate of the population variance.

Introducing the basic notation

We introduce E, called the expectation operator, meaning that $E[Y]$ represents the expected value of Y (in other words the average value of the random variable Y). There are five important properties of E:

1. $E[k] = k$ for constant k.
2. $E[k + Y] = k + E[Y]$ for constant k and random variable Y.
3. $E[X + Y] = E[X] + E[Y]$ for random variables X and Y.
4. $E[kY] = k\, E[Y]$ for constant k and random variable Y.
5. $E[XY] = E[X]\, E[Y]$ for independent random variables X and Y.

These properties are largely intuitive, and we will now make use of them to prove the two statistical truths mentioned above. First of all, however, we use the notation to define the variance and the sample mean.

Using the notation to define the variance of a sample

The variance of a sample, by definition, is the average squared deviation from the mean. Using our new notation, this translates into $E[(Y - \mu)^2]$. How else can this definition be expressed?

$$\begin{aligned} E[(Y-\mu)^2] &= E[Y^2 - 2\mu Y + \mu^2]. && \text{\{just by multiplying out the square\}} \\ &= E[Y^2] - 2\mu E[Y] + \mu^2 && \text{\{using properties 1, 3 \& 4 above\}} \\ &= E[Y^2] - 2\mu^2 + \mu^2 && \text{\{because } E[Y]=\mu\text{\}} \\ &= E[Y^2] - \mu^2 && \text{\{by algebra\}} \\ &= E[Y^2] - (E[Y])^2. && \text{\{using } E[Y]=\mu \text{ again for a pretty conclusion\}}. \end{aligned}$$

We can now define the variance operator, V for a random variable X.

$$V[X] = E[X^2] - (E[X])^2 \quad (1)$$

If we already know the variance, the following rearrangement is often useful.

$$E[X^2] = V[X] + (E[X])^2 \quad (2)$$

Using the notation to define the mean of a sample

Let Y_i, for $i = 1, 2, 3. \ldots n$, represent independent identically distributed random variables, with their mean and variance defined as:

$$E[Y_i] = \mu \text{ and } V[Y_i] = \sigma^2$$

and then let

$$\bar{Y} = \frac{\Sigma Y_i}{n}.$$

$\bar{Y}$ is a random variable representing the mean of a sample of size n. Deriving the obvious statement that the expected value of the sample mean is μ shows how we use properties 3 and 4 in a very simple case:

$$E[\bar{Y}] = E\left[\frac{\Sigma Y_i}{n}\right] = \frac{E[\Sigma Y_i]}{n} = \frac{\Sigma E[Y_i]}{n} = \frac{\Sigma \mu}{n} = \frac{n\mu}{n} = \mu \quad (3)$$

Defining the variance of the sample mean

Now we can move on to proving our first statistical truth: that is, the variance of the sample mean is the population variance divided by n, or in symbols

$$V[\bar{Y}] = \frac{\sigma^2}{n}.$$

Using the results we illustrated in equations (1) and (3) for $\bar{Y}$, we know that

$$V[\bar{Y}] = E[\bar{Y}^2] - \mu^2$$

and so by definition of $\bar{Y}$,

$$V[\bar{Y}] = E\left[\left(\frac{\sum Y_i}{n}\right)^2\right] - \mu^2 = \frac{E[(\sum Y_i)^2]}{n^2} - \mu^2. \qquad (4)$$

How can we simplify the term $E[(\sum Y_i)^2]$? Well

$$(\sum Y_i)^2 = (\sum Y_i)(\sum Y_i)$$

and when we multiply out the right hand side, there are n terms like Y_1^2, and $n(n-1)$ terms like $Y_1 Y_2$ (because, for each of the n values of Y_i, there are $n-1$ other values of Y_i to multiply with which don't have the same subscript). Formally

$$\left(\sum Y_i\right)^2 = \sum Y_i^2 + \sum_{i \neq j} Y_i Y_j$$

and so

$$E\left[\left(\sum Y_i\right)^2\right] = E\left[\sum Y_i^2 + \sum_{i \neq j} Y_i Y_j\right] = \sum E[Y_i^2] + \sum_{i \neq j} E[Y_i Y_j].$$

Now using equation (2) with Y_i instead of X gives us

$$E[Y_i^2] = \mu^2 + \sigma^2 \text{ so that } \sum E[Y_i^2] = \sum(\mu^2 + \sigma^2) = n(\mu^2 + \sigma^2). \qquad (5)$$

Then using Property 5, given that in the cases where $i \neq j$ the random variables are independent, we can say that $E[Y_i Y_j] = \mu^2$ so that

$$\sum_{i \neq j} E[Y_i Y_j] = \sum_{i \neq j} \mu^2 = n(n-1)\mu^2$$

because there are $n(n-1)$ terms in the sum. Hence

$$E[(\sum Y_i)^2] = n(\mu^2 + \sigma^2) + n(n-1)\mu^2 = n\sigma^2 + n^2\mu^2.$$

So we have now finally simplified this term, and can substitute it back into equation (4) to give

$$V[\bar{Y}] = \frac{n\sigma^2 + n^2\mu^2}{n^2} - \mu^2 = \frac{\sigma^2}{n} \qquad (6)$$

proving the required result.

To illustrate why the sample variance must be calculated with *n* – 1 in its denominator (rather than *n*) to be an unbiased estimate of the population variance

Using the notation, the result we need to show is that:

$$E\left[\frac{\sum(Y_i - \bar{Y})^2}{n-1}\right] = \sigma^2.$$

Taking the numerator, we can manipulate it as follows:

$$E[\Sigma(Y_i - \bar{Y})^2] = E[\Sigma(Y_i^2 - 2Y_i\bar{Y} + \bar{Y}^2)] = E[\Sigma Y_i^2 - 2\bar{Y}\Sigma Y_i + n\bar{Y}^2]$$

and given that $E[\Sigma Y_i] = n\bar{Y}$ then

$$= E[\Sigma Y_i^2 - 2n\bar{Y}^2 + n\bar{Y}^2] = E[\Sigma Y_i^2 - n\bar{Y}^2].$$

Hence

$$E[\Sigma(Y_i - \bar{Y})^2] = E[\Sigma Y_i^2] - nE[\bar{Y}^2]. \qquad (7)$$

Now we saw in equation (5) above that $E[\Sigma Y_i^2]$ equals $n\mu^2 + n\sigma^2$. Then we employ equation (2) with $\bar{Y}$ in place of X, and use results about the mean and variance of $\bar{Y}$ from equations (3) and (6), to find that $E[\bar{Y}^2]$ equals $\mu^2 + \frac{\sigma^2}{n}$. Substituting all this information into equation (7), we obtain

$$E[\Sigma(Y_i - \bar{Y})^2] = n\mu^2 + n\sigma^2 - n\left(\mu^2 + \frac{\sigma^2}{n}\right) = (n-1)\sigma^2.$$

Dividing through by $n - 1$ gives us

$$E\left[\frac{\Sigma(Y_i - \bar{Y})^2}{n-1}\right] = \sigma^2$$

which was the desired result. Interestingly, if we divide through by n instead, we get

$$E\left[\frac{\Sigma(Y_i - \bar{Y})^2}{n}\right] = \left(\frac{n-1}{n}\right)\sigma^2$$

which means that by using n rather than $n - 1$ as the denominator, we would get a systematic underestimation of σ^2.

Finally, note that this use of $n - 1$ rather than n arises because we have to use $\bar{Y}$ as an estimate of μ. This could be illustrated by proving the following result:

$$E\left[\frac{\Sigma(Y_i - \mu)^2}{n}\right] = \sigma^2.$$

You could try this as an exercise, or just read the simple proof laid out below:

$$E[\Sigma(Y_i - \mu)^2] = E[\Sigma(Y_i^2 - \mu^2)] = E[\Sigma(Y_i^2) - n\mu^2].$$

Now using equation (5)

$$E[\Sigma(Y_i - \mu)^2] = n(\mu^2 + \sigma^2) - n\mu^2 = n\sigma^2.$$

So $E\left[\frac{\Sigma(Y_i - \mu)^2}{n}\right] = \sigma^2$ as desired. So if μ were known, we would use n in the denominator of the variance estimate.

APPENDIX THREE Probability distributions

Some gentle theory

We meet the following probability distributions in this book: the Normal and Standard Normal distributions, *t*-distributions, chi-square distributions, and *F*-distributions. We also meet the Poisson distribution, but will not consider it further here. In this appendix, we show how to explore the distributions concerned with continuous *y* variables, and explain some of the relationships between them.

The concept of a **random variable** will be needed. A random variable is a quantity whose value is not known, but which takes a range of values with specified probabilities. We will write $X \sim N(\mu, \sigma)$ to indicate the probability that X lies within an interval (a, b). This is the area between the interval (a, b) under the curve of a Normal distribution with mean μ and standard deviation σ.

We first meet the concept of **standardising** in the Revision section. If $X \sim N(\mu, \sigma)$, and $Y = (X - \mu)/\sigma$, then it follows that $Y \sim N(0, 1)$, which is the Standard Normal Distribution. If we take a sample of size n from $N(\mu, \sigma)$, then we can define the mean and variance as random variables M and V. First, we ask 'what is the distribution of the M?'. The answer is $M \sim N(\mu, \sigma/\sqrt{n})$, from which it follows that $Z = (M - \mu)/(\sigma/\sqrt{n})$ and has a Standard Normal Distribution.

Second, we ask 'what is the distribution of V?'. Before answering, we must take two diversions. First, suppose that $Z \sim N(0, 1)$, and ask what is the distribution of Z^2? Z^2 must be positive. Its average must be one, because the average of Z^2 equals the variance of Z. The name given to this distribution is a chi-square distribution with 1 degree of freedom. So, $C = Z^2$, then we will write $C \sim \chi^2_1$. The second diversion is to suppose that we have altogether k variables with χ^2_1 distribution, called $C_1, C_2 \ldots C_k$. If we add them together to obtain $D = C_1 + C_2 + C_3 + \ldots + C_k$, and we assume that all k variables are independent of each other, then the name for the distribution of D is a chi-square distribution with k degrees of freedom. Formally, $D \sim \chi^2_k$. We add the means to find the mean of a sum, and so χ^2_k has a mean equal to k.

We are now ready to answer our question about the distribution of V. It is not V itself whose distribution has a name. Rather, $(n-1)V/\sigma^2 \sim \chi^2_{n-1}$. This is quite good news, because we hope that V is an unbiased estimate of σ^2, and if so, then the average value of $(n-1)V/\sigma^2$ is 1, and so the average value of $(n-1)V/\sigma^2$ is $n-1$. But we know that the average of χ^2_{n-1} is its degrees of freedom, which is indeed $n-1$.

Thus the sampling distributions of the mean and variance of a sample from a Normal distribution are related to Normal and chi-square distributions.

When we standardised X above, we did so with the standard deviation, but this will usually not be known. What happens when we use the estimated standard deviation instead, as we have to in practice when standardising M? The estimated standard deviation is a random variable calculated from the variance as $S = \sqrt{V}$, and so the estimated standard error of the mean, i.e. the standard deviation of M, is given by $S/\sqrt{n} = \sqrt{(V/n)}$. Let us clairvoyantly notate $(M - \mu)/(S/\sqrt{n})$ as T. Then divide top and bottom of the fraction by $\sigma/\sqrt{n}$, and we find that $T = \dfrac{(M-\mu)/(\sigma/\sqrt{n})}{S/(\sigma/\sqrt{n})}$. But this is a standard Normal variable, divided by the square root of a chi-square variable divided by its degrees of freedom:

$$T = \frac{\dfrac{M-\mu}{\sigma/\sqrt{n}}}{\sqrt{\dfrac{V}{\sigma^2}}} \sim \frac{\mathrm{N}(0, 1)}{\sqrt{\chi^2_{n-1}/(n-1)}}.$$

The name given to such a distribution is a t-distribution with $n-1$ degrees of freedom.

The interesting feature of $\chi^2_{n-1}/(n-1)$ is that we know its average is one. We now turn to looking at a ratio of two independent forms of this kind, namely

$$\frac{\chi^2_m/m}{\chi^2_n/n}.$$

This distribution is called the F-distribution, after RA Fisher, who pioneered its use in the course of inventing most useful statistics (though he worked initially with its logarithm). It is more precisely the F-distribution with m and n degrees of freedom, where m and n are called the numerator and denominator degrees of freedom, respectively, for obvious reasons.

Thus F-ratios are constructed as the ratio of two mean squares. A mean square is a variance estimate, and the F-ratio is also called the variance ratio. We choose numerator and denominator in an F-test so that, under the null hypothesis, the numerator and denominator are independent estimates of the same variance. So each mean square has a distribution of $\sigma^2\ \chi^2_k/k$ for its own degrees of freedom k. When we divide, the σ^2s cancel out, leaving the ratio of χ^2_k/k that we see in the formula for the F-distribution.

We end this technical section by noting that if we square a t-distribution on k degrees of freedom, we obtain an F-distribution with 1 and k degrees of freedom. For

$$\left(\frac{\mathrm{N}(0, 1)}{\sqrt{\chi^2_k/k}}\right)^2 = \frac{(\mathrm{N}(0, 1))^2}{\chi^2_k/k} = \frac{\chi^2_1/1}{\chi^2_k/k}.$$

This is connected to an important fact about terms with one degree of freedom in the analysis of variance. The t-ratio in the coefficients table, when squared, equals the F-ratio in the analysis of variance table. Thus, the p-values will be identical, and we have just one test, and not two different tests of the same null hypothesis.

Confirming simulations

In a mathematics textbook, we would prove the various assertions in the preceding section using the formulae for the probability distributions. Here instead we show how with some simple simulations we can satisfy ourselves of the assertions, and gain some familiarity with the ideas. The computing details are given in the package specific supplements.

Standardising a Normal variable

This simple case is performed to illustrate the method. We begin by constructing a variable called X. A random variable is a single object whose value is uncertain, but we represent it by a variable in a dataset, which has lots of rows in it. If there are 1000 rows, then the histogram of the variable will look reasonably like the probability distribution of the random variable; and the estimated mean given by your package will be very close to the true mean a mathematician would have derived analytically. Depending on various capacities of your computer and program, and your patience, you may be able to have 10 000 rows or even many more. Give X a Normal distribution, with a mean and standard deviation of your choice. Here they will be referred to as μ and σ^2, but of course in instructing your program, you must substitute the numerical values. Then we derive a further variable Y by using the formula $(X - \mu)/\sigma$. We can draw a histogram of X and Y, and also obtain basic statistics. The mean and standard deviation of Y should be 0 and 1.

Sampling distribution of the estimated mean and estimated variance of a sample from a Normal distribution

Begin by choosing your sample size. Let us suppose it is 6. Create six variables $X1$, $X2$, up to $X6$, and fill them all with the same Normal distribution, $N(\mu, \sigma)$. Create the mean in a variable called M by computing M from the formula $(X1 + X2 + X3 + X4 + X5 + X6)/6$. Create the variance estimate by calculating a variable V from the formula $((X1 - M)^2 + (X2 - M)^2 + (X3 - M)^2 + (X4 - M)^2 + (X5 - M)^2 + (X6 - M)^2)/5$. You should find when you plot a histogram that M has a Normal distribution with mean μ and standard deviation $\sigma/\sqrt{6}$. V should have a strongly right-skewed distribution that is proportional to a chi-square distribution on 5 degrees of freedom. Check this out. Your package will create for you a variable with random numbers taken from a χ^2_5 distribution. Make one and plot the histogram to compare with the histogram of V.

Sampling distribution of *T*

If we were going to test whether the mean of the population was indeed μ, we would construct a t-ratio as $M/\sqrt{(V/6)}$. So do this calculation and put in a variable called T. Now theory tells us that the variance of a t-distribution on k degrees of freedom is $k/(k - 2)$, which in our case is $5/3 = 1.667$. If the standardising of M had been performed with the true variance of M, namely $\sigma/\sqrt{6}$, then the variance would have been 1. Thus the effect of having to estimate the variance can be seen by obtaining summary statistics of T and comparing the variance to 1 and to 1.667. (You can even construct a variable $Z = (M - \mu)/(\sigma/\sqrt{6})$, and confirm that variance is 1.) If you cast your eyes over the values of V, you will see why the variance of T is so high: the variance is very poorly estimated, with occasional very high values, and lots of rather low ones. Of course a really low variance estimate will tend to give a very high T, and so the null hypothesis distribution of T contains all these high values.

Comment on the sampling distribution of *T* and sample size

With 1000 datapoints in the variables, we repeated the above exercise 50 times. The lowest and highest variances of T were 1.36 and 1.97; the mean of the 50 variances was 1.66437. So 1000 is enough to see that the variance of T is not 1, and 50 000 is enough to get a very close approximation to the theoretical value of 1.667.

T^2 has an *F*-distribution

Calculate a variable called `T2` with the formula `T2` = T^2. Then generate random numbers with an F-distribution on 1 and 5 degrees of freedom into a comparison column, draw two histograms and obtain summary statistics. You should find that the two are very similar. We noted before that the variance of a t-distribution is $k/(k-2)$. As its average is zero, the mean of the square of T must be $k/(k-2)$. Indeed in general the mean of an F-distribution with k degrees of freedom in the denominator is $k/(k-2)$. In this case the mean of `T2` should be $5/3 = 1.667$.

F is equal to a ratio of chi-squares divided by their degrees of freedom

It is straightforward to instruct your computer (see package specific supplements) to create one variable, `C1`, with a chi-square distribution on 3 degrees of freedom, and another, `C2`, with a chi-square distribution on 6 degrees of freedom. Then calculate `F12` with the formula (`C1`/3)/(`C2`/6). Finally construct a variable with an F-distribution with 3 and 6 degrees of freedom, and compare. The histograms and summary statistics should be similar. The mean should be close to $6/(6-2) = 1.5$. A sample of 10 000 is enough to get pretty close to this theoretical value.

In general, modern packages make it easy to play with distributions in a way that was previously accessible only to mathematicians. These simulations can also give quite a concrete feeling for the processes of statistical inference.

Bibliography

This book is an accessible introduction to General Linear Models, and it is a fair bet that readers will occasionally want to consult other books on statistics for various purposes. We recommend some in the lists below. Statistics is used in a wide variety of disciplines, and our selections are inevitably biased towards biology.

Elementary textbooks. Each different subject will have its own list from what must be hundreds of introductory statistics books.

TA Watt (1997) *Introductory statistics for biology students* (2nd edition), CRC Press. A simple and accessible introduction to statistics.

R Mead, RN Curnow and AM Hasted (1993) *Statistical methods in agriculture and experimental biology* (2nd edition), Chapman and Hall. This is a good elementary text.

ML Samuels (1989) *Statistics for the life sciences*, Maxwell–Macmillan International. This is an extremely thorough introductory text, which spells out every step of each argument. It also has a good chapter on 'multiplicity'.

H Motulsky (1995) *Intuitive biostatistics*. Oxford University Press. An appealing introduction that emphasises interpretation and de-emphasises mathematics.

MG Bulmer (1979) *Principle of statistics* (2nd edition), Dover Publications, New York. This book provides an introduction for the mathematically minded, including derivations of the formulae for the *t*-distribution, *F*-distribution, and so on. Very accessible to those with a modicum of mathematical ambition.

Large and comprehensive but not at all computerised. Likely to be of help to teachers and gradulate students.

GW Snedecor and WG Cochran (1989) *Statistical methods* (8th edition), Ames: Iowa University Press. The classic authoritative textbook. Although it is pre-GLM, the statistical wisdom and detail in this book make it an important source for statistical advisors.

P Armitage and G Berry (1987) *Statistical methods in medical research* (3rd edition), Blackwell Scientific. This gives extremely lucid and concise accounts of statistical tests and principles, and starts from scratch. It is statistically authoritative, more modern than Snedecor and Cochran, and very sensible.

RR Sokal and FJ Rohlf (1981) *Biometry* (2nd edition), W.H. Freeman. This is a large, comprehensive and expensive book that is a standard text for many biology researchers.

RR Sokal and FJ Rohlf (1987) *Introduction to biostatistics* (2nd edition), W.H. Freeman. This is a slimmed down version of *Biometry*.

Advanced books—for those who wish to check up the details and derivations of the results we present in this book, and who wish to look further.

DJ Saville and GR Wood (1991) *Statistical methods: the geometrical approach*, Springer. This book applies the geometric approach, and a computer package (Minitab), but is rather mathematical, and doesn't use model formulae.

AJ Dobson (1990) *An introduction to generalised linear models*, Chapman and Hall. This is an introduction to the mathematical theory of general linear models.

P McCullagh and JA Nelder (1989) *Generalized Linear Models* (2nd edition), Chapman and Hall. The authoritative text on Generalized Linear Models. Not for the faint-hearted, but if you use these methods a lot, this is essential reading for either you or your statistical advisor!

MJ Crawley (1993) *GLIM for ecologists*, Blackwell Scientific Publications. An introduction for biologists to the package and the methods.

M Aitkin, D Anderson, B Francis and J Hinde (1989) *Statistical modelling in GLIM*. Many worked examples of Generalised Linear Models in a wide variety of applications.

R Mead (1988) *The design of experiments*, Cambridge University Press. This is a thorough and excellent modern treatment. Mead shows that the availability of computer software has changed the importance of balance and orthogonality in design. Very much a biologist's book, rather than a social scientist's, in its view of 'random effects'.

Index